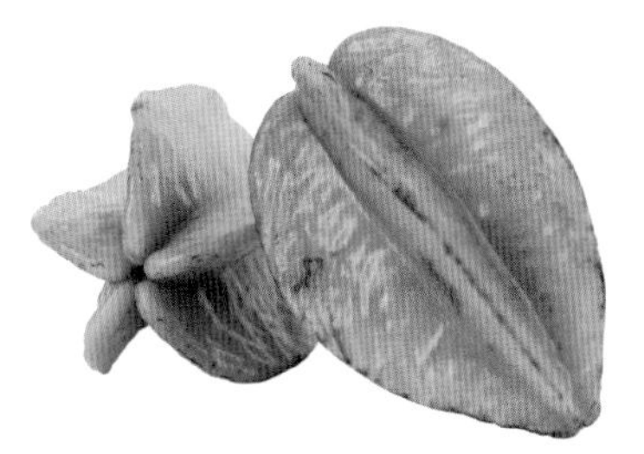

Edible

An Illustrated Guide to the World's Food Plants

Edible

An Illustrated Guide to the World's Food Plants

Edible: An Illustrated Guide to the World's Food Plants

PUBLISHED BY THE NATIONAL GEOGRAPHIC SOCIETY
John M. Fahey, Jr., President and Chief Executive
Gilbert M. Grosvenor, Chairman of the Board
Tim T. Kelly, President, Global Media Group
John Q. Griffin, President, Publishing
Nina D. Hoffman, Executive Vice President;
President, Book Publishing Group

PREPARED BY THE BOOK DIVISION
Kevin Mulroy, Senior Vice President and Publisher
Leah Bendavid-Val, Director of Photography Publishing and Illustrations
Marianne R. Koszorus, Director of Design
Barbara Brownell Grogan, Executive Editor
Elizabeth Newhouse, Director of Travel Publishing
Carl Mehler, Director of Maps
Barbara Levitt, Project Editor

MANUFACTURING AND QUALITY MANAGEMENT
Christopher A. Liedel, Chief Finanacial Officer
Phillip L. Schlosser, Vice President

Conceived and produced by
Global Book Publishing
Level 8, 15 Orion Road
Lane Cove, NSW, 2066
Australia
Ph: (612) 9425 5800 Fax: (612) 9425 5804
Email: rightsmanager@globalpub.com.au

MANAGING DIRECTOR	Chryl Campbell
PUBLISHING DIRECTOR	Sarah Anderson
ART DIRECTOR	Kylie Mulquin
PROJECT MANAGER	John Mapps
CONSULTANTS	Barbara Santich (Culinary), Geoff Bryant (Horticulture)
CONTRIBUTORS	Josephine Bacon, Claire Clifton, David Connor, Amy Felder, Steven Foster, Jennifer Graue, Jessica Loyer, Mary Moorachian, Catherine Rabb, Philippa Sandall, Stephanie Santich, Karl Stybe, Robin Stybe
COMMISSIONING EDITOR	Dannielle Doggett
EDITORS	Helen Bateman, Monica Berton, Fiona Doig, Emma Driver, John Mapps, Selena Quintrell
COVER DESIGN	Sanaa Akkach
DESIGNERS	Jo Buckley, Susanne Geppert, Kerry Klinner, Jacqueline Richards,
JUNIOR DESIGNER	Althea Aseoche
DESIGN CONCEPT	Kerry Klinner, Kylie Mulquin
GRAPHICS	Althea Aseoche
PICTURE RESEARCH	Tracey Gibson
INDEX	Jon Jermey
PROOFREADER	Elizabeth Connolly
PRODUCTION	Ian Coles
CONTRACTS	Alan Edwards
FOREIGN RIGHTS	Kate Hill
PUBLISHING ASSISTANT	Christine Leonards

Founded in 1888, the National Geographic Society is one of the largest nonprofit scientific and educational organizations in the world. It reaches more than 285 million people worldwide each month through its official journal, *National Geographic*, and its four other magazines; the National Geographic Channel; television documentaries; radio programs; films; books; videos and DVDs; maps; and interactive media. National Geographic has funded more than 8,000 scientific research projects and supports an education program combating geographic illiteracy.

For more information, please call 1-800-NGS LINE
(647-5463) or write to the following address:
National Geographic Society
1145 17th Street N.W.
Washington, D.C. 20036-4688 U.S.A.

Visit us online at www.nationalgeographic.com/books

For information about special discounts for bulk purchases, please contact National Geographic Books Special Sales: ngspecsales@ngs.org

For rights or permissions inquiries, please contact National Geographic Books Subsidiary Rights: ngbookrights@ngs.org

ISBN 978-1-4262-0372-5 (trade)

Printed in China by SNP Leefung Printers Limited
Color separation Pica Digital Pte Ltd, Singapore

COVER: *Solanum melongena* eggplant and *Capsicum sp.* chili

Consultants

BARBARA SANTICH is a Professor at the University of Adelaide, Australia, where she designed the curriculum for the Graduate Program in Gastronomy and developed its core courses in the history and culture of food and drink. She also teaches in the university's Food Writing course. Her interest in food and eating was stimulated by her early studies in biochemistry. She has been a food writer for almost 30 years, and has written six books as well as contributing to numerous publications, including *Australian Gourmet Traveller*, *The Journal of Gastronomy*, and *Petits Propos Culinaires*.

GEOFF BRYANT is a New Zealand-based horticultural writer and photographer and was a plant propagator and nurseryman for some 10 years. He has written a total of 12 books, including several widely sold plant propagation handbooks, as well as numerous magazine articles illustrated with his own photographs. He has been a technical editor, writer, and/or photographic contributor to a large number of major plant and gardening titles. Geoff's photographs can be seen at his image library website, www.cfgphoto.com, and he is also represented by several stock photo libraries.

Contributors

JOSEPHINE BACON is the author of 14 books on cookery, history, and travel. She is also a food historian and the author of monographs and articles, including a biography of Dorothy Hartley. She has lived in three citrus-growing countries and been involved in agricultural research. Her book, *The Citrus Cookbook* (1982) contains much citrus lore and legend. Bacon's main occupation is as a translator; her latest book translation, *A Sprig of Dill*, won the Gourmand Prize 2008 for cookbook translation.

CLAIRE CLIFTON is an author, journalist, and artist. Born in the United States but a long-time resident of the United Kingdom, she and her husband Colin Spencer run a bed and breakfast in a nineteenth-century vicarage in Icklesham, East Sussex. As a Type I diabetic she has a particular interest in what constitutes an enjoyable healthy diet—essentially fresh, local, and seasonal ingredients prepared with a minimum of fuss.

DAVID CONNOR is Emeritus Professor of Agriculture at the University of Melbourne, Australia, where he taught and researched in crop production from 1984 to 2002. Before that he held posts at the University of Queensland, Brisbane, and at LaTrobe University, Melbourne. For five years until 2002 he led a major Australian aid project on rice and wheat production in Bangladesh. He has published research papers and scientific reviews on many aspects of crop production and is the author (with R. S. Loomis, University of California) of *Crop Ecology*, published by Cambridge University Press.

A chef for more than 20 years, **AMY FELDER** has worked in small fine-dining restaurants and large resort hotels throughout the United States as well as in the Austrian Alps. She is currently a chef instructor at Johnson & Wales University in Charlotte, North Carolina, where she teaches in both the Baking and Pastry and Culinary programs. Her specialty is plated desserts with an emphasis on flavor. In 2007 she published *Savory Sweets*, a resource for anyone interested in learning how to analyze taste and flavor and how to use that knowledge to create flavorful desserts.

STEVEN FOSTER, one of the most internationally respected names in the herbal field, started his career at the Sabbathday Lake Shaker's Herb Department, the oldest herb business in the United States, dating to 1799. He is senior author of three Peterson field guides, and a dozen more books, including *Tyler's Honest Herbal* and the award-winning *101 Medicinal Herbs*. Foster's photographs appear in hundreds of publications. He is associate editor of *HerbalGram*, and a Trustee of the American Botanical Council. Foster is senior author of National Geographic's *A Desk Reference to Nature's Medicine* (with Rebecca Johnson).

JENNIFER GRAUE is a freelance journalist and food writer who splits her time between South Australia and the United States. In 2007 she graduated with a Master of Arts in Gastronomy from the University of Adelaide, Australia. Prior to moving to Australia, she was a television news producer in Phoenix, Arizona, between 1995 and 2005. Since finishing the Gastronomy program, she has carried out research for chef David Thompson and is currently working with chef Cheong Liew on a television project. She also contributes to a weekly radio program about food and works on a sensory panel for a wine consultancy firm.

JESSICA LOYER holds a Master of Arts degree in Gastronomy from the University of Adelaide, Australia, where she continues to work as a research assistant and tutor. She also holds a Bachelor of Arts degree in History from Barnard College, Columbia University. Her areas of interest include Jewish cuisine and its adaptations in secular New World communities, local food production and markets in South Australia, and interactive children's food education. In addition to gastronomic research and writing, she also conducts tours of the Adelaide Central Market and works in the hospitality industry.

MARY MOORACHIAN is a Professor at Johnson & Wales University, Charlotte, North Carolina, where she teaches personalized nutrition management in the Culinary Arts program. Dr. Moorachian has served as an instructor in higher education for 20 years, previously serving as instructor and program director for accredited Dietetic Education programs. She is actively engaged in numerous professional association activities, and has presented regionally and nationally on topics of healthy cuisine. She has served as a reviewer for numerous nutrition books. Dr. Moorachian is a licensed Registered Dietitian, a Certified Culinary Professional, and is certified by the American Association of Family and Consumer Science.

CATHERINE RABB is an Associate Instructor at Johnson & Wales University, Charlotte, North Carolina, where she teaches about beverages. She is also an instructor for certification classes for the Wine and Spirits Education Trust, and the International Sommelier Guild. Rabb writes the wine column for the *Charlotte Observer*. Over the years, she has owned and operated several restaurants, including Fenwick's on Providence, which has been operating in Charlotte since 1984.

PHILIPPA SANDALL is a freelance writer and editor of GI News (http://ginews.blogspot.com), the online monthly newsletter for the Human Nutrition Unit of the University of Sydney, Australia. She has co-authored seven books on food, health, and nutrition, including *Herbaceous*, *Spicery*, and *Sticks, Seeds, Pods & Leaves* (with Ian and Liz Hemphill). Since 1995 she been an integral part of the research and writing team involved in the best-selling *New Glucose Revolution* series of books on the glycemic index, with Professor Jennie Brand-Miller. Her most recent co-authored book is *The Low GI Family Cookbook*.

STEPHANIE SANTICH is a writer of fiction and non-fiction. She has worked in the publishing industry for a number of years as a reader and editor. Stephanie has a particular interest in food and food history, especially Australian food from the early twentieth century to the present day, and also in the application of herbs and other food plants in literature, folklore, and medicine. She is currently undertaking post-graduate studies in literature and creative writing at the University of Adelaide, Australia.

KARL STYBE is an Associate Professor at the College of Culinary Arts at Johnson & Wales University, Charlotte, North Carolina. He has been active in the industry for 15 years and has worked throughout the United States and the Caribbean as a chef and consultant. He has appeared on television and radio, and was a leading recipe editor for the James Beard Foundation's *Winning Styles Cookbook*. He holds multiple certifications through the American Culinary Federation, the American Hotel & Lodging Educational Institute, and Foodservice Educators Network International.

ROBIN STYBE is an Associate Professor at the College of Culinary Arts at Johnson & Wales University, Charlotte, North Carolina. She worked with the Rhode Island Department of Education as a Team Nutrition Culinary Instructor. She also owned and operated The Organic Gourmet, and is a Certified Master Gardener with an emphasis on organic farming. A contributing food writer for newspapers and magazines, she also makes regular appearances on local television and radio shows. She serves as food stylist, recipe tester, and editor for many publications and companies. She is a Certified Culinary Instructor through Foodservice Educators Network International and the National Restaurant Association.

Contents

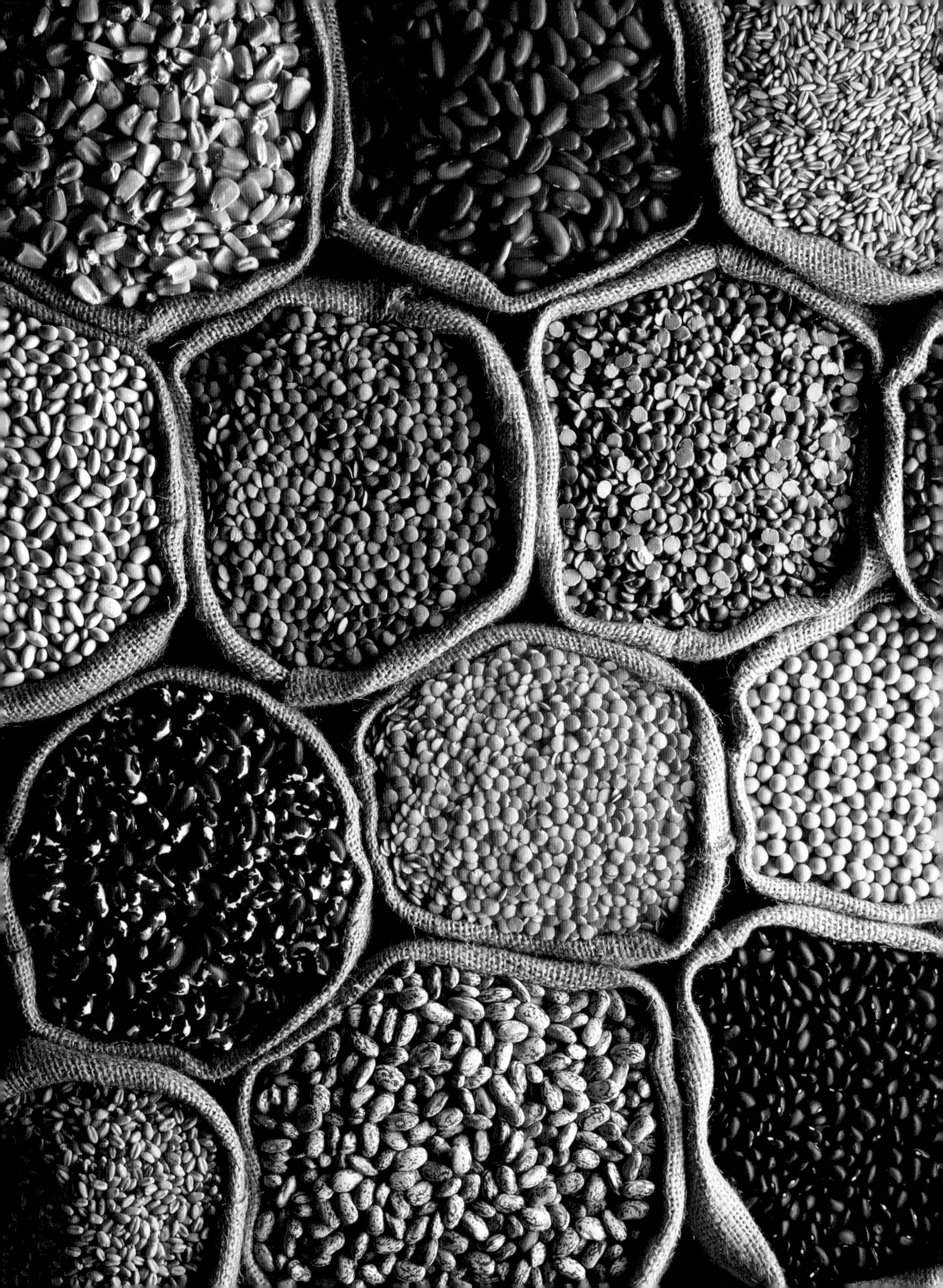

Foreword

Fewer than a hundred years ago, the New York Agricultural Experiment Station in Geneva, New York, published a series of monographs on the fruits of New York State. Each of these volumes contains hundreds of entries—600 pears in the pear volume alone—cultivar upon cultivar of cherry, peach, plum, grape, apple, and berry. This stunning portrait of diversity in a single state stands in stark contrast with the lack of variety encountered today by any supermarket shopper. For each fruit, vegetable, herb, or grain on the shelf, very many are absent. We don't know what we're missing.

Our ignorance goes hand in hand with our long-distance agricultural systems, extensive food processing, and a way of life that threatens to eliminate home cooking. We no longer pause to consider the merits of this cultivar over that; instead, we simply read the label on the can or packet. Our distance from the foods we eat can be measured not only in miles, but also in our diminishing knowledge and physical intimacy with plant foods. We know the shank of a leek, but seldom see its long banners of leaves.

But change is in the air. The international Slow Food movement argues for a slower pace of life that includes both the pleasures of the table and a keen awareness of the foods we eat. More farmers markets are opening worldwide. People discuss the pros and cons of eating seasonal produce from their own area. Gardens are cropping up everywhere. Local foods are celebrated in homes and restaurants, and on college campuses. And organizations such as the Seed Savers Exchange, which has been promoting heirloom vegetables for 30 years, are no longer the preserve of a few eccentric gardeners.

All of this strongly suggests that we are starting to participate in our food worlds in ways that are more active and direct. And certainly, the publication of *Edible: The Illustrated Guide to the World's Food Plants* is an indicator that there is a hunger to move from disconnection and ignorance to connectivity and knowledge.

That there's lovage among this book's herbs, and shagbark hickories among the nuts, gives me confidence that *Edible* goes beyond the territory of most books on food. It's an enormous undertaking to make a portrait of the world's plant foods, and yet the book is truly for the ordinary person who is curious about what plants, climates, continents, and conditions give rise to our foods. And in presenting this food world to us, the book does not shy away from botany and its language. I've always felt that everyone could benefit from a little botanical knowledge. Why shouldn't we be familiar with words like drupe, pome, and pericarp? Just knowing a few plant families can give someone the kind of knowledge to help him or her move about the kitchen intuitively and with confidence, which finally translates to pleasure and appreciation.

Plant diversity is essential to our survival. Without it, we are in a very dangerous position as eaters. But as *Edible* so handsomely illustrates, the world's food plants are a rich source of delight, an adventure in taste, form, and color that connects to history and culture as well as to the table. A timely corrective to the narrow food choices we now have, *Edible* invites us to expand our food world, explore it, and delight in the extraordinary range of food plants. And it *is* delightful. I mean, who knew that when the shell of the pistachio is split, it's said to be laughing?

Deborah Madison

Deborah Madison, founding chef of San Francisco's Greens restaurant, has long been committed to local and sustainable approaches to growing food, and has, through cooking, writing, and teaching, helped chefs, home cooks, and young people make that connection for themselves. She is the author of nine cookbooks, including *Local Flavors: Cooking and Eating from America's Farmers' Markets*.

PART ONE

From Plants to Food

Gatherers to Growers

RIGHT: *Illawarra plums, riberries, and acacia pods are among the traditional nuts, fruits, and seeds collected by Australian Aboriginal peoples, who lived as hunter–gatherers.*

Since the beginning of time, all organisms have lived by two basic rules: survival of self and survival of species. The tumultuous rise to dominance of early humans followed this pattern, and began with an instinctive quest for food. Through experimentation and observation, the ability to distinguish between edible and inedible plants has helped determine the destiny of humans. Adequate intake of calories and nutrition paved the way for human reproduction, therefore ensuring survival of the species.

Many scientists believe the world's first modern humans, *Homo sapiens*, inhabited Africa in nomadic groups about 200,000 BCE. Controversy remains as to the true origin of anatomically modern humans, but one thing is unchallenged: Our prehistoric ancestors used hunting and gathering as a means of providing food. This way of life continued until the end of the Mesolithic era, between 9000 and 8000 BCE.

BELOW: *Flint and other tools helped early humans hunt more meat, and the benefits of flint in harnessing fire enhanced their diets considerably. This flint tool is dated c. 6000–2000 BCE.*

THE QUEST FOR EDIBLE PLANTS: HUNTER–GATHERERS

A hunter–gatherer society is defined as one that primarily obtains sustenance from edible plants and animals in the wild, strictly using foraging and hunting techniques without resorting to any sort of domestication. Early hunter–gatherers were subsistence-based nomads, normally moving on a seasonal basis, if not more frequently. They followed herds and searched for favorable environments that provided sufficient resources to sustain the population. In the beginning, humans undoubtedly gathered and scavenged far more than they hunted. Roots, tubers, fruits, leaves, seeds, and nuts provided the bulk of their diet. Their remaining diet comprised small, easy to catch animals, fish, shellfish, insects, and abandoned animal kills.

As they refined their tool- and weapon-making skills with the creation of flint points, axes, snares, hooks, spears, bows, and arrows, the early humans began to develop and improve their fishing and hunting techniques; the use of domesticated dogs to hunt larger game ensured additional dietary protein. Simple and efficient techniques for preserving food were developed: Drying, salting, and utilizing cold temperatures were among the earliest methods. This use of short-term storage to help supply calories during lean times, and the increase in animal proteins in the diet, are thought to have helped produce healthier, longer-living populations.

AND THEN THERE WAS FIRE

A monumental event increasing humans' food choices and overall vigor was the discovery of fire. Fire was undoubtedly first observed as a brush fire from a lightning strike. Humans probably noticed wild animals feasting on charred remains left in burned fields, imitated the act, and found cooked meat superior to raw. The effort to successfully harness this highly regarded resource eluded humans for many millennia. Once captured, live embers produced permanent hearth fires that were maintained with the utmost care. Employing simple methods of combustion followed, leading to the discovery of the spark-inducing properties of flint. Fire's ability to provide light and heat created safer and more comfortable living environments for early humans, improved their overall health, and helped the movement toward developing more structured human societies.

Safety was probably the initial use for fire. Flames helped keep nocturnal predators at bay while everyone rested. The warmth from fire also helped protect humans from drastic temperature fluctuations. This was particularly significant for older and younger members of the population. Longevity in older populations added experience to the groups. The healthier youngsters were, the better their chances of surviving to breeding age and adulthood, therefore ensuring the survival of the species.

Fire also served a key social function. Communal fires became the nucleus of human grouping, and light from evening fires led to more social interaction between tribe members. This interaction promoted better communication skills through storytelling, the precursor to recording historical events. Organizational skills, used for social purposes such as planning events, were also important in developing a modern civilization.

Another fire-related health benefit was fire's ability to render food more palatable, nutritious, and safe through cooking. Cooking meat helped to kill bacteria associated with food-borne illness, leading to improved health. The preservation technique of smoking followed cooking. Cooking also expanded man's plant-based food supply. The difference in texture between cooked and raw foods in the diet is responsible for the evolutionary change in the skull structure in modern humans. Cooking food made its texture softer—massive jaws and large molars were no longer essential. The addition of cooked food made people healthier, stronger, and smarter. Meat eating became more common, a wider variety of food was available, and the consumption of higher quality nutrients led to increased brain power.

The nutritional improvements triggered by the use of fire, as well as climatic changes brought on by the end of the Ice Age, altered the dynamic of emerging societies. Migration slowed and populations grew, stressing food supplies. To compensate for inconsistent food supplies, societies began experimenting with herding animals and the process that was to revolutionize the world—agriculture.

ABOVE: *The benefits of fire to the diets of early humans were incalculable. Roots and tubers too tough or toxic to eat raw were rendered palatable by cooking over hearths similar to the reconstructed Stone Age one, pictured.*

LEFT: *In this fresco by an unknown artist, hunter-gatherers from Asia cross the Bering Strait to Alaska during the first Ice Age. Scientists believe these migrating Siberians were the ancestors of the Native Americans whose agricultural innovations would change the world.*

Settling Down: The First Farms

The first permanent farm settlements most likely formed in the Middle East, in the area now known as the Fertile Crescent, so-called because of the presence of an abundant water source and the diversity of plant and animal life suitable for domestication. The founding crops were wheat, barley, peas, lentils, chickpeas, and vetch. Rice cultivation followed closely behind in China; some argue, however, that rice cultivation in Korea preceded farming in the Fertile Crescent by thousands of years.

RIGHT: *Evidence of rice paddies dates back to the Neolithic era. This nineteenth-century painting shows paddies being watered by hand.*

THE NEOLITHIC REVOLUTION: THE AGE OF AGRICULTURE

Most historians agree that the gradual transition from hunting and gathering to agriculture and animal domestication began about 12,000 years ago with the Neolithic Revolution. There is strong evidence that weather patterns of the earlier Mesolithic era were the major environmental catalyst for change. With the end of the last Ice Age, temperatures and rainfall rose, as did the levels of oceans and rivers. Plant life and game flourished, enticing nomadic groups to linger rather than move along.

Although migration was not uncommon, the practice slowed in some societies and almost stopped completely in others. Birth rates increased in response to slower migration patterns, but the expanding societies were vulnerable to lack of food when they remained in an area once seasonally abandoned for greener pastures. Sedentary societies saw large game supplies and plant life dwindle from overhunting and over-gathering within limited areas. Additionally, a brief return of colder drier weather further hindered the ability to gather sufficient quantities of wild edible plants on a regular basis. Hunter–gatherers learned the importance of storing grain for lean periods, which led to increased efforts to renew plant resources through cultivation. Replanting wild plants and sowing harvested seeds yielded routine supplements to the food gathered in the wild.

BELOW: *Food storage jars are among the oldest ceramic artifacts surviving from early human societies. This one was found in Ganshu province, China, and is dated c. 2500 BCE.*

Sturdier homes and structures followed the emergence of crop cultivation. Where hunters and gatherers once depended on mobile or temporary living quarters, these sedentary communities needed stronger and more permanent dwellings, storage facilities, and communal buildings.

Agriculture emerged in other areas around the world after the communities in the Fertile Crescent settled in. Agricultural areas outside the Middle East included northern China, southern China, Central Mexico, the south-central Andes, eastern North America, and West and central Africa. This produced a commonality of skills among early peoples. Whether their agricultural knowledge was gained independently or was a result of migration from the original source is still a hotly debated topic among historians.

THE CHALLENGES OF AGRICULTURAL SOCIETY

Evolving human societies understandably faced considerable growing pains during their development. All these challenges, however, simply spurred them on toward the development of more advanced innovations in the areas of technology, religion, and government.

As agriculture evolved, societies began to depend upon mass plantings of fewer plant species, so were left vulnerable to weather variations, pests, and soil fertility problems. Previously, seasonal migration and a diverse diet were the answers to limited food supplies. Sedentary settlements, however, had to gamble on favorable weather conditions and other environmental factors. Animal and insect pests added to the struggle. Although "slash and burn" techniques for clearing and enriching soil were familiar, other practices to increase crop yields, including irrigation and crop rotation, took some time to be adopted.

One interesting result of the unreliable nature of agriculture was the dependence on a higher being for good fortune. Religion became the central driving force in early agricultural societies, as people looked to particular deities for the assurance of a successful harvest. Members of the society's clergy took offerings or payments of food and stored it in granaries.

Nutrition was also a significant issue in early societies. Migration allowed humans to gather different plant varieties and develop a well-rounded nutrient base. When more permanent settlements were established and populations grew, supplies of localized wild crops were soon exhausted, leaving these societies dependent upon farmed crops. The first of these were cereal grains, which did not provide all of the necessary nutrients for growth and good health. The addition of protein-rich legumes helped, but was not quite enough. Domestic livestock—goats, sheep, pigs, and cattle—increased protein in the diet, but meat was still not plentiful. Evidence suggests that animals may have been regarded more highly as religious sacrifices in the early days of animal domestication, though their meat was typically eaten.

"When tillage begins, other arts follow. The farmers therefore are the founders of human civilization."
Daniel Webster (1820–1905)

ABOVE: *Broad beans, also called fava or horse beans, are believed to be one of the oldest cultivated legume plants. Legumes added protein to the diets of ancient humans, but did not prevent malnutrition on their own.*

LEFT: *Storing grain in central storage areas made it much easier for societies to distribute excess product in times of hardship. Religious institutions often controlled this distribution. In this fresco, c. 2000 BCE, Egyptian workers transport grain to granaries.*

ABOVE: *Historians cannot agree on exactly when the potter's wheel was invented, but many agree that by 5000 BCE potters were creating wheel-thrown ceramic pots for storing food. This limestone model, c. 2500 BCE, is from Egypt.*

With agricultural societies living in closer quarters, however, came sanitation- and disease-related problems humans had not encountered in their nomadic lifestyle. Improper disposal of the dead and inadequate waste removal led to epidemics of infectious diseases, including those spread from animals to people. As time passed, humans developed immunity, which reduced the severity of disease and decreased mortality rates in communities.

The challenges faced by early agricultural societies and the evolving needs of their people helped usher in formal systems of social organization, religion, and government responsible for spiritual guidance and social order. Farming required strict observation of planting and harvest times. Organized work schedules, full group participation, and the intense physical labor required for agriculture were little-known experiences for hunter–gatherer societies. Strict discipline became necessary on an individual and social level. Planning for the future became even more essential. Stores of food and seeds needed to be protected from moisture, pests, and thieves. Controlled distribution of food to ensure a steady supply between harvest times quickly became a crucial part of each community's basic survival.

BUILDERS AND INNOVATORS

One of the greatest benefits offered by this new way of life was the opportunity for societies to develop new technologies to improve food production. With this came the chance to develop social organization, giving these new societies some cohesion.

Permanent housing became one of the top priorities of early settled societies. The properties of sun-hardened clay proved suitable for building sturdy houses and granaries. Builders then discovered how to make lime plaster to coat the clay structures, increasing their strength and aesthetic appeal. Sun-drying clay and making plaster were the Middle East's first steps toward producing harder, more versatile, pit-fired pottery. Fired pottery appeared after 6000 BCE (earlier in Japan), with the invention of the potter's wheel coming shortly afterward in Sumer.

RIGHT: *Animal-drawn plows have been used in agriculture for millennia, and are still used today. Adding large animals to the labor force meant that more agricultural products needed to be set aside for animal feed.*

Does Size Really Matter?

In many ways, hunter–gatherers had a dietary advantage over the early sedentary agricultural societies, because of lower group numbers, diets high in animal protein, and ready access to a wide assortment of plant foods. Studies show hunter–gatherers were bigger, more muscular, and better fed than those who came along after the Neolithic era began (a Neolithic skull is pictured at left). As humans shifted to a grain-based diet, they became less dependent on animal proteins and suffered some nutritional imbalances and periods of famine. Anthropologists have found that human brain size has reduced by 11 percent in that time. However, the technological advances of the Industrial Revolution all but eliminated this in developed countries, and our brains are now smaller but have an increased capacity. Smaller brains require fewer nutrients, so this can be seen as evolutionary efficiency.

The potter's wheel is believed to be the first recorded use of a wheel in history. Sumerian wheeled carts followed soon after.

Once farming became the primary method of food production, the need to develop new technology was essential. Digging sticks, stone axes, grinding stones, hoes, and sickles were among the first farm tools. Threshing boards to separate grain from its husk appeared after 6000 BCE, while animal-powered plows simplified and sped up the processes of soil preparation and planting. Metalworking skills began to be developed after about 3500 BCE, during the Bronze Age. This introduced stronger sharper tools more suited to large-scale agriculture.

Advances in irrigation technology, first recorded in the Middle East in 6000 BCE, enabled farmers to compensate for rainfall variances. Farmers were well acquainted with the benefit of the flooding and draining cycle of rivers. Harnessing this resource required the digging of canals to channel water where it was needed. Irrigation made fields more productive, which in turn allowed surplus grains to be stored for future use.

One of the by-products of domesticating animals was manure, probably one of the first forms of applied fertilizer. Leaving fields fallow for a season (or longer) and rotating crops also enriched the soil. The benefits of these practices were discovered by accident. Groups of nomadic peoples would leave fields they thought were depleted, but later return to find the land renewed and fertile. They discovered that if grains were planted in a field following a crop of legumes, plant growth was more vigorous. Once these beneficial practices were understood, they were regularly used to manage farmland.

DIVIDING LINES: ECONOMY AND STRATIFICATION

Increased populations in sedentary societies resulted in a larger potential workforce, so some chose to forgo agriculture to develop specialized skills. The choice to pursue a nonfarming profession produced a new segment of society no longer dependent on self-sufficiency. Food surpluses made this possible in conjunction with an organized means for exchange; enter the government, organized religion, and taxes.

As settled communities grew, strong leaders emerged to channel the efforts of their subjects. The government and religious organizations gained control of most of the land, now a valuable, finite asset within their realm of power. Land ownership was highly regulated and limited to those deemed worthy either by wealth, birth, or status. Taxes on food and valuable items provided revenue, and food distribution was decreed by the government or religious organization based on perceived class worth. Class stratification took a firm hold in societies.

Acquisition of land, wealth, and food gave rise to threats of theft and raiding from local or neighboring inhabitants. New threats generated the need to build walls around settlements, develop weapons, and form armies for the purpose of defense.

BELOW: *Lentils—seen here growing in a field—have been a popular food crop for 10,000 years. They were used by the Ancient Greek and Middle Eastern societies, and archaeologists have unearthed evidence of them in ancient tombs.*

The Fertile Crescent and the Origin of Agriculture

RIGHT: *The Euphrates River, seen here in Turkey, was a calmer, more predictable, and more stable source of water for farming in ancient times than it is today. Damming projects and environmental changes have made it less reliable for farmers.*

The Fertile Crescent is a region of 193,000 square miles (500,000 square km) that arcs from the Persian Gulf to the Mediterranean Sea. The Tigris and Euphrates Rivers run a roughly parallel course through the center of the crescent. Some maps also include Egypt and the Nile River basin in the defined area of the Fertile Crescent. Many experts attribute the origin of modern agriculture to the Fertile Crescent because of strong archaeological evidence, geographic location, and a climate that predisposed the area to successful agricultural activity.

AGRICULTURE'S IDEAL ENVIRONMENT

The Fertile Crescent was known for its hot dry summers and wet mild winters. This created the perfect growing environment for wild grains able to withstand harsh summer conditions. Natural evolution produced annual wild cereal species that wasted little energy on thick stems or leaves and put all their effort into creating large seeds.

Many wild grains found in the Fertile Crescent were self-pollinating—meaning that each flower contains everything necessary for pollination—with occasional instances of cross-pollination. Self-pollinating grains were dependable because the plant's product was always true to its original form. If cross-pollination did occur, the results were often beneficial, as in the creation of bread wheat.

BELOW: *Emmer wheat is an ancient grain, and was the most important wheat crop until c. 4000 BCE. Today, emmer is still grown in countries such as India, Ethiopia, and Italy.*

The variety of altitudes and terrains in the Fertile Crescent produced a level of biodiversity rarely seen in other developing areas. Plants were not the only valuable asset resulting from this diversity. Goats, sheep, cows, and pigs—four of the world's most important mammals—made their wild home in the Fertile Crescent. Humans' first attempts at animal domestication and herding took place before, or very shortly after, they took their first steps toward farming.

By 8000 BCE, plentiful game and plant life prompted nomadic groups to form semipermanent settlements near the rivers within the crescent. Settlers would plant seeds in soil still moist from winter rain or enriched with alluvial silt from recent floods. Soft, rich soil made planting easy and crops fruitful. The arid atmosphere helped prevent molds and disease from harming the grain during the growth season. It was a natural progression for these settlers to find suitable permanent locations close to these fertile lands. The close proximity to rivers also gave settlers the means to fish and irrigate, and eventually to transport goods for trade.

Two key societies emerged from these early settlers in the Fertile Crescent region: Egypt and Sumer, which developed at a similar rate, both benefiting from and expanding upon each other's knowledge.

EGYPT AND SUMER

The Sumerians (flourished c. 5000–1730 BCE) were responsible for many significant firsts that spurred the development of civilization. Although Egyptian society preceded the Sumerian civilization, the Egyptians made use of Sumerian technology, capitalizing on its innovations and adapting it to become one of the ancient world's central powers from c. 6000 to 30 BCE. The cultures of these ancient regions brought remarkable change to the world. They both developed well-organized agricultural systems, governments,

religion, resources, and technologies. As a result, both became strong societies with all of the defining characteristics of civilization.

Ancient Egypt was divided into an upper and lower region along the Nile River, which empties into the Mediterranean Sea. The Nile gave life to an otherwise hostile environment and, as trade developed in the region, it provided a convenient mode of transport for food and goods. Just before 5000 BCE, Nile settlements adopted farming as their main economic activity, although pottery and other crafts supplied additional products for trade both within Egypt and across the Mediterranean into the Middle East.

Nile farming used irrigation in the form of catch basins, dykes, and the "shaduf," a counterweighted delivery system used to convey water to higher elevations. Egypt, however, was also extremely dependent on flooding and the rich alluvial silt left behind after the water receded. Without this floodwater, desert conditions would have made farming virtually impossible. Despite these hardships, Egypt went on to become one of the greatest centers of culture and civilization in the ancient and modern worlds.

Sumer was located in Mesopotamia between the Tigris and Euphrates rivers. The region closely followed Egypt's path to civilization and culture, significantly improving many of the basic concepts learned from the Egyptians. By 5000 BCE, improved technology, organized labor forces, fertile farming lands, and access to water for irrigation allowed Sumerian farmers to employ some of the earliest large-scale production methods for the cultivation of grains and legumes. Plentiful supplies of grain enabled the enterprising Sumerians to use almost 40 percent of their harvest to produce some 19 types of beer.

Periods of abundant food supplies also allowed tradespeople time to practice their skills and create excess goods to sell and trade. With the advent of the potter's wheel and utility wheel, a strong economy based on surplus food, trade, and social order followed. Clear distinctions between working classes detailed each group's specific duties. Not everyone produced their own food, but everyone needed it, so a structured system of controlled distribution, measurement, and exchange was developed. Because of the need to keep agricultural records, track food distribution, codify laws, and preserve religious theology, archaeologists believe the first written language and arithmetic were developed in Sumer, and the ancient artifacts found in the region support this theory.

ABOVE: *Today, agriculture is still important along the Nile River's fertile banks. Because of their reliance on flooding, the early Nile farmers were often the victims of droughts that caused extended periods of famine in the region.*

BELOW: *This ziggurat, a temple dating back to 2100 BCE, still stands in the Sumerian city of Ur. Sumerian temples owned large tracts of agricultural land and large granaries.*

Staple Foods

ABOVE: *Wheat, one of the world's most popular staple grains, is grown primarily in the United States, China, and Russia. Other wheat-growing regions include India, Pakistan, western Europe, Canada, Argentina, and Australia.*

RIGHT: *This Egyptian figure, c. 2500 BCE, shows a brewer kneading beer dough. The heavy yeasted dough was partially baked, crushed, mixed with water and some extra moistened grain, then fermented and sieved to create beer.*

As hunter–gatherers made the gradual transition to an agrarian lifestyle, the foods they most frequently gathered became their crops of choice when farming took a firm hold. People sought out high-yielding varieties that gave them the most return for their planting efforts. These original crops became staple foods and formed the foundation for the dietary needs of each society.

FUNDAMENTAL SUSTENANCE

Staple foods are the dominant foods that provide the majority of the energy and nutrition needed for a basic diet. They typically make up one-half or more of the caloric needs of a population. Staple foods were originally based on the wild forms of plants first cultivated in fledgling farm societies. They are typically starchy, nutrient-dense, and fairly easy to obtain, and must also be able to be stored for long periods of time. Although staple foods are generally wholesome and add bulk to diets, they rarely provide all the essential nutrients required for growth and maintenance on their own.

The first domesticated crops were native to the areas in which they were grown, so they were ideally suited to local environments. Although some wild food plants were common in more than one area, they generally varied from region to region, so each agricultural center hosted different staple crops for its population.

THE EARLY CROPS THAT FED THE WORLD

Some of the most important staple food plants were indigenous to the Middle East, and are referred to as "founder crops." Cereal grains were primarily used to make porridges, breads, and fermented beverages. Legumes helped add protein to the average diet. This was especially important as lack of resources and social stratification denied the lower classes access to sufficient amounts of meat and left them dependent on plant foods. Some of the earliest staple food crops are discussed in more detail below.

Emmer is a low-yielding, bristly wheat that was one of the dominant food crops both before and during the Neolithic era (c. 8500–4500 BCE). Once primarily gathered, it became prized for cultivation when farming became established. Emmer wheat is self-sowing, and was the preferred wheat grown in Ancient Egypt. It is cultivated less often as a food crop today, but still grows in the wild.

Einkorn is a higher yielding wheat that was gathered before the Neolithic era, and widely grown during and after this period. It still grows in the wild and as a local plant. It is harvested to make cracked wheat and is used for animal feed. An interesting fact about this grain is that protein from the gluten in einkorn may not be as toxic to those with gluten sensitivities as standard wheat proteins are.

> *"Rice is the best, the most nutritive and unquestionably the most widespread staple in the world."*
> Auguste Escoffier (1846–1935)

Barley is a large cereal grain very widely gathered and cultivated before and during the Neolithic era. It was used to make bread and beer. Ancient forms of barley were higher in protein and therefore more nutritious than many modern forms of the grain. It remains in fourth place on the list of the world's most important grains, the first three of which are wheat, rice, and maize.

Lentils, also called pulses, are low-growing, scrubby plants in the legume family. They are grown for their small, lens-shaped, protein-rich seeds. Their excellent nutritional content made lentils one of the most important cultivated crops of ancient times.

The first cultivated peas were probably more like field peas which resemble dried beans, rather than the typical green pea we are familiar with. Not only were these peas an important food crop, but they also helped replenish the soil after harvesting grains because of the nitrogen-fixing properties of their roots.

Chickpeas are a large seeded plant in the legume family. They require more water than lentils so they were probably first grown in areas with more rain or greater access to

ABOVE: *Peas are one of the oldest cultivated crops in the world. Fossilized peas have been found at ancient Swiss lake sites and dated to c. 6000 BCE.*

Food for Trade and Travel

As technology advanced, farming communities became more productive and managed to generate surplus products for trade. Food was a highly regarded resource for trade. Since trade involved lengthy travel, it was essential to select foods able to withstand harsh conditions. Hardy grains, nuts, legumes, and tubers were chosen for this reason. Trade brought new plant species from afar; this introduced new staple foods. One example is taro, pictured, which originated in Asia but was taken up by the Ancient Egyptians and Romans. Later, the exploration of the New World of the Americas and the Pacific brought a host of new staple foods to the populations of Europe and Asia, and with them a population explosion. As people migrated from their rapidly overcrowded homelands, they spread their staple crops across the world.

Taro is now a staple food on many Pacific Islands.

RIGHT: *Sorghum is a tall grass that probably originated in Africa. While today it is often grown for animal feed, it is also a valuable food crop that withstands heat and aridity. Its grains can be ground for porridges and flatbreads.*

water. They contain one of the highest levels of protein found in plants.

Bitter vetch is a small-seeded member of the legume family—the protein-packed seed resembles a red lentil. As the name implies, bitter vetch has an unpleasant taste, and requires several soakings in hot water before cooking to make it palatable. It may have been a human food of last resort during times of famine, but was probably used more for medicinal purposes and as animal feed.

SIGNIFICANT STAPLES

Staple food production may have started in the Middle East, but people quickly adopted staples from other areas as trade introduced new food plants throughout the world.

Rice, taro, and chestnuts were initially cultivated in Asia. Rice competed closely with wheat in its value as a staple cereal grain but was able to feed more people for much longer periods than did any other grain. Taro is the starchy corm from a dry cultivated or semiaquatic plant and is toxic in its raw state. Chestnuts are borne on a large deciduous tree and contain twice as much starch as potatoes—and a fraction of the fat found in most nuts.

Africa first produced finger millet, pearl millet, sorghum, and yams. Millets are annual grasses that tolerate unfavorable growing conditions. Some varieties native to Asia became more important than rice in areas of Korea and China. Sorghum is a wild grass grown for food, beverages, and animal fodder. It can withstand drought and acid soils. Yams are large, starchy, nutritious tubers.

Although all staple foods served to feed the masses, some stood out as catalysts to monumental change across the world. The New World foods from the Americas and the Pacific are believed to have been responsible for huge population explosions in China and Africa, eventually spreading to Europe. Migration caused by growing populations further expanded the staple food base where new settlements formed.

BELOW: *Native to tropical South America, the peanut is a highly concentrated source of nutrients, which makes it a prized crop in areas threatened by famine.*

These new foods included maize (corn), potatoes, cassava, beans, sweet potatoes, and peanuts. Maize is a productive cereal grain high in starch and sugar, which became a primary staple in some countries. Potatoes are an extremely productive root crop and are high in starch and many nutrients. They provided famine relief wherever they were adopted, but early forms were toxic when raw. Cassava (manioc) is a woody shrub that is grown for its starchy, low-protein tuber and protein-rich leaves. Most cultivars are toxic in their uncooked state. Beans are members of the legume family that produce large, high-protein seeds. Sweet potatoes are members of the morning glory family that produce large tubers high in nutrients and sugar. Peanuts are actually members of the legume family, and they store fat instead of starch. They contain a higher percentage of protein by dry weight than meat.

THE NUTRITIONAL COST OF STAPLE FOODS

So, why did certain plants become staple crops? Plants that were adaptable and produced large yields were prime candidates. One of the most serious drawbacks of this, however, was that plants that fit these criteria were not necessarily able to supply people with enough nutrients for healthy survival. Most staples provided basic sustenance only if consumed in sufficient amounts with other foods entering into the mix. Protein-rich meat was a luxury for most, so plants had to provide the bulk of the nutrients. Dairy products could be a little easier to come by, but were seldom a plentiful resource. When food supplies fell too low, or were not varied enough, populations experienced chronic bouts of malnutrition leading to skeletal deformities, stunted growth in children, blindness, and anemia.

Another important consideration with a vegetable-based diet involves the processing of nutrients in the body. Many important nutrients can remain chemically bound in grains and vegetables, making it difficult to assimilate the vitamins and minerals found in those foods. Accidental remedies for this problem were sometimes found in food preparation methods. Soaking, grating, rinsing, or cooking releases locked nutrients. These processing methods also help to neutralize some of the toxic properties of many staple foods.

Most grains and vegetables are very nutritious when consumed as part of a healthy diet. However, to get the proper amount of dietary proteins, common grains must be combined with legumes, nuts, or seeds to produce complementary proteins in order to supply all of the essential amino acids necessary for human maintenance and growth. Many cultures were able to address some of these nutritional challenges by serving foods in specific combinations, such as chickpeas (*hummus*) with flatbread, beans with rice, or maize with beans. Whether these practices were first directed by flavor, nutrition, convenience, or accident is difficult to say, but regardless of the reason, these combinations now define numerous cuisines around the world.

If poor nutrition was not enough, dental troubles frequently accompanied grain-based diets. To produce breads and gruel, grains were generally ground. Stone was the most common material used for this process, and grinding always left tiny rock particles in the food. Over time, these particles wore down tooth enamel, causing cavities and infection. Maize had the added disadvantage of a sticky consistency and high sugar content, so worn enamel plus sugar-laden deposits resulted in more tooth damage. Interestingly, wealthy classes experienced more dental problems than lower classes because they could afford more highly processed grains. These grains introduced more rock particles into their diets, and consequently eroded their teeth.

ABOVE: *This miller in Corum, Turkey, grinds grain using a millstone powered by a donkey. Grinding grain is still a labor-intensive task in many countries, and is often performed by hand.*

LEFT: *Soaking and roasting chickpeas is one way of releasing this staple food's nutrients—including folate, fiber, protein, and iron. Here a street vendor sells roasted chickpeas at his stall in Istanbul, Turkey.*

RIGHT: *Mommina women in West Papua clean the marrow of the sago palm, before filtering it and using it to make flour. Sago palm is indigenous to Indonesia, and is a staple food of the Mommina people.*

BELOW: *In Tunisia, North Africa, grain and other foodstuffs were stored in high-ceilinged rooms called* ghorfas. *Foods such as chickpeas, wheat, and barley were dried in the sun, then stored in the* ghorfas *for up to 2 years.*

FAMINE FOODS

In times of famine, alternative food choices frequently appear as a last-resort effort to stave off starvation, but many offer sustenance at a price. Bad taste, reduced nutrition, low yields for high effort, and deadly toxin levels are all associated with some "famine foods." Others can be quite tasty and nutritious, but when people associate them with times of hardship they are less inclined to consume them unless absolutely necessary.

Greater yams, for example, are a starchy tuber that hold well in the ground, which makes them a good insurance crop, but they contain toxic alkaloids and must be handled with care. Bitter cassava is a very toxic form of cassava—another tuber—that must be soaked or cooked thoroughly before it is eaten. Sago palm is a tree that produces a large quantity of starch in the trunk, and is often used for food when rice crops fail. Grass peas are weedy invaders of cereal grain fields. Overconsumption of these peas is associated with paralysis, caused by a neurotoxic amino acid found in the seeds.

FOOD STORAGE AND PROCESSING

When large-scale farming techniques were developed, it became possible to produce larger quantities of staple foods than ever before, so storage and processing facilities had to keep up with the growing demand.

Previously, skins, baskets, caves, and crudely dug pits served to store what few surplus items nomadic groups maintained, but this was primarily for convenience in the short term rather than long-term use. Sedentary farm communities, on the other hand, began to rely on stores of food to see them through less productive months. The need to safely store grains and tubers led to the use of clay vessels; when sealed, they successfully kept out insects and vermin and helped keep grains dry. Roots were still stored in grass-lined pits, which maintained even humidity levels and temperatures.

Some drier roots and other foods were stored in small mud domes or houses.

These methods were ideal for individual families, but food storage for communities required more planning. Wooden and brick buildings served as granaries, which were controlled by the local administration or religious organization. Wealthier families constructed smaller granaries on their own land. The Chinese built elaborate storage pits, the sides of which were burned, oiled, and lined with wood or bricks before being sealed up with dirt. Because the pits were airtight, grain stored in them was kept free from insect, bacteria, and fungus damage.

Processing staple foods before and after storage was essential. Roots had to be brushed clean and dried before storage to toughen the skin and improve their holding capabilities. Later, waxes were applied to some roots to prevent excess moisture loss. Before grains could be stored they were parched or lightly roasted. This served to prevent premature sprouting, loosen the outer husk for easier removal, and lower the moisture content to allow for longer storage times and aid in the grinding process.

With more people reliant on grains for a large part of their diet, bread became an important staple by-product. Grinding enough grain to produce the daily bread requirement was an all-day task when small, hand-powered stone mills called querns were used. Large mills, owned by the community's government or religious organization, were soon necessary. These were run by slaves or animal labor and, later, by water or wind power. Grinding services were paid for by community members with a percentage of their grain. In some of the communities, the use of small querns was even outlawed to ensure regular grain payments to official mills.

ABOVE: *In Yamouneh, Lebanon, a villager grinds maize using a rotary quern. The kernels are spread over the grindstone, the top stone is replaced, and a handle turns the stone to grind the maize.*

Food preservation practices were also advancing. Salting, smoking, fermenting, drying, and pickling extended the shelf life of many perishable foods. Fermented beverages had the additional benefit of rendering water safe for drinking, while the process of fermenting dairy products—a significant protein source—allowed the creation of longer-lived yogurts and cheeses.

Despite the improvements brought by better food storage and processing capabilities, populations still experienced devastating famines brought on by crop failure, overpopulation, and war. This was further complicated by class stratification, which in times of crisis kept the meager stores of food in the hands of the privileged few. Unfortunately, this age-old problem still plagues our modern societies.

Staples of the Ancient World: Where Are They Now?

The top three staple foods of ancient times—wheat, rice, and maize—have retained their dominant status throughout history. Roots, legumes, and nuts were also major players in early agriculture, providing energy and sustenance to countless populations, and many are just as important today. Chestnuts, lentils, peas, chickpeas, peanuts, potatoes, beans, and sweet potatoes have been integrated into most modern cuisines. Taro, millet, yams, and cassava still appeal as food crops. Sorghum and barley are primarily used as livestock feed and serve a secondary purpose in human food and beverages. Vetch is no longer a popular food crop, although it is still used as animal feed.

Roasted chestnuts are found on menus and street corners across the globe.

Food Trade, Exploration, and Conquest

ABOVE: *The exotic foods of the New World fascinated Europeans. This image from a French tapesty of the eighteenth century shows a West Indian carrying a basket of tropical fruits.*

In the recent decade or so there has been a craze for television cooking shows, celebrity cookbooks, and kitchen "reality" programs that has captured the world's interest in food in a way that has seldom been seen before. Chefs are now given the exalted status of celebrities, and consumers are demanding more exotic and esoteric ingredients from their grocers. Part of this trend in the last 20 years has been a fascination with the cultural side of food, including "fusion cuisine"—the joining of cuisines or techniques from different regions of the world, with varying levels of success.

If today's "foodies" look with a long enough view to the past, they may be quite surprised to find that virtually all modern cuisines are "fused" in some way. Humankind has always sought the next gastronomic thrill, no matter where it might take them. It led our ancestors over the next hill, or across the next river or ocean, to find out what was to eat on the other side. A succession of traders, explorers, marauding invaders, steely eyed conquerors, and persevering agriculturists continued to find the next new thing that was good to eat and spread the word across the globe.

TRADERS: FOOD ON THE MOVE

As discussed earlier, the shift from nomadic hunter–gatherer societies to more settled farming communities about 12,000 years ago brought about humans' first forays into agriculture and organized food storage. Archaeological evidence suggests animals began to be domesticated soon after, which allowed people to harvest and transport their products more efficiently. The invention of the wheel continued to enhance early traders' ability to travel longer distances with their wares. As communities grew larger, trade routes and roads opened up. On their journeys, traders obtained foods from other regions and, through trial and error, were able to see which of these products would thrive back home. The introduction of the sail also brought a whole new dimension to the concept of trade.

Middle Eastern traders made early contact with their counterparts in India, China, and Southeast Asia, who brought them the spices that created such a stir in the early culinary world, along with refreshingly sweet additions such as bananas and coconuts from

"Food is about agriculture, about ecology, about man's relationship with nature, about the climate, about nation-building, cultural struggles, friends and enemies, alliances, wars, religion."
Mark Kurlansky (1948–), *Choice Cuts*

the South Pacific. Demand for these wonders created entire industries and economies based on their exchange, and created one of the greatest drives to explore farther and farther over the next horizon.

EXPLORING THE (EDIBLE) WORLD

When transport became easier, traders turned into explorers and, over time, came into contact with cultures having their own agricultural revolutions, thereby introducing new products far and wide. Wheat and barley made it to Europe and Egypt, as well as India and China to the east, and in turn foods such as rice and onions were brought across to the Middle East, and eventually became standard in the local cuisine.

Exploration and the distribution of food often occur hand in hand. For instance, when European explorers set out for the New World they brought along wheat, and grape cultivars that had not been known in the region prior to that time. Both have become integral to the diet of the Americas. New migrants to the Americas then experimented with viticulture to produce a greatly missed beverage from home, and in recent years South American and North American wines have garnered many awards.

THE CONQUEROR'S PALATE

Contact between different groups is not always benign, however. Aggressive conquest and war, however abhorrent, was a very important part of the process of food exchange. An army on the march through a foreign territory still needs to be fed; a victorious leader in a conquered land may wish to celebrate victory with a lavish feast.

When Europeans conquered the New World, or Mongols invaded China, or the Ancient Persians marched across vast tracts of the Middle East, or the French made inroads into Africa, they gained more than territory—they hit a mother lode of new ingredients that changed their own cuisines forever. So, while warmongering brought unspeakable devastation to the conquered peoples, it also accelerated the adoption of new agricultural and food products.

ABOVE LEFT: A *Phoenician trading ship reaches Egypt in this painting by Albert Sebille. The Phoenicians created a mercantile empire that stretched across the ancient Mediterranean.*

BELOW: *Armies have always been instrumental in spreading new foods across the globe. This Assyrian relief from the seventh century BCE shows soldiers sharing a meal.*

The Americas

The impact of the New World larder is almost too big to contemplate in so small a space. It is impossible to imagine a world without Hungarian paprika, or curries without the punch that the chili pepper provides. It would not be a proper German or English meal without potatoes to add a level of comfort to the table. Italian food would just not be the same without the juicy tomato to highlight its pastas. What would any of these meals be without a decadent dessert such as chocolate mousse, or any of a host of other delicious possibilities to round out the menu? Centuries of trade, exploration, and, unfortunately, conquest have made it possible for the world to enjoy the bounties of the Americas.

ABOVE: *Hungarian paprika is made from finely ground dried red peppers, which made their way from the Americas, through the Balkans, to Hungary in the sixteenth century.*

CONQUERING THE AMERICAS

When Christopher Columbus, sponsored by the Spanish, first set out to the west in 1492, he was hoping to find a sea route to the spice-rich lands of India and beyond. The spice trade was a lucrative one, and by finding a new route east, Spain hoped to gain an advantage over the other European powers. While Columbus did not find the fabled route to the Spice Islands, the new lands proved to be even more valuable, and more influential on the world's diet, than anyone could have foreseen. Maize, potatoes, tomatoes, sweet potatoes, chili peppers, chocolate, pineapples, beans, and pumpkins were all native to the Americas, and their introduction across the world had a huge impact on global culinary culture.

Columbus may have been the first, but he was certainly not the last of his kind to have an impact on the region. By 1511, Spanish colonies were founded in Jamaica, Puerto Rico, and Cuba, and the stage was set for even greater penetration to the west. In 1519 Hernán Cortés set out with some 600 men from Cuba to conquer the Aztec peoples in Mexico. By 1521 the Spanish had conquered Mexico and in the process destroyed the great city of Tenochtitlan, which today is Mexico City. Francisco Pizarro

BELOW: *An undated painting shows Christopher Columbus arriving at a Caribbean island on his voyage to find the Spice Islands. On later voyages, Columbus introduced new plants, including the all-important sugar cane, to the Caribbean.*

conquered the Incas of Peru in a similar fashion, then moved on to exploit the people and the Incas' caches of silver and gold, which he sent back to Spain. The other conquistadores to the north did not initially have this much luck. These lands would prove their value in time, but they did not quench the Spanish conquerors' immediate thirst for gold. Although initially overlooked by the Spanish, the greatest reward this New World had to offer was food, not precious metals.

In their relentless quest for riches, these foreign conquerors discovered much about the agriculture of the areas they invaded, and began sending specimens back to Europe and to the countless other regions of the globe where their trading ships docked. The inhabitants of Central and South America had been building elaborate civilizations around the rivers, lakes, and swamps of the Americas as early as 2600 BCE. As in virtually all newly flourishing societies, this process had begun when agricultural methods had allowed the storage of surplus foods, so people could look beyond their day-to-day survival and build monuments, cities, and roads, and develop arts and sciences.

MAIZE: THE VERSATILE CROP

Maize is a crop that is indigenous to the Americas, and wherever it flourished, the local civilization flourished, too. It became such an important crop that the Olmec people, their successors the Maya, and the Aztec cultures developed religions and calendars to better chart the growing seasons of maize. As well as developing better strains of maize, these early farmers also soaked the kernels in ashes or lime, making the maize softer and easier to eat. Although the Aztecs did not know it, this also increased the grain's protein availability and released niacin, an important nutrient.

Maize was used as a thickening agent, ground to make breads, and formed an important component of beverages; it was even used with chocolate. The crop worked its way north into the southwestern United States and by CE 200 had finally reached the Eastern Woodlands region (east of the Mississippi River to the Atlantic Ocean). It was eventually shipped back to Europe where it spread throughout Spain, Portugal, Italy, and the Near East. Traders brought it to ports in India and China, and the Manila galleons—Spanish trading ships—brought it to the Philippines.

ABOVE: *Europeans in the New World noted the Native Americans' agricultural methods and food storage practices. In this 1591 illustration, fruit is transported by canoe.*

Genetic Engineering—Ancient Style

One of the first examples of genetic engineering—and some say the most important one—occurred when the farmers of the Americas started to develop better strains of a grass called teosinte (pictured) approximately 7,000 years ago. Teosinte is very possibly the forerunner of modern maize, or corn, and the cultivars developed were adapted to different climates with the aim of increasing yield per cob. This allowed the rise of the Olmec people, who lived in the area around modern-day Veracruz, in southeast Mexico. Scientists may disagree about the relationship between modern maize and teosinte, but there is no doubt its cultivation played a central role in the growth of early American civilizations.

THE POPULAR POTATO

The humble potato has been cultivated for approximately 5,000 years, and has definitely had an impact on the diets of people the world over. Pizarro first came upon them during his conquest of the Incas, but they were slow to catch on in Europe (see feature box, below). The Spanish used them for long sea voyages because they were inexpensive, nutritious, and held up rather well. Potatoes were being eaten in the Netherlands in the sixteenth century, and they started to gain popularity in the rest of Europe after the Thirty Years' War in the mid-seventeenth century, as people discovered them to be easier to farm and a more durable crop than the grain crops popular at the time.

Potatoes were often looked down upon as food for the poor, but they found acceptance in the American colonies, and became a staple food source in Ireland—so much so that when the Irish potato crops failed in the 1840s and 1850s, the result was the Potato Famine, which reduced the Irish population by as much as 25 percent. They were first brought to India and China by European traders, and now figure prominently in virtually all the cuisines of the world.

ABOVE: *Potatoes have come a long way in global popularity since being initially shunned in Europe. Their importance as a food crop was highlighted when the United Nations declared 2008 as the International Year of the Potato (IYP).*

MODERN MUST-HAVES: CHOCOLATE AND SUGAR

When Hernán Cortés first reached Mexico in 1519, it is said that the local inhabitants brought him some "odd-looking almonds," which they seemed to use for currency. It took some time for him to understand that these odd beans, called *xocolatl*, were cocoa beans and had been cultivated by the local inhabitants for millennia. They used them to make a stimulating bitter beverage that could be mixed with chilies, spices, maize, and even honey. The Aztec emperor Moctezuma is said to have quaffed gallons of it daily in his palace. The drink became popular with the Spanish, who sent it back to the court of Castille where it was a jealously guarded secret for many years. Like so many secrets, however, it became public, and sweetened and flavored with vanilla it created a real sensation and became chocolate, the drink of those "in the know." It was also partly responsible for the desire for sugar; this, in turn, spurred on the trade in human cargo.

ABOVE: *Cocoa trees were first cultivated in what is now Mexico, and the beans were even used as currency in some Native American societies.*

Sugar cane, believed to have originated in the South Pacific, was introduced to the Caribbean by European conquerors, who recognized that the Caribbean climate was

The Potato Catches On

The Inca domesticated the potato approximately 5,000–6,000 years ago. The "spud" was so important to these inhabitants of the Andes that they cultivated over 3,000 varieties to adapt to different altitudes and soils. These became so crucial to the survival of the empire that they were freeze-dried (called *chu no*) to preserve them for years. The Spanish conquistadores brought them back to Europe, but it took time for them to catch on. In France they had long been held to be food for the poor until they were popularized by the eminent scientist Antoine-Auguste Parmentier (1737–1813), who it is said gave potato flowers to Louis XVI and Marie Antoinette. It is also claimed that he used to keep his experimental potato fields outside Paris heavily guarded in an effort to entice interest by the local folk.

suitable for the crop. At that time, sugar was in huge demand in Europe for sweetening the new beverages of chocolate, coffee, and tea. Sugar is difficult to process and a huge labor force was needed. Portuguese and Spanish traders saw the opportunity to bring goods from the New World and exchange them for African slaves to work the sugar plantations. One result of the abhorrent slave trade was that many crops discovered in the west were brought to Africa and from there spread across the world.

Cassava and sweet potatoes, for example, were cheap and durable crops, and flourished on both sides of the Atlantic. From Africa they made their way to Malaysia and the South Pacific, where they are still popular today. Peanuts were another nutritious and drought-tolerant New World food that caught on in Africa, and maize, pineapples, pumpkins, beans, and chilies were also adopted.

> *"I observed a number of jars, above fifty, brought in, filled with foaming chocolate of which he took some."*
>
> Bernal Díaz del Castillo (c. 1492–1584), describing a meal of the emperor Moctezuma (1519)

CHILI AND TOMATO: UNLIKELY COUSINS

ABOVE: *Red Habanero chili peppers are very hot members of the* Capsicum *genus. Chilies have been domesticated for at least 5,500 years.*

The chili pepper was domesticated in Mexico, and it certainly caught the Spanish by surprise. The fiery fruit caught on in Portugal and Spain, and was highly prized in India when the Portuguese took it to trade. Parts of China and Southeast Asia also took a liking to the crop. Chilies were first domesticated along with tomatoes, which were initially viewed with some suspicion and feared to be poisonous. But the plant grew easily in the warm Mediterranean climate, and began to be cultivated there from the mid-sixteenth century. Interestingly, tomatoes did not become a driving force in the cuisine of southern Italy until almost 200 years later.

BELOW: *In this color plate from* Ten Views in the Island of Antigua *(1823) by William Clark, slaves plant sugar cane on an Antiguan plantation.*

Oceania, India, and Southeast Asia

No study of world cuisine is complete without a look at these most influential regions of the world. Without the food plants indigenous to the South Pacific region, we would not now have coconut, sugar cane, bananas, and many other tropical delights playing such a central role in the cuisine we know. India, with its fusion of the best things the world had to eat, found itself perfectly positioned between the creative gastronomes of the Middle East and the exotic offerings of the Spice Islands in Southeast Asia. As in other parts of the world, the combination of food, technology, and curiosity resulted in something new to add to the pantheon of world cuisine.

ABOVE: *Coconuts are a valuable food in the Pacific Islands, containing important vitamins, minerals, and energy. Here a coconut-picker scales a tree in American Samoa.*

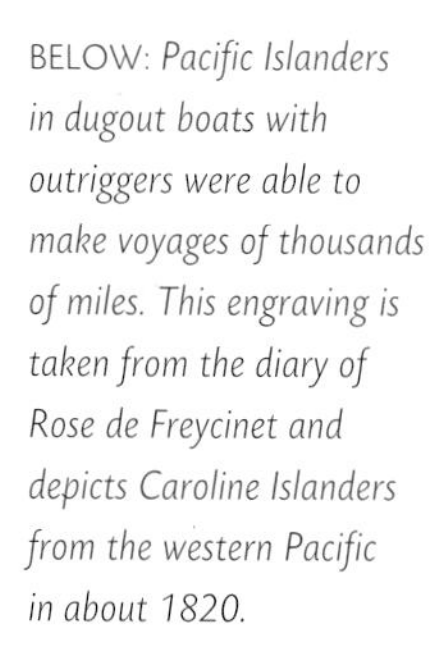

BELOW: *Pacific Islanders in dugout boats with outriggers were able to make voyages of thousands of miles. This engraving is taken from the diary of Rose de Freycinet and depicts Caroline Islanders from the western Pacific in about 1820.*

THE SOUTH PACIFIC'S INTREPID TRAVELERS

The archaeological record shows that the first human settlers made their way into Australia and New Guinea some 50,000 years ago. The Polynesian islands of the South Pacific are thought to have been first populated by settlers from southern China or Southeast Asia via Indonesia, some 5,000–5,500 years ago, beginning in the Solomon Islands. These intrepid explorers did not remain complacent for long; over time, it appears that they ventured farther and farther out into the Pacific, settling more and more of the uninhabited islands they encountered in the region.

Early Islander explorers and their small crews journeyed for thousands of miles, accompanied by durable foodstuffs to sustain them on their voyages into the unknown. It is probable that as they touched different islands they introduced these food items, which then thrived throughout the region and were a boon for later explorers and settlers. To this day, the seafood-based diet of Pacific Islanders is supplemented by breadfruit, taro, yams, plantains, bananas, sugar cane, and nutritious coconuts—the very same edible plants they disseminated across the islands thousands of years ago.

Exploration was not always from within; over the following centuries, many contacts were made with far-flung cultures.

Several European nations made forays into the region. Magellan, the famous Portuguese navigator, visited the area as he attempted to complete a circumnavigation of the globe. The British Royal Navy opened up the region with James Cook's expeditions into this section of the world as well. His early landings on the Australian continent led to the first British settlement at Sydney in 1788.

The famous episode of the mutiny on HMS *Bounty* was a direct consequence of the "bounties" of these tropical islands. William Bligh, a Royal Navy lieutenant, was tasked with collecting and shipping approximately 1,500 breadfruit plants from Tahiti to Jamaica. This was in an effort to supply a cheap, nutritious food source to one of England's most important sugar-growing colonies. A mutiny led by Fletcher Christian set Bligh adrift at sea in a ship's launch, which he piloted over 3,000 miles (4,800 km) of open ocean to safety (the Polynesians would have been proud). Bligh led a second Tahitian expedition in 1791, and eventually delivered the promised breadfruit plants to the West Indies. Initially, however, the slaves refused to eat the fruit, so the mission was useless. It took a while to catch on, but breadfruit is now one of the staple foods in the West Indian diet.

ABOVE: *This nutmeg seed and mace (the red outer covering) was grown on Bandaneira Island in Indonesia. Some say it was this island Christopher Columbus hoped to find when he set sail in 1492, only to land in America.*

INDIA: THE CULINARY CROSSROADS

India was perfectly situated as a crossroads for all cultures and foods of the ancient world. The Harappan culture of the Indus River Valley was trading with the Persians in the Middle East as early as 3500 BCE. These Indus Valley peoples had developed a system of standard weights and measures, as well as marking tabs (sometimes in the shape of animals) to differentiate various cargoes and their origins. This opened up an exchange of fruits, nuts, grains, vegetables, and spices across the ancient Middle East. The fabled Spice Islands to the east (today known as the Maluku or Molucca Islands) were, until only a few centuries ago, the only known source of highly valued spices such as cloves, nutmeg, and mace. Such rare commodities naturally fetched a very high price for those who traded them. By the Middle Ages, this hunger for spices created a worldwide trade that made some countries rich, crushed others, and forced some to take to the sea.

Before Ancient Greece or Rome were powers to be reckoned with, Indian ships laden with spices, textiles, and perfumes delivered their precious cargo to the Middle East, where it found its way to the agricultural heartlands of Mesopotamia and the burgeoning cities of Egypt. Spices including cinnamon, turmeric, cardamom, and black

LEFT: *Breadfruit, on the right-hand side of this 1855 illustration by Elizabeth Twining, is an important food in the Pacific, with a high starch content and useful amounts of protein. It can be eaten like a vegetable, or used in desserts when ripe.*

pepper were native to the Indian region, and were in great demand from traders for thousands of years.

In the first century BCE, the markets of Alexandria in Egypt were some of the busiest in the world, and were well stocked with Indian spices on their way to markets across Europe. Over time, trade was opened with the Romans, who came to India looking for exotic foods and brought with them their grapes and the skills of viticulture. Early trade with the Greeks probably introduced fenugreek, with its distinctly sweet musky smell, to the Indian spice rack; today, it is considered a must in many garam masalas. Saffron was funneled from Arab lands through Tibet and China to the Indian subcontinent. It would later become a mainstay in Persian cuisine, and make its way to the Iberian Peninsula in modern-day Spain and Portugal.

As the years wore on and trade leap-frogged around the world, food was in the wake of virtually every ship. The exotic spices of India and Southeast Asia found themselves featured in cuisines of far-flung areas of the globe. Cardamom found its way northward to Scandinavia, where it became a must in stews and breads. Nutmeg, cinnamon, and ginger made serious inroads into the food of the Caribbean, where it is perfectly right and proper to sample roti and curried goat from one stall or egg rolls laced with soy sauce from a neighboring stall, and to finish off with a refreshing beverage infused with tamarind.

Trade with Europeans had its own impact on the cuisine of India, too; the fiery vindaloo, a meat stew flavored with vinegar and garlic, is said to have its roots deep in Portuguese cooking. It was these same trade routes that brought chilies, pineapples, and tomatoes to Europe from the Americas.

ABOVE: *Native to the Mediterranean, fenugreek is now cultivated in Asia as well as Europe. The seeds are roasted and ground to reduce their bitterness.*

BELOW: *The emperor of India, Ashoka, embraced Buddhism and, while not strictly vegetarian, he did enforce a law banning the slaughter of certain animals, and encouraging "abstention from killing living beings."*

INVASIONS AND EXCHANGES

Due to its centralized location and wealth of natural resources, India has been the target of many external invasions. The early Aryan peoples were pastoral shepherds with a sideline in military prowess who first pushed down from the northwest through the Hindu Kush around 2000 BCE. They were meat-eaters by nature, but they may have been one of the first to declare that cows should not be slaughtered and eaten. This was not in deference to their sacred status, but rather because they were worth more alive, as they could provide milk to make ghee—cooking butter—and yogurt. According to legend, in the third century BCE the Indian emperor Ashoka embraced Buddhism after witnessing one very violent battle, and set out to disseminate its teachings throughout India. With its emphasis on a vegetable-based rather than a meat-oriented diet, the spread of Buddhism had a long-lasting effect on the landscape of Indian food and culture.

The list of other invaders in the region is long and distinguished. In the second half of the fourth century BCE, the Macedonian leader Alexander the Great made forays into the northwest region, in what is now called the Punjab, but was only partly successful, as his troops mutinied and he was forced to pull back; he died shortly afterward. The Persian Sassanid Empire (CE 224–651) was more successful pushing into the region and spreading its gastronomic love for complex dishes across the area. The Persian love of nuts and sugar candy has lasted to this day in India. The spread of Islam led to the fall of the Persians and the taking of Delhi, bringing further complexity to the food. When Babur, the Muslim Mughal emperor, pushed home his attack on Hindustan in 1526, he maligned the local food, but that was to change with the use of local produce, and a Middle Eastern love of excellent cooking.

INDIA AND EUROPE: A TWO-WAY STREET

European usurpers from the sixteenth century onward also began to exert a huge influence on the agriculture and agronomy of India. The Dutch East India Company did much to organize the growth of rice in the Ganges Valley, and the British began the process of modernizing their tea plantations. Many of the dishes that are so common to "curry houses" in the UK today are distant cousins of the original dishes found in India. Mulligatawny soup, a popular Anglo-Indian dish, is said to be derived from the Tamil word for "pepper water." It was once a wholly vegetarian dish, in keeping with the Indian diet, but today's Anglicized version usually contains chicken or other meat.

ABOVE: *The Persian walnut is grown in many regions of the world, but is especially important in India, where it is grown in the Himalayan regions, and its nuts are exported around the world.*

Curry or *Kari*?

The word "curry" has been the subject of endless confusion over the centuries. In the West, it refers to a plethora of Indian and Southeast Asian stews that have become popular in the last few decades. In a grocery store you can purchase curry powder, a blend of spices that a curry purist would view with a jaundiced eye. The word "curry" is said to derive from the Tamil word *kari*, which refers to a sauce or gravy. The Tamils are inhabitants of southern India and Sri Lanka, and have one of the oldest vegetable-based diets in the world. They have become experts in the use of rice and legumes, as well as the blending of flavors such as tamarind, curry leaves, cloves, nutmeg, cardamom, and coconut to tempt the taste buds. Outsiders, such as Western colonialists, tended to simplify these complex cuisines and in so doing created the curry confusion.

Europe and the Mediterranean

In ancient times, continental Europe and the Mediterranean were home to some of the world's most renowned explorers. The maritime cultures of Crete and Mycenae made contact with all the major cultures known at the time, and acted as conduits for foods from the east and south. Celtic tribes from Europe ventured down the Atlantic coast to Spain and over to Ireland, spreading their culture and influence. Later, the Viking traders of Scandinavia left their mark far and wide, from Russia to the coast of North America and wherever their ships would take them. With a food culture based on agriculture, but a limited amount of land available in their home regions, they sought other fertile areas to raid or conquer for their expanding populations.

> *"Triptolemus, one of the principal figures in Greek religion, is said to be the inventor of the plow and of agriculture, and therefore the real father of what we call civilization."*
>
> M. F. K. Fisher (1908–1992), in her 1949 translation of *The Physiology of Taste*

Much later, in the fifteenth and sixteenth centuries, came the "Age of Exploration," when some of the best known voyages and discoveries took place. Columbus set off for the Spice Islands of the Indies and found a whole new world. Da Gama sought a way to India and China by sea, aiming to break the lucrative monopoly of the Arab spice traders. Magellan set out to circumnavigate the globe. The world seemed to be getting smaller at a shocking rate, and its foods would soon reflect this change. Conquest, exploration, and trade led to a diversification of European food on a scale that would not have been possible had Europeans relied only on indigenous agricultural products.

TRADING DIETS: FROM MINOANS TO ROMANS

Trade began in the Mediterranean region about 2000 BCE with the Minoans of Crete, and was followed by the Mycenaeans who migrated from the Balkans. They were both

BELOW: *Viking ships on trading expeditions bought silver, silk, spices, wine, jewelry, glass, and pottery. In return, they sold items such as honey, tin, wheat, wool, wood, iron, fur, fish, leather, and walrus ivory.*

FAR LEFT: *In Ancient Rome, ginger was regarded as such a specialty that a duty was levied for its trade through the port of Alexandria in Egypt.*

LEFT: *Roman grapes (like these white grapes from Emilia-Romagna, Italy) and viticulture spread through Europe, but the Greeks had been making wine as early as 1600 BCE.*

excellent shipbuilders and navigators, and traded with Greece and Egypt for commodities such as olives, wine, and grain. Around 1000 BCE the Phoenicians of coastal Lebanon and Syria had picked up the baton, rapidly expanding trade and setting up outposts in North Africa, Spain, Sicily, Persia, and the Indus. They were thought to have traded as far north as England, and established a central trading colony around Carthage on Africa's north coast. It is thought they brought the date palm of the Middle East to Spain and North Africa, where it is today grown widely for its fruit, sap, and wood.

The emergence of the Romans in Europe brought about a change in the trading powers of the region. The Roman Republic defeated Carthage in the Punic Wars in the second century BCE and took over trading dominance, rapidly expanding its interests. Rome's tentacles reached out to India and China to obtain luxury spices such as ginger, cinnamon, and pepper. They also found it necessary to import many basic foodstuffs, such as olives and grain, to feed their ever-growing and hungry populace.

The Romans' need for lumber to build warships and trading vessels deforested Roman provinces such as Britannia (England), which led to a shift to agriculture there. As Romans set up trading outposts in different areas they also brought along some of their most beloved foods, so olives soon flourished in Spain and southern France. Viticulture also followed closely in the tracks of the Romans, with vines being introduced into the Rhine, Mosel, and Rhône River valleys. From Roman times until around the 1400s, even England had a thriving wine industry, which upset the French so much that they enacted laws to prohibit the importation of those excellent wines. Unfortunately the onset of the Little Ice Age dropped the temperatures too low for England to continue with grapes. This is why, in the northern climes of Europe, grain-based beverages such as aquavit, beer, and whiskey are produced.

As with other conquests throughout history, the Roman occupation of Europe was not always welcome, but the Roman centurions stationed across the continent had a tremendous influence on the diets of the local people. Soldiers were solidly, if not elegantly, fed and wherever they went so too did their lentils, beans, cabbages, and mustard seeds. This seemed to be the case especially in northern and central Europe, where it would be difficult now to imagine a meal without mustard and tangy sauerkraut.

BELOW: *Mustard, seen here flowering in a Czech field, was a popular addition to Ancient Greek and Roman food and medicinal preparations, later becoming an integral part of European cuisine.*

RIGHT: *Saffron is a key ingredient in Spanish paella, and the La Mancha saffron crocus, pictured here, is especially prized. The stigmas of the flower are harvested painstakingly by hand.*

NEW PLAYERS IN FOOD TRADE

The decline of the Roman Empire by the fifth century CE left a power vacuum which was initially filled by the expansion of the Muslim Arabs. Their abilities in trade and navigation saw them spread rapidly across North Africa and up into the Balkans, and they soon took control of the Iberian Peninsula. One crucially important outcome of this occupation was an intensive use of hydrology for irrigation, and a newfound interest in gastronomy on the part of those subjugated. Sugar cane, originally sourced from the South Pacific, together with the technology of turning it into granulated sugar, were both introduced to the Iberian region, and fueled a love of sweets that has gone on undiminished. Dates, lemons, limes, spinach, bitter oranges, and aubergines (eggplants) also made inroads into the local pantry, especially in Sicily and Spain. Rice became important in the cuisines of the region, along with saffron. Today, we would be hard-pressed to think of Valencia without paella, or Milan without its risotto. The food of the Balkans was so heavily influenced from the East that we can still see delicate phyllo pastries of honey and nuts, and the addition of dried fruits to savory dishes to give them a satisfying complexity.

BELOW: *First arriving in Europe in the early part of the seventeenth century, coffee was embraced by the Europeans, who soon began to open coffee houses in which politics, philosophy, and current issues of the day were discussed.*

After the fall of the Roman Empire, other groups also set out to make new commercial and in many cases military gains in the Mediterranean and Europe, with varying levels of success. The fearsome Vikings became shrewd traders, ferrying wine, spices, and grain up the Rhine. In their successful vanquishing and colonization of parts of north England and Normandy, they introduced pears, apples, cherries, and plums. In time they were supplanted by the unified Hanseatic League, which took up where the Vikings left off in supplying goods to the northern European regions. Threats came from the east during the thirteenth century, from the Mongols in Asia, and later in the sixteenth century from the Ottoman Turks, who actually reached the outskirts of Vienna, where it is reported they left mountains of coffee behind. This began a whole new intellectual movement: the coffee house, which in turn promoted a new search for sugar as demand for the popular beverage continued to escalate.

EUROPE EXTENDS ITS INFLUENCE

It would be incorrect to view Europe as only being on the receiving end of conquest, however. In the centuries after the fall of the Roman Empire, Europe went through a period of fierce infighting followed by a period of unification under the influence of the Christian Church. Pope Urban II called for a holy war to wrest sacred biblical lands from the grip of the Muslims. He promised glory and forgiveness to knights for their past sins, and the promise worked; the "crusades" (*c.* 1100–1300) took more than 50,000 people to the Holy Land.

Some of the culinary fallout from all this activity was an increased trade in foods from the Middle East—luxuries such as spices, dried fruits, and, later, sugar. Venice became an influential commercial power, and enjoyed a lucrative trade supplying equipment and transport to the crusaders. Other Italian cities such as Genoa and Pisa also became important trading hubs. At the same time, recipes began to circulate in manuscript form.

NEW WORLD EATING

While the crusades failed to remove the Muslims from Palestine, the Reconquista succeeded in pushing them out of Europe. By 1492 the era of Muslim dominance of the Iberian Peninsula had ended. The nations of Europe had been seeking a better way to get the spices they loved from the East without going via the middle men in the Byzantine Empire. As discussed earlier, this is when Christopher Columbus sailed west to find a sea route to the Spice Islands, and instead realized the significance of the Americas for wealth, land, and food. The Spanish and other Europeans swiftly conquered the indigenous peoples of the Caribbean, and of Central and South America, and before long the flavors of cuisines across the world were changed forever.

ABOVE: *Cardamom originated in India and was embraced by northern Europe as a flavouring for sweet dishes. Europeans also mixed it with coffee for a spiced beverage.*

Chili peppers made their fiery way over to Hungary via the Turks, and to India and China by way of the Spanish and Portuguese traders. Tomatoes were paired with bread and olive oil in Spain to make gazpacho, and with cardamom in India to make chutneys. Potatoes quickly became the staple food of Ireland. Maize arrived in China in the holds of European trading ships and was instrumental in feeding a hungry populace, leading to a new population explosion.

Our Daily Bread

During the Middle Ages, bread had become such a part of daily life in Europe and the Mediterranean that many laws were passed to prohibit dishonest bakers from selling underweight loaves, or bread that had been spoiled in any way. One English baker had an apprentice conceal himself in front of the oven, and when the baker would prepare to cook the loaves the lad pinched bits of dough from the bottom. These bits of dough were combined so the baker could bake and sell extra loaves at no cost. Offending bakers were punished severely. The baker could have the loaf tied around his neck, and then be pilloried (put in stocks) in the street. Grain that was contaminated with ergot could cause dry gangrene, and even hallucinations. Some say that the myth of the werewolf is tied to some of the bizarre behavior and horrible visions this form of poisoning brought to unwary sufferers.

RIGHT: *The ergot fungus, seen here on a rye plant, decimates grain crops by replacing grains of the host plant with black masses of branching filaments. Currently, there are no commercial cultivars of wheat, rye, or barley that are resistant to ergot.*

China and East Asia

BELOW: *Bean crops flourish at this bend in the Yellow River near Shapotou, China. Across the river are dry dunes, forming the edge of the Gobi Desert.*

In China, as in other parts of the world, the location of the first permanent settlements (c. 5000 BCE) was determined by the suitability of the land for cultivating food crops. Millet had been growing wild for several millennia along the fertile banks of the Yellow River, so it is believed the first permanent villages emerged here. By 2200 BCE the first kingdom was formed under the Shang Dynasty. While other civilizations elsewhere in the world were starting to vie for supremacy, Chinese communities and their agricultural products were developing in relative obscurity, protected by rivers, oceans and mountain ranges. This was not to be the case forever, and the global food trade was soon to benefit from East Asian influence and products.

BELOW: *Soybeans are native to China, but the majority of their production occurs in the Americas. The USA has over 60 million acres (24 million hectares) of soybean plantations.*

BEYOND CHINA'S BORDERS

It is said that the Han Dynasty sent out explorers as early as 200 BCE to make contact with the kingdoms of the West—the Persians, Indians, and possibly the Romans in Syria. They then turned their attention to the sea and, by mastering the monsoon winds and designing a stern-mounted rudder instead of an oar, were able to sail into the Pacific to such places as Japan, Taiwan, Tahiti, and Fiji, all the while finding edible plants and introducing them to each new area they colonized. Yams, bananas, and breadfruit spread across the Pacific Islands in this way. Chinese traders also ventured far out into the Indian Ocean where it is said they made it as far as the Arabian Peninsula and the east coast of Africa.

Having established contact with the West, China began a lucrative and diverse trade in many commodities. Pepper, nutmeg, cinnamon, cassia, and ginger, as well as apricots, many types of citrus, apples, and even peaches found a ready market. Soybeans, which had been well known in China for millennia, were especially popular with the Buddhists, who very possibly brought them to Japan; in Korea they became a popular snack food and a major source of protein, and yielded oil and the ubiquitous soy sauce. Buckwheat from Japan made it back to China where it was well received in the country's northern grain-growing regions. Rice was one of the most important exports from the southern regions of the Yangtze, where it funneled through the trade ports of the Middle East to Africa and was eventually cultivated in the New World to feed the slaves.

The Chinese traders did not miss an opportunity to gather new things to please the palates of their masters and their fellow citizens. They relished the grapes, cucumbers, pomegranates, coriander, and peas from the Middle East, and the sesame from India.

TAKING TEA TO THE WORLD

Tea became one of China's most sought-after exports. There is some debate concerning the origins of the tea plant, but it is thought to have originated in China. Some believe that the brewing of tea occurred with the accidental mixing of hot water and a wind-blown tea leaf. Whatever its source as a beverage, the Chinese have been enjoying tea for thousands of years. It has been used as a medicinal preparation, and became popular as a stimulant among Buddhists and Muslims, who eschewed the use of alcohol. By land it spread through China to Korea, Turkey, Afghanistan, and Egypt. By sea it spread to Japan and Java, and by the early sixteenth century it was introduced by the Portuguese to Europe. The English took to it with enthusiasm, and touted its health benefits.

> *"If you do not interfere with the busy season in the fields, then there will be more grain than people can eat."*
> Mencius (372–289 BCE), Chinese philosopher

CONQUER OR BE CONQUERED

During its long history, China has been both the conqueror and the conquered. Its influence in eastern Asia—over Taiwan, the northern parts of Vietnam, and the Korean Peninsula—is evident in the cooking technologies and food of these regions. Even in the Philippines, far into the Pacific, China's culinary presence is felt, where stir-frying and *lumpia* (fried pastries similar to spring rolls) are popular. The Chinese have known the bitterness of defeat as well. The Mongols under Genghis Khan invaded China's northern regions in the early thirteenth century, sacking villages and causing havoc until Genghis turned his attention toward enemies to the west. China had not seen the last of the Mongols, however. Kublai Khan, Genghis's grandson, later defeated the Song Dynasty and crowned himself emperor of the Yuan Dynasty.

During this time, the Mongols' connections with the West meant that overland trade routes, such as the famous Silk Road, were kept open and improved between China and western Asia through to Europe. Granaries were built across the Mongols' empire, and many new food products were introduced. A Flemish monk named William of Rubruck traveled to the Mongolian empire in the thirteenth century and described the Mongols' love of mares' milk, from which they produced an alcoholic beverage called *kumiss*.

BELOW: *This nineteenth-century watercolor by an unknown Chinese artist, shows workers watering a tea crop with buckets. The tea plant requires a climate with a high annual rainfall to thrive.*

The Middle East and Africa

The Middle East has long been held to be the cradle of civilization, and, as we have already seen, archaeological evidence shows semipermanent settlements in the region, and some early agriculture by 8000 BCE. The people of the ancient Middle East proved to be excellent mathematicians and astronomers, and ventured forth in all directions to make contact with other cultures to see what foods and goods they had to offer.

The culinary cultures of Africa are possibly the most diverse in the world. The people of this massive continent, especially in the north, had close contact with the early Middle Eastern pioneers, so the food of the north and the Mediterranean coast tends to emulate the food of the Middle East, while the cuisine of the east coast encompasses many of the best offerings of India and the Middle East, the product of a long mercantile history. Today, the food of the west coast is so closely intertwined with the cuisine of the Caribbean and the southeastern United States as to be virtually inseparable.

BELOW: *The date palm (pictured here with ripe dates) has been cultivated in North Africa and the Middle East for thousands of years. One story suggests the tree was introduced into northern India by Alexander the Great, whose soldiers spat date stones around their camp when campaigning there.*

ABOVE: *Watermelons in their wild form are indigenous to Africa and have been an important source of water in dry regions since ancient times.*

LOOKING EAST

Geographically, the Middle East was perfectly situated for outside contact as it sat between the Mediterranean on the west and Asia to the east. It was also crossed by the principal trade routes of the ancient world, including the Silk Road, which extended from Europe to China. From as early as 3000 BCE the Sumerians established great trade centers in modern Iraq that had regular contact with the outside world. The important city of Ur on the Persian Gulf enjoyed a lucrative trade agreement with the Harappan civilization in India. The Indians brought in such items as sesame oil, millet, and rice from China, as well as spices such as cinnamon and pepper. It is likely that onions and leeks were brought into the region in the same way. This exchange was by no means one-way. Middle Eastern traders exported foods such as wheat, barley, grapes, olives, dates, figs, pine nuts, almonds, pistachio nuts, and chickpeas.

Africa has historically been a site of "give and take" food exchange for the world. All manner of traders from the established east and the burgeoning west have passed around

Islam and Culinary Culture

When the Prophet Muhammad founded the religion of Islam in CE 610, it sparked a movement of adherents who set out in the Arabic lands to convert as many new followers as they could. The religion spread to Persia, North Africa, and Spain, and even made inroads into northern India, Russia, and the fringes of China and the Pacific. Along with their teachings, Islamic peoples brought their dietary laws and a love of good food. Cucumbers, aubergine, sugar cane, spinach, and other greens accompanied them, as well as rice, saffron, and new spices. The prohibition of alcohol for Muslims led to an increased demand for alternative beverages such as coffee (pictured). The 200 years of the Crusades also helped to spread Middle Eastern foods through Europe. When European knights returned to their villages, European cuisine was changed forever.

and through the great continent. Traders brought joy to the thirsty inhabitants of the Middle East with luscious watermelons. The okra and melegueta pepper of West Africa were well liked by the Persian gastronomes, and complemented the pilaus and biryanis of India. The coffee of Ethiopia swept the world with a jittery fanfare, and, in turn, spurred on a craving for sugar. This demand also led to the bitterness of the slave trade, and to the sugar plantations of the Caribbean. The transportation of slaves also led to the export of foods such as akee, rice, okra, and kola nuts to the Caribbean. In return, Africans adopted the maize, cassava, yams, pumpkins, peanuts, and pineapples of the New World.

As trade developed, more items began passing through the Middle East on their way to the west and south. Rice became wildly popular in what was to become Persian cuisine, and found its way to the west coast of Africa and the Caribbean to feed a hungry slave population. Sturdy crops such as yams and taro were adopted in Africa and found a ready market with the local people. The continent's cuisine was further diversified with the introduction of apricots, oranges, lemons, bananas, and limes from the east.

Spices, of course, played an enormous role in diversifying food in the Middle East and Africa. Overland trade routes such as the ancient Persian Royal Road and the far-reaching Silk Road ran through the Middle East, and exotic goods were available to the civilizations in the region. As we have seen, the quest for new routes to the Spice Islands led to the full exploration of the Americas, which, in turn, produced even more new agricultural products for the Middle East, Africa, and other areas.

LEFT: *Explorers of the "Age of Exploration" such as Vasco da Gama, pictured, were important conduits for new foods and culinary experiences. Da Gama stopped at many African ports on his journey to India in 1497–1498.*

High-Flying Foods

ABOVE: *Exotic fruits are sold fresh, frozen, and canned in regions far from where they were originally grown. Transporting fresh fruits across the globe by sea, air, rail, or refrigerated truck requires enormous planning to maintain quality and condition.*

When we go grocery shopping, many of us take for granted the sheer variety of fresh foods available to us. The produce section has dozens—if not hundreds—of different species on display, most of them perfectly formed and virtually unblemished. The colorful fruits and vegetables may come from all corners of the globe: grapes from Chile, kiwi fruit from New Zealand, and mangoes from India. This global cornucopia does not end when you leave the produce section, either. The coffees, spices, and chocolates were likely grown in places thousands of miles from where you stand holding your shopping basket. This is an agricultural and culinary utopia that our ancestors only dreamed about.

FOOD GOES GLOBAL

Exotic foods are grown and delivered to us across national and geographical boundaries via amazing feats of logistics, technology, and transportation. Our ability and desire to have this extensive array of food at our fingertips, however, is not without its consequences. Environmental concerns and the exploitation of some workers in the developing world are just two consequences of our globalized food system. How we deal with these issues shapes much of our modern food culture, as we try to find a balance between our desire for variety and its impact on our planet.

Until fairly recently, only the most powerful and wealthy people could afford to have exotic foods and spices brought to them from distant lands. The rest had no choice but to eat only seasonal and local foods that left them at nature's mercy. They had to find ways to conquer the seasons in order to sustain themselves through long winters when plants were dormant, or during years when frost, drought, or other disasters cut into their already precarious food supply. For centuries, the best—and often only—way

The Cool Change

One of the very first cool chains was used in the United States lettuce industry in the first half of the twentieth century. Lettuces were picked and taken to a shed where they were packed in ice and readied for transport. In the 1950s, vacuum-cooling was introduced, revolutionizing the cool chain system. It was more effective than ice and allowed the cooling process to move into the field at the point of harvest. Cool chains finally began to be widely used on a global level in the 1970s and 1980s, giving consumers year-long access to fresh fruits and vegetables that had previously been available only seasonally or priced beyond the reach of most people.

to achieve this was by drying and preserving foods. But with the Industrial Revolution came many technological changes that brought about a new era in food.

GETTING IT FROM A TO B

With little doubt, faster and better means of transportation have had the greatest impact upon the supply, variety, and affordability of the food available to us. Railways, automobiles, ships, and airplanes have the ability to move food across countries and around the globe in a matter of days or even hours. Because such large quantities of food can be transported at any one time, the food becomes much more affordable for consumers.

One of the greatest challenges involved in transporting food around the world is keeping it fresh. This is achieved through the use of "cool chains," which are complex systems for keeping food cool, preventing further ripening and spoilage between harvest and consumption (see feature box). Fresh produce is refrigerated shortly after harvest, transported in refrigerated vehicles or containers, off-loaded into specialised climate-controlled warehouses, delivered to retailers via refrigerated trucks, and stored in cool temperatures at our well-stocked neighborhood markets until they are needed on shelves.

STORAGE INNOVATIONS

The manufacturing sector plays a significant role in ensuring the availability of fresh produce by extending the shelf life of some fruits and vegetables. Storage warehouses keep the temperature and oxygen levels low, and carbon dioxide levels are raised to discourage further ripening. Just before the food is sold, it is exposed to ethylene gas to activate ripening enzymes. Plastic wrapping is extremely common because it reduces the amount of oxygen the produce receives and slows moisture loss. For some bagged produce, such as prewashed lettuces, oxygen is removed, and nitrogen and carbon dioxide added to slow spoilage. Waxing produce, which deters oxygen intake and water loss, is also common. Irradiation is a more controversial manufacturing process. To prolong the product's shelf life, irradiation renders ripening enzymes inactive, but can change the taste of the food.

ADVANCING TECHNOLOGIES

Agricultural technology has also increased the availability of fresh plant foods. The use of greenhouses, along with hydroponics, has enabled producers to grow food year-round in virtually any climate. Hybridization involves breeding two species or cultivars of plants to develop a new cultivar with desired characteristics from each parent plant. This type of technology has led to increased crop yields, and has created cultivars of fresh fruit and vegetables that may taste better or are better able to withstand early harvesting and the stresses of transport and handling.

More controversial are genetically modified organisms (GMOs). GM plants are different from hybrids in that the DNA

LEFT: *Dried fruits—such as these apricots, pears, pineapples, and prunes—have a much longer shelf life than their fresh counterparts, but sulfur dioxide and other preservatives are often added to prevent spoilage.*

BELOW: *Glass or plastic greenhouses trap the sun's warmth, providing ideal conditions for plant growth. Evidence suggests similar structures were used in Ancient Rome for growing foods such as cucumbers.*

of a GMO is altered in some way, whereas hybrids still go through sexual reproduction. GMOs are bred for a variety of purposes, as discussed in the next chapter: higher crop yields; resistance to herbicides, pests, and diseases; and delayed ripening, to name a few. GMO supporters say these new organisms will not only help combat world hunger, but are better for the overall health of the environment, since they can reduce the need for chemical sprays, cut down on water usage, and decrease the area of land under cultivation because crops can become more productive. Others object to GMOs because of the potential for loss of natural species, their possible effects on human health, and the monopolies that may be created by companies that sell GM seeds.

> *"With the introduction of agriculture mankind entered upon a long period of meanness, misery, and madness, from which they are only now being freed by the beneficent operation of the machine."*
>
> Bertrand Russell (1872–1970)

DOWNSIDES TO FOOD GLOBALIZATION

Although having almost any variety of fresh produce available to us year-round is a great convenience, there are some downsides to having our demands for it fulfilled. Some of these drawbacks directly affect the food we eat and our relationship to it. Other problems are less immediately tangible, but still have a tremendous impact upon our world in terms of the cost to people and the environment.

Our ability to move food around the globe so easily, along with the development of storage and ripening technologies, has resulted in a great deal of seasonal confusion. Previous generations knew that leafy greens were best in the spring, peaches and tomatoes in the summer, and apples and pears in autumn. Now that these foods are available to us all year long, many of us have become completely out of sync with the cycles of nature. This seasonal confusion can lead to disappointment in matters of freshness or taste. If you have ever bought a fresh peach in the middle of winter only to be dissatisfied with its aroma, taste, or texture, you have experienced this consequence firsthand.

RIGHT: *A worker at Monsanto, a high-profile agricultural development company, collects pollen from genetically modified maize in a lab. Monsanto has developed controversial cultivars of maize that resist pests and diseases.*

Some produce, such as pears and bananas, withstand early harvesting, lengthy transport, and ripening by gases quite well. Others, such as peaches and citrus fruits, need to stay on the tree until they are ripe, or nearly ripe. Although peaches will soften and their aromas will intensify somewhat after harvest, they will not get any sweeter, which is why you can have trouble finding peaches in winter that taste as good as the ones you can buy in the summer at a local farmers market or grocer.

Even foods that respond well to cold storage or artificial ripening occasionally disappoint our palates, especially those that keep well for long periods of time such as apples and pears. These foods maintain their outward appearance for months under cold storage, but studies have shown their texture can begin to deteriorate, resulting in mealy or soft fruit.

BIOLOGICAL AND HUMAN COSTS

There are also more serious issues that confront the global food system. Ensuring food safety presents us with a tremendous challenge. Chemicals used in growing or preserving produce, or fruits and vegetables harboring pests or food-borne illnesses, all pose potential safety threats. One country's pesticide or preservative may be another country's poison, and although most countries have strict regulations on the food they allow to be imported, it is not possible to regulate exactly how another nation grows the food it exports, nor is it feasible for a country to inspect each individual piece of produce that crosses its borders.

The enormity of the global food industry adds another layer to the safety issue. Often, freshly picked fruits and vegetables from dozens or even hundreds of plantations or farms are sent to a central processing facility to be readied for shipment. If produce from even one of the farms is contaminated with an illness-causing bacteria or microorganism,

ABOVE LEFT: *When destined for local markets, peaches can be picked when nearly ripe, but for export they must be picked earlier, increasing the chance that their sugars will not have developed sufficiently.*

ABOVE RIGHT: *In this refrigerated warehouse in California, USA, boxes of produce are stacked ready for distribution.*

LEFT: *Crop dusting with fertilizers and pesticides is routinely performed on food crops. Many early pesticides were made from naturally occurring but hazardous compounds, such as arsenic and lead.*

Pioneering Tomato Technology

Tomatoes were the world's first commercially available GM food, but the success of this trailblazing fruit was decidedly short-lived. The Flavr Savr™ tomato was first sold in the United States in 1994. Scientists found a way to turn off the tomato's "ripening" gene, which gave the Flavr Savr™ an extended shelf life. In the UK, the same technology was developed to create a more flavorful tomato for tomato puree. Beginning in 1996, this canned puree, clearly labeled as genetically modified, was sold in two UK grocery chains and easily outsold the competing brands made with conventional tomatoes.

Both products lasted just a little more than 2 years in stores. The company that designed the Flavr Savr™ tomato was prevented from being profitable by its industrial inexperience. In the UK, the tomato puree fell out of favor after increasing resistance to GM foods, which have famously been called "Frankenfoods" since the early 1990s.

it is possible for it to contaminate anything that moves through the facility or that is processed on its equipment.

Perhaps the most compelling argument against globalization is the exploitation of workers in underdeveloped and developing nations. Some are bound to contracts that ultimately benefit a produce company. Workers can toil endlessly on a crop, but may be paid nothing if weather or plague destroys it, if labor unrest prevents it from being harvested, or if the company merely feels the crop is unacceptable. Another problem for indigenous workers is that farmers are often required to take native crops, which are frequently dietary staples, out of production to grow fruits and vegetables that are destined for overseas markets. Some argue, however, that this type of food production in the developing world provides a much-needed economic boost for these communities.

BELOW: *Vegetables and fruit are transported via sea freight in refrigerated containers, then may be loaded onto barges for delivery into ports. Sea freight suits hardier types of produce; more delicate foods are exported by air.*

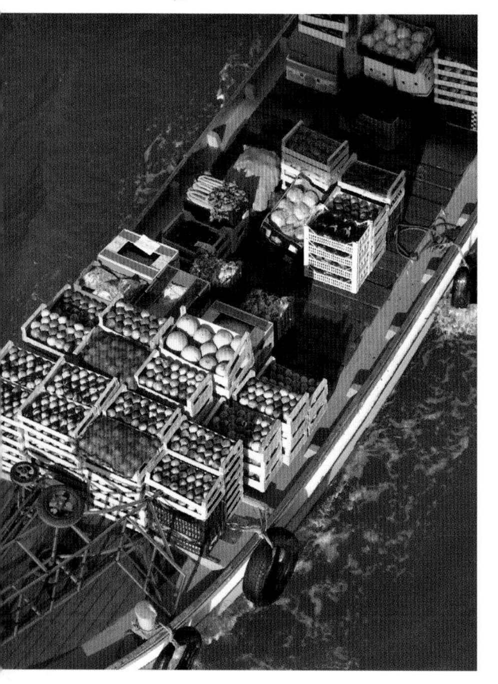

Another pressing issue surrounding the globalization of food, and an increasingly popular topic of discussion, deals with the "carbon footprint" of food—that is, the amount of fossil fuels required to grow and then ship food from the point of harvest to the point of consumption. The concern about "food miles"—how far the food has traveled between its harvest and its eventual point of consumption—has intensified due to concerns that the world's supply of fossil fuels could soon be depleted, and also because of increasing evidence of global warming, which scientists theorize is due in part to the burning of fossil fuels.

THE CONSUMERS WEIGH IN

Given the benefits and the downsides to the globalization of produce, the public has responded to each in positive ways that are not completely at odds with one another. On the one hand, people like having a variety of fresh fruits and vegetables to choose from at any time of year. In fact, since the rise of global cool chains in the 1970s, many more products have been made available, and fresh fruit and vegetable consumption in countries such as the United States, Australia, and the United Kingdom has increased substantially. Consumer demand has even turned some fruits and vegetables that were at one time considered rare and exotic into standard daily fare across the world.

But even as the global market for plant foods grew, so too did markets for locally grown produce. The United States was the first to experience exponential growth in farmers markets. In 1970, there were just over 300 farmers markets in the US. Thirty years later that number had grown to more

than 3,000. Farmers markets did not start in the United Kingdom or Australia until the late 1990s, but both countries have seen phenomenal growth since then.

People give many explanations as to why they prefer to shop at farmers markets, and several reasons seem to be a response to the downsides of globalization. Some shop at farmers markets because they prefer doing business directly with the person who grew the food—a near impossibility in today's global marketplace. Other reasons people give for shopping at farmers markets are to keep money in the local economy, to access food that is fresher and has a less significant carbon footprint (due to the reduced distance it has traveled to the point of purchase), and to support the awareness of seasonal eating that farmers markets encourage.

Public concerns about food safety pose perhaps the biggest challenge to the global food system. Reports of contaminated food have made people wary of some foods or of buying any food from certain countries. These food scares have led some consumers to seek out organic or locally grown produce, but these types of options are not always affordable, nor available, to everyone.

MAKING INFORMED CHOICES

Despite the downsides to high-flying food, there is little chance that our global food system will go the way of the dinosaurs, namely because we have come to depend on it for the delightful range of foods it provides, not to mention the convenience. But perhaps as we enjoy the benefits of this system, if we keep in mind the drawbacks they will inspire us to make more thoughtful food choices every day.

For instance, there are certain products, such as coffee, tea, sugar, and some spices, for which most of us depend upon global trade. Because these foods often come from developing nations, some consumers who have concerns about the exploitation of workers and the environment in these countries seek out "fair trade" products. Fair trade means the producers have received a fair price for their product, working conditions are safe and reasonable, and production methods are environmentally responsible, among other things. These kinds of choices take advantage of the global food economy, but at the same time aim to minimize its human cost. It is these decisions that, in the long term, could make the future of our global food network better for everyone involved.

LEFT: *Farmers markets, such as this one in Pula, Croatia, are an important outlet for local growers to sell their produce direct to consumers. Fresh food markets are particularly important in parts of the developing world where large supermarket chains do not exist.*

BELOW: *These ripe coffee beans are harvested by coffee farmer Patrick Kajjura and one of his sons, in the Kamuli region of Uganda. The beans are then dried and sold to a coffee company in a fair trade arrangement.*

Green Future

ABOVE: *Barley plants, seen here in their early germinating stage, were first domesticated in the Middle East at least 9,000 years ago. The crop has been continually bred and improved as a food staple ever since.*

In the history of life on planet earth, the green plants came first—and everything else followed. All life on earth is underpinned by the marvel of the newly germinating plant that "plugs in" to the sun and accesses solar energy. By taking in sunlight, absorbing carbon dioxide from the atmosphere, and taking up water (which is split into hydrogen and oxygen) and necessary inorganic nutrients from the soil, green plants produce carbohydrates, a core element in human food. One group, the legumes, in a bilateral deal with some bacteria, evolved to access the vast nitrogen store of the atmosphere and enhanced the production of protein, another vital component for human nutrition. Our very lives depend on these natural processes, so it is no wonder that for thousands of years humans have sought ways to improve them to produce their food.

> *"If people let government decide what foods they eat and what medicines they take, their bodies will soon be in as sorry a state as are the souls of those who live under tyranny."*
>
> Thomas Jefferson (1743–1826)

THE EVOLUTION OF FOOD PLANTS

Over time—hundreds of millions of years—plants have evolved a huge variety of forms, structures, and storage organs such as seeds, bulbs, and tubers. These storage structures are often essential for the plants' long-term survival. We humans, very late but clever arrivals, have used these as our food base. First, we selected them from nature; then we began deliberately improving their size, shape, and taste. This process of selection and improvement was launched as soon as humankind began experimenting with agriculture, and accelerated rapidly as societies turned from hunting and gathering toward more settled farming communities.

Inevitably, the plants we grew for food also became attractive as a food source to

RIGHT: *This field of swedes in Somerset, UK, is covered with netting to protect it from the cabbage root fly. The larvae of the fly eat through the swedes and destroy the crop.*

animals, birds, insects, and microorganisms. When we did not wish to share our crops with them, we devised methods of crop protection. Physical separation, such as netting, was used to keep out undesirable organisms; chemical products were applied to kill the creature or organism; and now genetically modified (GM) or transgenic plants are developed to be physically or chemically unattractive to pests. Over time, as the size of our communities grew, we also invented methods of increasing crop yield to feed more and more people. All these choices and innovations are central to our "green future"—the future of the plants we use to feed our population.

HARNESSING PLANT GROWTH

Because living plants are the key to the earth's energy supply, it is no surprise that the long processes of evolution have led to a vast array of plants on earth that have adapted to most habitats, including huge areas apparently inhospitable because of low moisture and infertile soils. However, even when the enormous range is studied, the basic metabolism of plants remains relatively constant. A plant's growth is tightly linked to its nutrient uptake from the soil, and the nutrient content in its organs: the more growth, the more nutrients needed. So, as our food is taken from fields in crop products, nutrients must be fully replaced in the soil to sustain the crop's yield over the longer term. These processes are necessary for any plant to grow, so even if technology allows us to speed up the process, there are unlikely to be new ways of dramatically increasing crop yields.

Nevertheless, the managers of a "green future" for the planet now have the task to build on these basic plant processes as they work to feed the now 6.5 billion people on the planet—one-quarter of whom are inadequately nourished—and plan for 8.5 billion people by 2050. While some of our food is produced from the sea, and some through hydroponic methods, most of it is grown in soil. It therefore requires land, so food production faces increasing competition for land—with urban and industrial use, national parks and wilderness, recreational activities, farm animals, and biofuels, to name a few. To feed the human population, then, our efficiency must be high and our crop yields maximized. Greenhouse gas production and its link to climate change is also a key issue, so gaining maximum crop yield per unit of fuel consumption is essential for the future survival of our planet and its people.

IMPROVING PLANT PROPERTIES AND PERFORMANCE

Our basic energy needs are supplied by a few main crops, called staples: we eat their grains, fruits, and roots. Many other fruits,

ABOVE: *Hydroponic methods of growing plants do not require soil. Rather, the plants are grown using a solution of water and nutrients, or in a material such as gravel, using light from the sun or a lamp.*

LEFT: *The range of habitats edible plants have adapted to is enormous. The pinyon pine, for example, survives in the rocky deserts of North America. Its nuts provided a crucial source of nutrition for Native American peoples.*

ABOVE: *The seed bank at the International Maize and Wheat Improvement Centre (CIMMYT) in Mexico City collects seeds from every native strain of maize in Mexico.*

vegetables, nuts, and herbs make our diets more interesting and add protein, minerals, and vitamins. These foods need to be safe and high quality, with only low residues of harmful chemicals, and no toxins or pathogens. To meet all these requirements for a safe, nutritious, stable, and adequate food supply, humans must continually work to improve the performance of plants to adapt them to different conditions and climates.

While it is true that the world depends on a relatively small number of staple crops, there is a huge range of known genetic material in each crop. Take the potato, for example. Although only a few cultivars are available to worldwide consumers, many local wild types still exist. The olive, one of the earliest domesticated food plants, has 4,000 or so recognized cultivars, even though breeding and production systems only concentrate on about 10 of them to meet the demand for nutritious and healthy oil. To protect the gene pool of all these cultivars and seeds, living plant collections are systematically and securely stored around the world. Agricultural scientists then use them to research the lesser-known cultivars and adapt commercial cultivars—via breeding or genetic technologies—to withstand water shortage, extreme temperature, salinity, pests, diseases, and other environmental challenges.

WHAT ARE GENETICALLY MODIFIED ORGANISMS (GMOs)?

Genetic modification (GM), or manipulation, is the process of taking a gene from one organism or species, and transferring it into another organism or species, thereby changing or enhancing the properties of the plant. New plants produced in this way are known as genetically modified organisms (GMOs) or transgenic organisms. Previously, new genes could only be introduced via cross-breeding, but now they can be isolated individually and transferred, even across species barriers, to new plants in the laboratory. Examples under development include cultivars with improved nutritive qualities, and with resistances to herbicides, insects, and harsh environmental

Fuel or Food?

What impact would the use of crops for biofuels have on food production and environmental conservation? A good human diet requires crop production equivalent to 1,100 lb (500 kg) of grain per year. Almost 53 gal (200 L) of bioethanol can be produced from the same amount of grain. If doubling food production by 2050 is already a significant challenge, doubling it again to contribute even a small proportion of fuel for transportation appears impossible. Nonfood crops compete with food crops for land, water, and fertilizers. Scientists are looking into other alternatives. Crop stubble, for example, returns essential organic matter to soil for the soil's continuing productivity; how much could be removed to produce the biofuel cellulosic ethanol remains unresolved. An alternative is making fuel from true organic (e.g. urban) wastes.

LEFT: *Drought is a persistent problem for farmers all over the world; this farm in Nebraska, USA, was one of many in the area affected by drought in 2002. Crops such as rice and corn are currently being genetically engineered to withstand drought conditions.*

conditions. Most breeding programs routinely use molecular (genetic) markers linked to favorable traits for selecting progeny. There is enormous potential for these types of transgenic crops to address the dire food shortages in the developing world. Drought-resistant cultivars of common staple foods, for example, could increase food production in countries where traditional crops regularly fail due to water shortage.

These new techniques that penetrate to the basic building blocks of life are raising concerns, and many people caution or argue against the use of GMOs in crop production. Biologically, there is the unknown potential for disruption to flora and fauna if the genes escape into the natural environment, and also the possibility that food produced in this way might be harmful to humans. One further concern is that these new techniques place the future of plant improvement in the hands of few, usually multinational, companies. This may disadvantage smaller companies and older methods of plant improvement, and further limit the diversity within individual crop species. If these large companies sell their transgenic seeds only to those who can afford them, it may leave those most in need without the benefits. Clearly, any introduction of transgenic or GM food plants needs to be carefully implemented, monitored, and evaluated to minimize these risks.

SUSTAINABLE PRODUCTION SYSTEMS

With a quarter of the current population of the world underfed, the challenge to properly and sustainably feed 8.5 billion people by 2050 cannot be underestimated. Food production must be doubled, and food storage must be significantly increased. For scientists, the key is productivity. During the last century, total food production generally kept pace with population growth, through the use of more land and greater yields from existing land. But in most countries land is now limited: clearing forests poses dangers for the environment, and land is needed for recreation, wildlife, biodiversity, and water supplies. Growing crops for biofuels—that is, fuel produced from renewable biological sources, such as grains—is also an emerging issue, because it is another fierce competitor for land and the use of crop products (see feature box, facing page).

Developing a sustainable system of food production therefore requires us to find ways to keep the land productive over time while feeding the greatest number of people. One key issue is the use of chemicals, which may eliminate pests and diseases, but which can potentially contaminate soil.

Chemical crop protectants are still widely used for pest control, but scientists aim to develop crops with resistance to pests and diseases to eliminate this need. Pest and disease resistance is the focus of most breeding effort, although breeding solutions are more successful against disease. Breeding for resistance to insects is more difficult, but insect control is a major concern in many food production systems requiring trapping, deterrence, or the application of insecticides. Some crops are especially attractive to insects. Insect-resistant GMOs, however, are becoming available for food crops such as maize, and are now grown widely, reducing insecticide use. Productivity also requires control of weeds that share land with food crops. Control can be achieved through rotations and tillage—plowing or cultivating the soil—but it can be costly, damaging to soil, and ineffective, so chemicals are also used. Supplying soil with all-important nitrogen can also be problematic. Legume crops are used as much as possible to "fix" it from the atmosphere, but in some cases nitrogen is applied as

BELOW: *Eliminating weeds from fields by tilling soil can reduce herbicide use, but can also contribute to soil erosion. Striking a balance between productive crops and a healthy environment is a key agricultural issue.*

RIGHT: *Delivery trucks transport food for many thousands of miles, which means the price of our fresh food is directly affected by the availability and cost of fuel. To make informed decisions, consumers need to understand all the factors which impact on the chains of food supply.*

applied as synthetic urea fertilizer or liquid nitrogen fertilizer, also captured from the atmosphere, to increase crop yields.

The transportation of food also feeds into the production process. As we have seen, large retailers offer fresh fruits and vegetables all year round, sourcing them from distant places, even overseas. The effect of food transport on the environment can vary greatly according to mode (e.g., a small truck, a long train, or a cargo ship), and raises the contentious question of "food miles," as discussed in the previous section.

To create a process of sustainable food production, therefore, many things have to be balanced, and many judgments made. When food producers fully comprehend the complexities of these systems of food production, they are able to make judgments about "Good Agricultural Practices" (GAP). Implementing a worldwide GAP system will allow food to be labeled so consumers know that it has been produced responsibly, and that the food itself is safe for humans to consume. For example, the GLOBALGAP scheme (formerly known as EUREPGAP), is an international certification system that requires its members—either food producers or retailers—to follow a series of specific guidelines and protocols certifying the use of sustainable processes.

BELOW: *These carrots were grown on an organic farm in the United States. For produce to be certified organic, it must not have been treated with chemicals, contaminated by heavy metals, or subjected to any genetic modification.*

THE ORGANIC REVOLUTION?

In recent decades, the organic, biological, biodynamic, and permaculture movements have argued that a more sustainable way of producing food is via a low-input, organic system of food production. The movements first gained momentum in Western countries, and now have increasing support in the developing world. About 0.5 percent of the world's food is currently produced by these methods. Proponents of organic farming point out the hazards of large-scale agriculture, such as the use of chemicals and water pollution, and argue that only legume plants should be used to add nitrogen to soil, not chemicals; that only tillage should be used for weed control; that only organic substances should be used to supply nutrients to soil; and that diseases and insects should be controlled using biocontrol (that is, non-chemical methods). Supporters of organic food are also opposed to the genetic manipulation of food plants.

Because of the low-input, low-yield nature of organic farming, its critics argue that its methods are not feasible in the light of current population (and predictions), the limits to cropland, and the concern over greenhouse gases that require maximizing yield per unit of land, energy, and water. When the world's population was much smaller, systems that relied on human and animal effort—and, later, fossil fuels—and natural cycles were easily able to feed communities. Yields were small per unit of land, but fertile land was not limited. Critics of the organic approach therefore argue that the expansion of organic methods would prevent food production keeping pace with general population growth and would encroach on land better reserved for nature. Current understanding concludes that a world using strictly organic food production could support only 3–4 billion people.

> *"The act of putting into your mouth what the earth has grown is perhaps your most direct interaction with the earth."*
> Frances Moore Lappé (1944–), *Diet for a Small Planet*

However, proponents of sustainable farming practices—and there are many outside the organic movement—have raised concerns that have resulted in changes to large-scale agricultural practices. Their calls to be more careful with chemicals in food production have been implemented in many areas. Rotating crops to control weeds, pests, and diseases; using legumes to fix atmospheric nitrogen; and composting and recycling nutrients are all common practices in agriculture today. Organic supporters argue that their produce has a superior taste; their opponents point out that this is because the products are harvested at maturity and enter a short, local-supply chain that is not available to all consumers, and further argue that by using genetic technologies, new crop cultivars can also have improved nutritional qualities—and flavor—without the higher price tag. Currently, organic agriculture is a niche activity, but some of the ideas advocated by the organic movement have influenced certification systems like GAP.

ESTABLISHING A GREEN FUTURE

Efficiently integrating all the factors that feed into the food production chain—land, crop choice, production method, processing, transport, and more—to improve our agricultural output is a key challenge for the future. It is a triumph that food quantity, quality, and safety have been maintained and delivered to so many people throughout the world, but there will always be work to be done. With strong research by governments, universities, and other organizations, passed on to farmers by intelligent communicators, and applied on the ground by skilled and understanding operatives, the work of global food production can be continually improved.

A "green future" means extracting better quality food for more people, and using less land and a minimum of energy. It means taking full advantage of our soil and water resources, and growing crops with the widest resistance to pests and diseases that is safely possible. It means considering all the scientific and technological options. It means carefully balancing the increasing demand for biofuels with the increasing demand for food. It means putting in place systems for certification and testing so people know the food they are eating is safe and was produced in an environmentally responsible way. It means all consumers understanding more about production systems and their own interaction with the environment. A "green future" means more choices, intelligently made.

ABOVE: *Agronomists and scientists are employed all over the world researching food crops, plant growth, soils, farming systems, and the environmental impact of agriculture.*

LEFT: *Crop rotation is widely practiced to prevent depletion of nutrients from the soil. Strip cropping–planting different crops in strips across the land's slope–reduces erosion and maximizes the use of rainfall on slopes.*

PART TWO

A Directory of Edible Plants

Fruits

Introduction to Fruits

ABOVE: *Small fruit stalls are a common sight in markets in many countries. This stall in Shanghai, China, offers a tempting array of fruits, including bananas, cantaloupes, oranges, and kiwi fruit.*

The English word "fruit" comes from the Latin verb *frui*, which means "to enjoy" or "take pleasure in." And from what we know of our hunter–gatherer ancestors, humans quickly discovered that sweet fruits and berries were not just enjoyable but safer to eat than plant foods whose bitter taste was a giveaway for the presence of toxins. Our innate sweet tooth seems to have had a very practical purpose. Fruits and berries were a delicious form of food that nourished us, and we and other animals and insects played our part, too, by helping to disperse their seeds. It was a two-way street.

The sugars in fruits and berries have provided energy in human diets for millions of years. Today we know that these foods not only provide glucose, our brain's essential fuel, but they are also important sources of dietary fiber, vitamins, and minerals. Fruits also contain nature's bodyguards, antioxidants such as beta-carotene and anthocyanins, which help to protect the body's cells from damage caused by pollutants and the natural process of aging. The message from health authorities around the world is very clear: Eat more fruits and vegetables. This chapter describes a cornucopia of fruits to help you take that important message to heart in a deliciously enjoyable way.

WHAT IS A FRUIT?

To a botanist, a "true fruit" is simply a mature ovary, a home for the seed or seeds inside. So in their scientific fruit basket, along with what are normally regarded as fruits and berries, you will find nuts and many foods we think of as vegetables—tomatoes, eggplants, capsicum, cucumber, and pumpkin and squash. And that is not a definitive list, by any means.

Cooks have a different attitude. In the kitchen, taste rules, and how a food is served tends to define the difference between fruits and vegetables. Certainly that is the approach the United States Supreme Court took in 1893. It ruled that a vegetable refers to a plant that is generally eaten as part of the main course, while a fruit refers to a plant part generally eaten as an appetizer, dessert, or out of hand.

And that is, generally, the approach we have taken in *Edible* for our fruit bowl. There are some exceptions of course (to avoid repetition). That is why you will find avocado (a stone fruit often used as a vegetable) in the Fruits chapter. Rhubarb is a tricky one as this popular pie "fruit" is no fruit at all, but a stalky cousin of sorrel and very much a leafy vegetable. So despite the decision of the United States Customs Court in Buffalo, New York, in 1947, to classify it as a fruit, we have placed it in the Vegetables chapter.

> *"In an orchard there should be enough to eat, enough to lay up, enough to be stolen, and enough to rot on the ground."*
>
> James Boswell (1740–1795)

WHAT TYPE OF FRUIT IS THAT?

To make it easy to find your favorite fruits, and to discover many that may be new to you, we have broken the Fruits chapter into six sections: Citrus, Stone and Drupe, Pome, Tropical Fruits, Berries, and Vine Fruits.

Citrus fruits, including oranges, lemons, limes, grapefruit, mandarins, pummelo, tangerines, and kumquats, all share a unique fruit structure. They have a tough skin, dotted with aromatic oil glands, that encloses a white pith of varying thickness. Inside that pith are the segments with seeds, which are embedded in large, juicy cells.

Stone and drupe fruits are fleshy fruits that contain a single stone or pit. The botanical term "drupe" may be unfamiliar, but the fruits will not be, as the section includes those seasonal summer favorites such as plums, peaches, nectarines, apricots, and cherries along with avocados and dates.

Pome fruits (named after the French word *pomme* for apple), such as apples, Asian (nashi) pears, and quince, are all members of the rose (Rosaceae) family and have a "core" of several small seeds, which are surrounded by a tough membrane that is encased in an edible layer of flesh.

Tropical fruits include drupes, pomes, and berries, and are from the tropical areas of the world (between the Tropics of Cancer and Capricorn), including highlands and lowlands, wet areas and dry. In this chapter you will discover exotic and familiar fruits and berries, including the very famous bromeliad that Christopher Columbus happened upon in the West Indies in 1493 (the pineapple); the sprawling, climbing cactus *Hylocereus undatus* (dragonfruit or pitaya); and that "false berry," the banana. What do they share? An intolerance of frost.

Berries have their seeds embedded in their juicy flesh, making them easy to eat, as there is no core, pit, or stone. The berry section includes all the familiar berries, such as strawberries, blackberries, raspberries, mulberries, and blueberries, along with figs, guavas, and feijoas.

Vine fruits come from climbing plants that need a supporting structure to grow on, such as a wall, trellis, pergola, or another plant. Some have tendrils with which to attach themselves to the structure; others twine up and around it. Compared with the large number of tree and bush fruit plants, there are relatively few vine fruits. The main ones are kiwi fruit, watermelon, passionfruits, and grapes. Popular for their flowers and fruit, vines have another very practical purpose in the garden, to cover up unsightly fences or outbuildings.

ABOVE: *The Surinam cherry, also known as the pitanga, bears small red fruits with a thin skin. Some trees fruit 2–3 times a year. Surinam cherries are eaten fresh, in desserts, or made into jellies and preserves.*

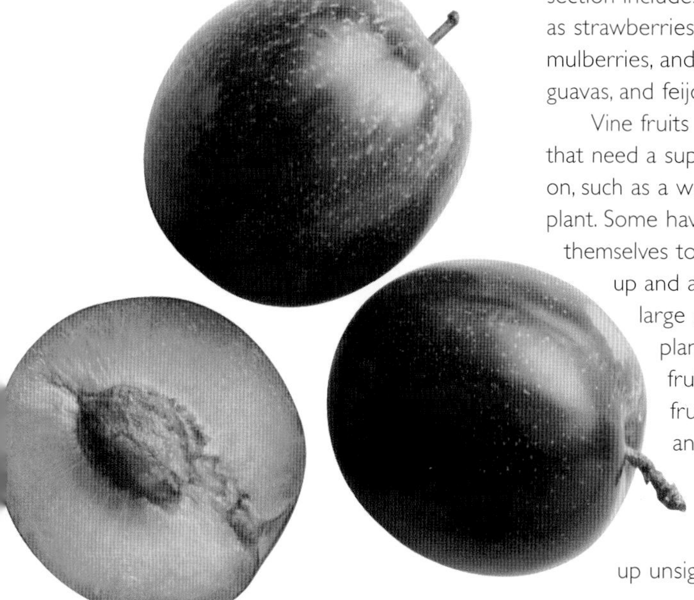

LEFT: *Japanese plums flower earlier than European plums and are more suited to warm, temperate climates. They are large and round, and quite juicy. They are not suitable for being dried for use as prunes.*

Calamondin, Limequat

Citrofortunella microcarpa Calamondin *Citrus x fortunella* Limequat

BELOW: *The calamondin is a small fruit that looks something like a tangerine, but is much more acidic. The tree can grow to 10 ft (3 m) high, and produces fruit all year round.*

The calamondin belongs to the *Fortunellas*, the same subfamily of citrus as the kumquat, whose thin skin adheres strongly to the pulp. The calamondin (*Citrofortunella microcarpa*) has been crossed with the key lime (*Citrus aurantifolia*) to produce the limequat (*Citrus x fortunella*). The limequat is yellow in color and is not generally available commercially.

Historical Origins The calamondin is probably native to Southeast Asia, since it is in this region that citrus originated. It is grown for food throughout the South Pacific, but elsewhere is mainly an ornamental.

Botanical Facts The calamondin belongs to the group of citrus whose members are extremely frost-sensitive. The fruit, which is usually about 1–1½ in (25–35 mm) in diameter, varies from orange to yellow in color. Its pale green oval leaves are about 3 in (8 cm) long. The flesh, though orange, is very sour.

Culinary Fare The strongly aromatic calamondin is an important ingredient in the cuisine of the Philippines. It is often served in a little saucer and used as a dipping sauce for meat and vegetable dishes, or it may be squeezed directly over food and made into drinks. The limequat is strongly flavored, and is an excellent fruit for marmalade.

Australian Finger Lime

Citrus australasica

Also called the Dooja lime or Gympie lime, this species of small cylindrical fruits is native to northern Australia. Like all native species of citrus, it grows within a monsoon belt that, in this case, runs across the northern tip of Australia, taking in Queensland, the Northern Territory, and Western Australia.

Historical Origins The Australian finger lime was a traditional food for Aboriginal Australians for thousands of years, but was not discovered by white settlers until the early twentieth century. The finger lime is now being cultivated commercially in Australia for general consumption.

Botanical Facts These limes and their relatives do not cluster like most citrus, but hang individually like miniature cucumbers from large trees. A number of cultivars have been developed, ranging in color from the original green to yellow, orange, purple, red, and almost black. Among these is the 'Alstonville', whose rather unattractive brown skin encloses a brilliant green pulp. One of the attributes of the finger lime is that it is almost seedless, and the juice sacs are encased in small, pearl-like membranes.

Culinary Fare The various species of Australian limes can be used in the same way as other limes, but the pearl-like juice sacs, separated from the skin of the fruit, look very attractive when dropped into a citrus-based drink or sprinkled over a fruit salad. The fruits can be sliced in half and used to surround a dish of fruit salad.

Limes

Citrus aurantifolia Lime, Key Lime *Citrus hystrix* Kaffir Lime *Citrus latifolia* Persian Lime, Tahitian Lime

LEFT: *Persian or Tahitian limes are harvested 8–12 times a year. The tree grows to 20 ft (6 m) and bears purplish colored flowers. A wedge of lime is often served with fish and avocado for extra tang.*

Limes are similar in shape to lemons, but the blossoms do not cluster in the same way as those of lemons. They are smaller than oranges, and the skin color varies between the brilliant green associated with their name, to a lemon-yellow color. The bright species, *Citrus aurantifolia*, known as the key lime, Mexican lime, or West Indian lime, is the same species provided by the British navy to its sailors for protection against scurvy. The role of citrus in preventing scurvy had been known in England since 1593, but it was not until 1747 that Dr. James Lind discovered that limes worked almost as well as lemons. So from that time, rations of lime juice were handed out with the rum, and the British abroad became known as "limeys."

Historical Origins Limes are believed to be native to southern India or the northern end of the Persian Gulf. *C. aurantifolia* was brought westward by Arab conquerors in the Middle Ages. Columbus included limes among the citrus fruits that he brought to the island of Hispaniola on his second voyage in 1493, and they quickly established themselves in the Caribbean. The Persian or Tahitian lime (*C. latifolia)* was introduced to the West from Persia in the nineteenth century.

Botanical Facts Limes belong to the group of citrus fruits that require plenty of water and year-round warmth. The trees can grow from 5 ft (1.5 m) to 20 ft (6 m) tall. The species most often seen in the supermarket is the Persian lime, the key lime being grown mainly for juice or for products using lime extracts. A Tahitian lime cultivar, 'Bearss', sometimes confusingly referred to as a Bearss lemon, is grown exlusively in California. The kaffir lime (*C. hystrix)* is grown mainly in Southeast Asia. Only the leaves of the kaffir lime are used in cooking, particularly in Thai cuisine; the juice of the fruits can be used as a hair rinse and insect repellent. The kaffir lime is also used as a hardy rootstock for other citrus species.

Culinary Fare The juice of the key lime is not as sour as that of lemons, which makes it an excellent ingredient in drinks such as "lime Rickey," and as a mixer for rum, gin, and Scotch whiskey. Its milder, sweeter flavor makes it more suitable for use in desserts, the most famous of which is key lime pie. Limes are added to other citrus fruits for use in marmalade and to make sherbets (water ices). In the Middle East, limes are dried and added to stews. In Indian cuisine, limes are used in chutneys and pickles. In French cooking (except in the French Antilles) they have had so little importance that the French name for them is merely *citron vert* (green lemon).

BELOW: *Key limes (shown here) are smaller than Persian limes, and have a thinner rind and more seeds. Limes are often used as an alternative to vinegar in dressings and sauces.*

Oranges

Citrus aurantium Seville Orange, Bitter Orange * *Citrus aurantium* subsp. *bergamia* Bergamot Orange
* *Citrus reticulata* x *Citrus sinensis* Florida Orange, Tangor * *Citrus sinensis* Sweet Orange

ABOVE: *Blood oranges are juicy and sweet with a dark red flesh. They are great for juicing, and their segments make an attractive addition to salads and sorbets. The juice adds a striking color to cocktail drinks.*

The most widely grown species of orange are the Seville or bitter orange (*Citrus aurantium*) and the sweet orange (*C. sinensis*), which is by far the more popular. It is not known which was the first of the two to reach Europe, but the Seville orange was the first to be widely planted. Seville oranges are considered to be the most beautiful of all citrus trees, cultivated mainly in Spain and southern Italy for their fragrant blossoms, fruits, and leaves.

Historical Origins The sweet orange, also called the China orange or Portuguese orange, reached Europe comparatively late, in about 1500, when Vasco da Gama brought the first ones from India. They may have been known earlier, but their keeping qualities are not as good as those of the sour oranges, and Europeans who first tasted them after their long voyage from the Tropics may have found the flesh dry and insipid. The sweet orange spread around Europe during the next century, but it was still an expensive luxury. The species was introduced to the New World in 1565 by the first Spanish colonists in St. Augustine, southern Florida, and Spanish missionaries planted sweet oranges in California in 1707. However, it was not until Florida became part of the United States in 1821 that sweet oranges became a major export. In 1824, the first sweet orange tree, imported from Bahia in Brazil, was planted in Australia, although exotic citrus had been introduced in 1788.

The tangor is a hybrid of the the sweet orange and the mandarin or tangerine. The name "tangor" is a combination of the "tang" of tangerine and the "or" of "orange."

The Seville orange is believed to have originated in Asia, on the lower slopes of the Himalayas. These oranges were introduced to Europe during the Muslim conquest of the Iberian peninsula. Descendants of the original sour orange trees planted by the Moorish conquerors at their palace, the Alhambra, in Granada, still line the courtyards. The Patio de los Naranjos, an orange grove next to the Grand Mosque of Cordoba, echoes the colonnades of the mosque in the lines of elegant trees, their fragrant white blossoms and brilliantly colored winter fruits contrasting starkly with the smooth, glossy green leaves that are typical of citrus. The Seville orange is no longer as widely grown as it once was, although large wild populations flourish in southern Florida and in South and Central America.

The bergamot orange is thought to be a subspecies of the Seville orange, and is cultivated in Italy mostly for its fragrance. It is used in the perfume industry and as an ingredient in Earl Grey tea.

RIGHT: *The most popular marmalade is made from Seville oranges. Both the juice of the fruit and its peel are boiled in water with sugar. The result is a tangy, jamlike preserve.*

ABOVE: *The 'Washington Navel' is one of the most popular cultivars of sweet orange. It is easy to peel, and the segments can be easily separated.*

In the seventeenth century, wealthy people in northern Europe were eager to possess orange trees, but even a relatively hardy citrus, such as the Seville orange, could not live outdoors in the cold, damp climate of northern Europe. So heated greenhouses that let in plenty of natural light, known as orangeries, were built to house the trees. The knowledge gained later served to transplant other exotic species.

Botanical Facts Although both Seville and sweet oranges (particularly the former) belong to the hardier types of citrus, exceptionally cold winters destroyed many early crops grown outside their native habitat. In Australia, for example, orange growing was concentrated in states such as Victoria and New South Wales, where the weather was not ideal. The most popular cultivars, due to their abundant fruiting and keeping qualities, continue to be the 'Washington Navel', 'Shamouti' ('Jaffa'), and 'Valencia'. A type of sweet orange is the blood orange, or Maltese orange, of which the best-known cultivar is 'Ruby'. The red streaks of color in blood oranges seem to have developed spontaneously during cultivation, since blood oranges are not known in the wild.

Culinary Fare Sweet oranges can be eaten fresh out of hand or squeezed for their juice, as can the tangor. The blood or Maltese orange is most popular in Spain, Scandinavia, and Germany as a juice. In France, it is used in the sauce that accompanies duck, which is consequently known as *sauce Maltaise*. The most popular use for the Seville orange is in drinks and, above all, in making marmalade, for the sharpness and bitterness of the skin and pulp. It is an important ingredient in the steak sauces favored by the British, and in orange-flavored liqueurs. Seville oranges are also an ingredient in the "bitters" first used by the Dutch to flavor beer and wine. Grated orange peel, or zest, is used in cooking to impart an orange flavor to dishes.

The Orange-Sellers of England

In England, sweet oranges were sold in the streets and to the audience inside theaters. The most famous orange-seller, Nell Gwyn (1650–1687), first encountered King Charles II, whose mistress she remained until his death, when she was selling oranges at the Theatre Royal, Drury Lane, in London. This is often thought of as somewhat ironic since the king's wife was Catherine of Braganza, whose Portuguese lineage had helped introduce the first oranges into England.

Lemons

Citrus ichangensis Ichang Lemon *Citrus limetta* Sweet Lemon *Citrus limon* Lemon *Citrus x meyeri* Meyer Lemon *Citrus ponderosa* Giant Lemon

ABOVE: *Lemon trees can grow to a height of 20 ft (6 m) and bear small, fragrant white flowers. Most of the fruits have seeds, but some are seedless. The juice is used to make lemonade.*

The lemon is unquestionably the most versatile of all fruits and has been cultivated for so long that wild species are unknown. Its attractive color, sharp odor, and acidic flavor have won it universal acclaim. There is not a single cuisine that does not include lemon as an essential flavoring. The lemon has supplanted verjuice (the juice of green wheat or unripe grapes) as the souring ingredient used in cooked foods. Its skin, grated into zest, is also used in many dishes.

The ichang lemon (*Citrus ichangensis*) is actually a papeda, a slow-growing subspecies of citrus, which is cultivated for its foliage and flowers, since the fruit itself is rather dry and unappealing. Most Florida oranges are grown on rough lemon (another papeda) rootstock, although lemons are not an important citrus crop in Florida because the climate is too mild. The sweet lemon (*C. limetta*) is not grown commercially, allegedly because the juice is insipid, but it is popular as a dooryard tree in Italy and Morocco. There are three cultivars, the 'Millsweet', 'Marakesh Limonette', and the 'Mediterranean Limonetta'.

Historical Origins The lemon first arrived in the Middle East from India and China in the twelfth century, and has been cultivated ever since in Palestine (Israel) and Persia (Iran). Columbus brought the fruit to the West Indies on his second voyage in 1493, and the seeds were introduced to Florida from Haiti in the early sixteenth century.

Botanical Facts Some botanists believe that the lemon is not actually a separate species of citrus, but a mutation of the citron (*C. medica*). The lemon belongs to the citrus types that are hardy. It needs an arid climate and a long, dry spell for the fruit to set, so it flourishes in drier, subtropical—rather than tropical—climates such as Israel, Cyprus, Spain, and the California–Arizona sunbelt of the United States. In Australia, lemons are grown mainly in the south of the country: Victoria and New South Wales. The main cultivars of lemon grown commercially are 'Eureka' and the 'Lisbon' (both *C. limon*). Both are thin-skinned, as is the delicious Meyer lemon (*C. x meyeri*), grown in California and elsewhere, which is very juicy. The giant lemon (*C. ponderosa*) is a thick-skinned species, a cross between a lemon and a citron, which has been introduced recently.

Lemons have attractive dark green foliage and the white blossoms are tinged with purple. Though the winter is the most productive period, the lemon tree can be encouraged to flower and fruit all year round, which is another reason why it is considered so important commercially.

Culinary Fare Every part of the lemon can be used in cuisine. Lemon juice is an essential ingredient in chicken soup, Chinese hot-and-sour soup, stews, and, of course, lemon meringue pie. The juice is sprinkled over fruits and vegetables that oxidize (turn brown) when cut, such as apples and jicama, to keep them white if they are to be eaten raw. The fragrant skin contains valuable oils that add flavor to drinks ("martini with a twist") and every kind of dessert. Grated and dried lemon rind makes a very acceptable salt substitute for people on a low-salt diet. Even the seeds and white parts are useful, because they are richer in pectin than other citrus fruits. These parts are tied in a cheesecloth bag and added to preserves and jellies during cooking to make them set. A useful chef's tip is to rub the hands with lemon juice while slicing meats or fish, so the odor of the food will not cling to the skin. The lemon is the citrus richest in vitamin C (a single lemon can supply 35 percent of the Recommended Daily Allowance) and also contains the antioxidants known as anthocyanins, as well as traces of copper and magnesium.

The Answer Is a Lemon

The lemon's uses are not confined to the kitchen. The citric acid in lemon juice is an important bleach, and when mixed with salt, it is used for cleaning copper (especially copper pans, because it is nontoxic and leaves no aftertaste). Lemon juice can remove stains from a kitchen counter—simply allow it to sit on the stain for a few minutes. It is a mild bleach, contained in shampoos for blonde hair, and the juice is also used in medicines, engraving compounds, dyes, plastics, and synthetic resins. Lemon oil is the most valuable of citrus oils and is used in all types of perfumes from room-fresheners to expensive French fragrances. It should not be added to furniture polish, however, as it has a drying effect. Fresh lemons can also be used as a room or refrigerator fragrance.

BELOW: *The Meyer lemon, slightly sweeter than other lemons, is very popular in home gardens. It is a very hardy tree–more tolerant of cold and frost than other species.*

Kumquat

Citrus japonica

BELOW: *The 'Nagami' kumquat is the most popular kumquat in the United States. The fruits make a sharp, delicious marmalade, or can be preserved, candied, or made into a sauce.*

The kumquat is the smallest of citrus fruits and the only one that is normally eaten without peeling. It originates in China, and the name is said to be a corruption of *chin kan* or *kin kan*, both Chinese and Japanese words for "golden orange." Miniature kumquat trees were sometimes given as ceremonial gifts in China, and even today such miniature ornamentals are popular in Florida.

Historical Origins The kumquat was brought to Europe by the British plant collector Robert Fortune, who discovered the plant in China in 1847. He sent it to England, from where it was soon exported to Florida, and has been grown there successfully ever since. Fortune's name has been used for this citrus genus, which has been classified separately because of the peculiarities of its fruits—a bitter flesh in contrast to the sweet, edible peel.

Botanical Facts There are three main cultivars of kumquat, depending on the shape of the fruit. The oval or 'Nagami', is the original kumquat brought to Europe by Fortune. The round-fruited 'Marumi' was introduced into Florida from Japan in 1885, and the 'Meiwa' between 1910 and 1912. The 'Meiwa' is the most popular kumquat now grown for the sweetness of its flesh. There is a fourth cultivar, the 'Hong Kong' kumquat, which is grown on that island as an ornamental, but is inedible due to its many large seeds that leave little room for the pulp. Despite its tiny fruits, the kumquat is not a miniature tree, and fully-grown specimens look magnificent—the deep green leaves dotted with little glowing orange fruits like lights. However, the kumquat is a slow-growing citrus, so it is suited to cultivation in pots, indoors or out. It fruits in winter. The kumquat is often crossed with other citrus species because it is very resistant to cold. It is believed that the calamondin, for example, is a cross between a kumquat and an orange.

Culinary Fare The kumquat can be eaten raw, but the pulp, especially of the 'Marumi' cultivar, is quite sour, and sometimes bitter. The 'Nagami' kumquat is sweet and has a sweet peel, and can be added to fruit salads. The kumquat goes well with savory dishes, such as meats, but it is mainly used in making marmalade, which has a lovely golden color. Also popular are brandied kumquats and kumquat liqueur.

Pummelo

Citrus maxima

The pummelo is the largest of the citrus fruits and almost certainly the "ancestor" of the grapefruit. The fruit is highly prized by the Chinese and Vietnamese for celebrating the New Year, and it can be found in oriental food stores all over the world at this time.

Historical Origins The name "pummelo" or "pomelo" originates from the Dutch word, *pompelmoes*, which means "big lemon." It has also been called "Adam's apple" and "forbidden fruit." The eighteenth-century poet James Grainger referred to it as "the golden Shaddock, the forbidden fruit" because it is known as a Shaddock in the West Indies—named for a Captain Shaddock who is supposed to have first brought the seeds to the Caribbean.

Botanical Facts The pummelo varies in shape from round to slightly pointed at the stem end, and it is greenish yellow in color. The white parts under the skin are soft, but the fibers are very tough around the segments, which are greenish yellow in color. The juice is contained in large sacs. The flavor is bittersweet, a bit like a grapefruit. The 'Sweetie' cultivar, sweeter than the original, has been developed in Israel, and there is also a pink-fleshed version.

Culinary Fare The pummelo is best eaten raw because the white parts between the segments are thick and hard to peel, too hard even for a grapefruit knife to scoop out. The individual juice sacs are so large that they can easily be separated out and incorporated into a fruit salad or mixed with seafood for an unusual appetizer.

ABOVE: *The pummelo is sweeter than the grapefruit. Slices of the fruit can be used in desserts and salads, and its skin is used in candies and preserves.*

Citron

Citrus medica

The citron has the most variable shape and size of all citrus fruits. When mature, it may be as small as a lemon or as large as a rugby ball and similar in shape. The white parts are very thick and the juice is sparse, but the fruit is strongly fragrant and is used in Southeast Asia as a natural air-freshener for rooms and cars. The main growing areas are Italy and Puerto Rico, serving Europe and the United States, respectively.

Historical Origins The citron was the first citrus fruit to be brought westward from its native habitat, believed to be the Arabian Gulf. It first reached Persia (Iran) in around 500 BCE and then spread westward, becoming the first citrus fruit to be known in the Western world. It is the only citrus fruit to be mentioned in the Bible, which makes reference to "goodly trees" (Leviticus 23:40).

Botanical Facts The citron is green, yellow, or yellow and green in color, and varies very widely in shape. The 'Etrog' cultivar is lemon-shaped, and may be knobbly, furrowed, or smooth. The cultivar known as 'Buddha's Hand' is the strangest of all citrus fruits—the segments are surrounded by rind and look like a bunch of fingers. It is highly prized in parts of China and Ceylon, but is not eaten.

Culinary Fare The citron is no longer cultivated for its juice, which is sparse and rather bitter, but for its thick peel, which is considered to make the best candied peel of any fruit.

BELOW: *The citron blooms all year round. Its fruit is dark green when young, and takes 3 months to ripen and turn yellow.*

Grapefruit

Citrus x paradisi

Citrus fruits cross very easily between species and the grapefruit is the best example of this. Grapefruits are believed to be hybrids of the pummelo with the sweet orange, and are the only citrus fruits native to the New World. They are a round, yellow fruit growing in huge, impressive clusters like a giant bunch of grapes, hence the name.

Historical Origins The grapefruit was "discovered" in the early nineteenth century. The main grapefruit cultivar now grown is 'Marsh', which is a seedless descendant of the 'Duncan', the oldest commercial grapefruit cultivar in the world. The parent tree, which is believed to be from Barbados, was planted in 1830 in Safety Harbor, Florida, and lived to be nearly 100 years old. In 1892, A. L. Duncan began propagating the tree near Dunedin (Tampa), Florida.

There is no point looking for grapefruit recipes in cookbooks dating from before the 1880s, because it was not until 1885 that Florida first shipped grapefruits to New York and Philadelphia, creating a flurry of culinary interest in this new kind of citrus fruit.

Botanical Facts The grapefruit prefers drier, cooler climates, and is easy to cultivate. It is hardy and very resistant to intense heat. The 'Duncan' grapefruit is the standard against which all other cultivars are measured. It is the most popular grapefruit for canning and the sweetest of all grapefruit cultivars. The 'Marsh' grapefruit also has pink types. Pink grapefruit are now even more popular than yellow. The 'Star Ruby' cultivar, grown in Texas, is a deeper red in color. There is a prevalent myth that pink grapefruits are sweeter than yellow-fleshed ones, but it is completely untrue, although pink grapefruit do contain higher amounts of vitamin A.

Culinary Fare While misguided attempts are made to cook grapefruit, as in broiled grapefruit, the fruits are far tastier when eaten raw, and this way they also retain all their valuable nutrients—especially vitamin C, which is easily destroyed by heat. Grapefruit juice is a breakfast staple. Roughly 40 percent of the grapefruits harvested each season in the United States are used for juice, while 60 percent are sold as fresh grapefruit. As soon as dieting became fashionable in the early 1920s, grapefruit was advocated as an important part of a weight-loss diet, due to its low calorie content. Recent studies have shown, however, that grapefruit juice significantly increases levels of estradiol, a sex hormone, in women. Diets that are based on grapefruits are no longer recommended.

LEFT: *Grapefruits, both pink and yellow, are a very popular breakfast fruit served sliced and chilled. They are also used to produce marmalade and jelly.*

Clementine, Mandarin, and Tangerine

Citrus reticulata

Citrus has an ability to produce hybrids and cultivars with exceptional ease, and as a result, its areas of classification are fluid and often in need of revision. The various types of tangerine share a thin-rind skin, which in some types is only loosely attached to the fruit pulp. The fruit itself is very juicy and the flavor is stronger than that of the orange.

Historical Origins Tachibana and mikan mandarins were brought to Japan from China in around CE 500 and brought to the United States in the mid-nineteenth century. The Clementine is an even more recent arrival in the West. It is so called because it was developed by Father Clément Rodier in Oran, Algeria, in 1902.

Botanical Facts The Satsuma is small and round with a thick skin and pale, seedless, orange-colored flesh. The 'Wilkins' or 'Wilkings' is much smaller, and the 'Honey' is a mandarin whose juice reputedly tastes of honey. The 'Dancy' has a loose skin. The Clementine has prominent oil glands and is round like an orange, rather than the flattish shape of a tangerine.

Culinary Fare The stronger flavor of the tangerine makes it more suitable for cooking than the orange. The one classic dish that features the tangerine is *crêpe suzette*. This is made of tangerine-flavored pancakes in a sauce of tangerine juice and curaçao.

ABOVE: *The Clementine is a small fruit with a skin that easily peels off. It is juicy and sweet, and not as acidic as an orange. It is mostly eaten fresh.*

Tangelo

Citrus x tangelo

The tangelo, a reddish-orange citrus, is a spontaneous hybrid of a tangerine and either a pummelo or a grapefruit. Ranging from the size of an orange to the size of a grapefruit, the tangelo is more popular in the United States than the tangerine.

Historical Origins The tangelo developed about 2,500 years ago in China, but has since been crossed and developed into a number of cultivars, chiefly in the late nineteenth and early twentieth centuries, with the spread of citrus-growing all over the subtropics and tropics.

Botanical Facts The tangelo requires a long, hot growing season and most cultivars ripen in the spring. Tangelo hybrids include the 'Minneola', a distinctive fruit with a deep orange color and a protruding "nipple" on the stem end, grown mainly in California. The 'Orlando' and the 'Seminole', as their names imply, are Florida tangelos. The Ugli fruit, developed in Jamaica, is a tangelo that is widely exported to Canada and Europe. The bumpy surface and blotchy color of the fruit make it live up to its name, but it has a good flavor.

Culinary Fare The tangelo produces copious juice and makes a better "orange" drink than oranges. The firmer-skinned cultivars are easier to juice. The tangelo can be used in the same way as all of the sweetest citrus, in drinks and fruit salads.

BELOW: *The tangelo has a sharp taste like a grapefruit, but is easy to peel, like a mandarin. The popular 'Minneola' cultivar is pictured here.*

Avocado

Persea americana

The creamy, buttery avocado is something of a culinary chameleon. Because avocados pair well with so many foods, they seemingly shift between the categories of fruit and vegetable. Although at first slow to catch on outside its Central American native habitat, once established the avocado quickly became a favorite fruit in kitchens around the globe.

Historical Origins Avocados have been cultivated in Central America for more than 7,000 years. The name comes from the Aztec word *ahuacatl*, which means "testicle" and referred to the fruit's shape and their tendency to hang in pairs. There is evidence avocados had spread as far south as Peru in the pre-Columbian era and were introduced to the Antilles in the Caribbean shortly after the Spanish conquest of South and Central America. Subsequent introductions occurred in 1750 in Indonesia and about 100 years later in California. Avocados were growing in Australia in the late nineteenth century but were not widely used until World War II, when American servicemen stationed there created a demand for them. In more recent years, the popularity of avocados has grown enormously, with world production tripling between 1961 and 1996.

Botanical Facts The avocado tree is an evergreen with elliptical leaves and clusters of yellow-green flowers. The trees can grow up to 60 ft (18 m) tall and produce 100 or more avocados each year. There are three original "races" of avocado—the Mexican, Guatemalan, and West Indian—from which the more than 500 cultivars are descended. Two of the most popular avocados are both Mexican–Guatemalan hybrids, 'Haas' and 'Fuerte'. Avocados can be harvested when mature but still "unripe," as they will ripen off the tree in a matter of days.

Culinary Fare A ripe avocado should be brown at the stem end and give slightly under gentle pressure. Avocados can be used in so many dishes because they go well with so many things—particularly fruits, vegetables, and poultry. Perhaps the best-known avocado dish is guacamole, which is derived from the Aztec sauce called *ahuaca-mulli*, and is now commonly used as a dip for tortilla chips. It is made using mashed avocados, lime juice, chillies, onion, garlic, tomato, salt, and cilantro (coriander leaf).

Avocados are good as sandwich fillings or in salads, and are served in some countries with sugar or in milk shakes. Avocado ice cream has long been a favorite in Brazil. Some think the creamy avocado is best eaten right out of its skin with a spoon, perhaps with a squirt of lemon juice. Avocados can become bitter when cooked, so if you are using them in hot dishes, add just before serving.

BELOW: *Avocados are a nutritional powerhouse. They are high in fat, but the fat is monounsaturated (the good fat). They are rich in antioxidants, potassium, and folate.*

Date

Phoenix dactylifera

This fruit of the date palm is so important to Middle Eastern cultures, there is said to be a different use for dates on every day of the year. In biblical references to the "land of milk and honey," the honey most likely refers to dates, as they were often a substitute for honey. Interestingly, Arabs commonly eat milk and dates together, and when they are combined, they provide a very wide range of necessary nutrients.

Historical Origins The name "date" comes from the Greek word *daktulos*, which means finger, a reference to the fruit's shape. Dates have been around since prehistoric times and likely originated in the region encompassing North Africa, the Middle East, and India. Desert dwellers ate dates fresh, dried, and pressed into cakes. They were also ground into flour and used to make wine. Dates were popular in Roman times, and *Apicius*, a Roman cookery book, had recipes for date sauces to go with meat and fish, stuffed sweetmeats, and dates fried in honey. The Chinese began importing dates from Persia during the Tang dynasty (CE 618–906).

Date production has spread into western Asia and as far north as southern Spain. North American date production in the hot, dry regions of California and Arizona began in the early twentieth century, but counts for just a fraction of world production. Iraq and Egypt are the world's top producers.

ABOVE: *Dried dates are eaten out of hand, or used on cereal, in cookies, or in candy bars. Young date leaves are also put to good use, cooked as a vegetable.*

Botanical Facts Date palms can grow over 100 ft (30 m) tall. They have extremely large leaves that are 9–16 ft (3–5 m) long and grow at the very top of the tree. Most of the species are grown as ornamentals, but *Phoenix dactylifera* is grown for the fruit. Palms are either male or female, but only the female bears fruit. The fruits grow in large clusters which can weigh 20 lb (10 kg). A mature date palm can produce 150 lb (70 kg) of fruit or more each year. Dates are typically around 2 in (5 cm) long. There are three types of date—hard, soft, and semidry—and the semidry is the most popular in the West.

Culinary Fare Outside the Middle East, fresh dates are a rare commodity. Because dried dates are so high in sugar (70 percent or more), they are frequently used in desserts such as date cakes, breads, and, of course, sticky date pudding. Dates are wonderful stuffed with blue cheese or goat's cheese and wrapped in bacon or prosciutto; the saltiness of the meat and piquancy of the cheese complements the sweet fruit. In Middle Eastern cuisine, dates are used in countless savory and sweet recipes. Dates also make delicious milk shakes.

LEFT: *The processing of dates has a long history, as this c. 1400–1390* BCE *wall painting shows. Here, villagers are pressing dates in a vat in order to extract the juice for use as a syrup or alcoholic beverage.*

Illawara Plum, Plum Pine

Podocarpus elatus

RIGHT: *The edible fruit of the Illawarra plum is attached to the branch, and the seed sits on the end of the fruit. The fruit can be eaten raw or cooked.*

This unusual fruit, in both form and flavor, is native to Australia, where it was part of the diet of Aboriginal Australians for thousands of years. Compared to other fruits, the Illawarra plum makes up only a small fraction of the commercial fruit market, and it is mostly gathered wild, although there have been a few attempts to grow it on plantations.

Historical Origins The Illawarra plum, a member of the podocarp family, is one of the world's most primitive fruits and has been around since dinosaurs roamed the earth. Fossil evidence shows that podocarps have existed for about 250 million years. Illawarra plums are native to rain forests in the eastern Australian states of New South Wales and Queensland. They were a favorite fruit of Aborigines in these areas, particularly in New South Wales, where other fruit choices were limited.

European settlers also enjoyed the fruits and found the wood suitable for making furniture, boats, and even pianos. Unfortunately, its usefulness as timber led to many Illawarra plum trees being cut down during the nineteenth century. Today, most Illawarra plums grow in public parks.

Botanical Facts Perhaps the most distinguishing feature of the Illawarra plum is that its seed grows outside the fruit. The seed is small, about ½ in (12 mm), and grows on the bottom of the fruit, which is about twice as large and is actually a stem for the seed. This species is a large shrub or fast-growing tree that can reach 50 ft (15 m) in height. The tree also has unusual leaves that are long and narrow and quite leathery, with sharp tips. The Illawarra plum tree is either male or female, but only the female trees bear fruits, which ripen between March and July just after the wet season.

Culinary Fare Some people find that the Illawarra plum is an acquired taste. It is similar to plum, but with pine characteristics and a resinous consistency. The fruits are most frequently used in jam making, but also in compotes, cheesecakes, and muffins. The Illawarra plum pairs nicely with chili and garlic to make savory chutneys and sauces to serve with meat. The plum should be chopped or blended before cooking, and should be cooked only in stainless steel cookware. It can easily become bitter if overcooked, and it is best to let sauces and jams cool a bit before tasting them, as they can taste bitter when they are still hot.

Apricot

Prunus armeniaca

One of the surest signs that summer has arrived is a tree full of almost-ripe apricots. The name apricot, which comes from the Latin word *praecocium* meaning precocious, is quite appropriate for one of the first stone fruits of the summer season.

Historical Origins The apricot is native to China, where it was cultivated more than 4,000 years ago, and likely moved west along the Silk Road to Iran where the Greeks and Romans found it in the first century CE. Despite the prospect of an inhospitable climate, King Henry VIII's gardener took the apricot to England in 1542, but it was not until some 200 years later that Lord Anson had great success with a cultivar he named 'Moorpark', after his estate north of London. Today apricots are grown from western Asia to Japan, and in southern Europe, northern and southern Africa, Australia, and California.

Botanical Facts The apricot is a small tree, typically about 25 ft (8 m) in height, with red-brown bark and a fairly wide canopy. Its serrated leaves are somewhat oval-shaped with a pointed tip, and it bears white or pale pink flowers. It grows best in a climate with cool but temperate winters with little danger of spring frosts. Apricots are generally the size of a golf ball or larger, and are golden to orange in color, often with a red blush. Their thin, velvety-soft skin feels like a baby's cheek to the touch. Apricots are quite delicate and damage easily.

Culinary Fare Fresh apricots are a wonderful treat, but because they are so delicate, they do not travel well. Early harvesting and refrigeration do this fruit no favors, and it is said that its flavor often diminishes in relation to its distance from the tree. You can eat apricots raw, of course, but they are also wonderful lightly poached or baked in crumbles or tarts. Apricots are rich in natural pectin and so make delicious jam, which is also useful as a pastry or confectionary glaze.

Dried apricots are usually preserved with sulfur dioxide to preserve color and flavor, and are quite versatile. They can be rehydrated and cooked into a filling for apricot bars, or chopped and added to bread or puddings. Dried apricots are also frequently used in Middle Eastern cuisine as an addition to stews and pilafs, and they pair well with lamb, in a tagine, or as an ingredient in stuffing for roast lamb.

ABOVE: *For dried apricots, only fruit picked when fully mature gives rise to a good shape, a bright orange color, and that full, strong, apricot flavor.*

BELOW: *Apricots are left on the tree for as long as possible. Their color and degree of softness indicate their level of maturity.*

Cherries

✳ *Prunus avium* Sweet Cherry ✳ *Prunus cerasus* Sour Cherry
✳ *Prunus padus* Bird Cherry ✳ *Prunus serotina* Black Cherry

RIGHT: *Ripe cherries are best picked and stored with their stems on. They can be plucked and eaten right away, gently poached and used as fillings, or served flambéed over ice cream.*

What this small stone fruit lacks in size, the succulent cherry makes up for in sheer beauty. Its lovely blossoms, resplendent skin, and refreshing sweetness have led to its celebration in art, literature, and song for centuries. It has a short fruiting season, and although it is one of the more expensive fruits, the cherry is consistently one of the most popular.

Celebrating the Cherry

Both cherry blossoms and their fruits have been cause for celebrations all over the world. In Japan, festivals for cherry blossom viewing, or *hanami*, have been going on since at least the seventh century. Today, there are hundreds of these festivals all over Japan, just like this one in Kyoto. Each spring, Washington, DC is awash with color for the National Cherry Blossom Festival, which commemorates the gift of 3,000 cherry trees to the city from the mayor of Tokyo in 1912.

During the Middle Ages in England, cherry fairs were held in the summer months. People would pick fruit, eat, drink, and even make love right in the middle of the orchards! In Surrey, the town of Chertsey still holds a Black Cherry Fair each July.

More than 500,000 people attend the National Cherry Festival each year in Traverse City, Michigan, which is the self-proclaimed cherry capital of the world. The week-long event features cherry pie-eating and pit-spitting contests. The cherry capital of Australia, Young, New South Wales, has its own cherry festival each spring.

Historical Origins The sweet and sour cherries we enjoy today have all descended from two wild species, *Prunus avium* and *P. cerasus*, which are believed to have originated in the region near the Caspian Sea. Archaeologists have found preserved cherry stones at prehistoric Swiss lake sites, but cherries were not actually cultivated until much later.

In 300 BCE, the philosopher Theophrastus indicated that Greeks were already cultivating cherries, and in the first century CE, Pliny the Elder wrote that cherries were prized in Italy and that the Romans planted them throughout their empire, including southeast England.

The cherry was reputedly one of King Henry VIII's favorite fruits, and he and his gardener, Richard Harris, are often credited with reinvigorating cherry production in England by improving the fruit's quality and introducing new cultivars.

Today there are well over 1,000 cultivars of cherries—about 900 sweet and 300 sour. Some of the better-known sweet cultivars are 'Bing', 'Lambert', 'Ranier', and 'Stella'; 'Montmorency' and 'Morello' are two of the most popular sour cultivars. Iran, Turkey, and the United States are the world's largest cherry-producing nations.

Botanical Facts The sweet cherry tree is deciduous, with clusters of white blossoms that bloom in the spring before the oval-shaped leaves appear. The trees grow quickly and can reach 50 ft (15 m) in height, but are usually kept smaller to make harvesting easier. The fruits range in color from red to black to creamy white with a red blush. Trees begin producing fruits in 5–6 years and can yield 50–100 lb (23–45 kg) of fruits each season.

Sour cherry trees are smaller, reaching 20 ft (6 m) in height, but have similar clusters of beautiful white blossoms that flower a bit later than the sweet cherry's. The fruit also looks much the same, but the sour cherry tree produces somewhat less, typically about 35–45 lb (16–20 kg) a season.

The bird cherry and the black cherry are both wild species that bear stalks of small white flowers in the spring and small fruits that are almost black in color. The bird cherry is native to northern Europe and Asia and reaches a height of about 50 ft (15 m). The black cherry is native to North America and can grow much higher, up to about 100 ft (30 m). Both these species are occasionally used in cooking.

Cherry trees should be planted in a sunny spot with good air circulation and well-drained, fertile soil. Trees should be pruned to make sure the branches get plenty of sunlight and air. Cherry trees need cool winter temperatures to ensure they flower and fruit, but spring frosts are damaging, as is rain near harvest time. Cherries are picked when fully ripe, as they will not ripen off the tree.

Culinary Fare The sweet-tartness of cherries is the perfect foil to many rich and creamy dishes. The classic example is Black Forest cherry cake (*Schwarzwälder Kirschtorte*), in which layers of rich chocolate cake and cream alternate with cooked sour cherries and kirsch liqueur. Cherries are also used to great effect in clafoutis, a custardlike, cakey dessert from central France. Cherry compotes and sauces are delicious on ice cream, cheesecake, and even served with duck or other rich meats.

Sour cherry soup is popular in northern Europe and Scandanavian countries. Cherries are wonderful in many baked goods, including pies, crumbles, and streusels. Cherries can be preserved by drying, glacéing, soaking in brandy, or as jams and jellies. Cherries are also used to make kirsch liqueur.

When choosing cherries, look for shiny, plump, firm, blemish-free fruits with their green stalks still intact.

BELOW: *Bursting with flavor, plump, sweet 'Bing' cherries fetch prime prices on markets worldwide. However, birds devour up to a third of these sweet treats before harvest.*

Plums

* *Prunus x domestica* European Plum * *Prunus institia* Damson * *Prunus nigra* Canadian Plum
* *Prunus salicina* Japanese Plum * *Prunus spinosa* Blackthorn, Sloe

ABOVE: *Japanese plums are commonly grown for the fresh market. They are suitable to be eaten out of hand. Plums are best picked when mature, with a good color, but not fully ripe.*

The plum is a humble fruit. It is not as flashy as the cherry, nor does it receive the praise that is so often lavished on the peach, although it is difficult to understand why. There are countless cultivars that come in a rainbow of colors—yellow, green, red, blue, and black—and many different sizes and tastes, all of which makes the plum an incredibly versatile fruit.

Historical Origins Plums have been around for a very long time, but their history is somewhat murky because so few people wrote about them. Damson stones have been discovered at prehistoric Swiss lake sites, but it was not until several thousand years later, in the first century CE, that the Roman writer Pliny the Elder became one of the first to describe plums. He listed a dozen cultivars growing in Italy, which indicates that they had been cultivated for some time. Pliny even noted how curious it was that others, such as Cato, failed to ever mention them.

What we do know is that damsons are older than European plums, and that they originated in eastern Europe or western Asia. *Apicius*, the Roman cookery book, includes recipes for a number of sauces made with them. European plums, or prune plums, are probably native to central Europe.

By the Middle Ages, plums were being cultivated in England, and cultivation became more important as time went on. Many new cultivars were produced in England in the nineteenth century. Plum growing gathered momentum in America in 1828 when the Prince Nursery in New York claimed to have 140 different kinds of plum trees for sale.

The Japanese plum is believed to be the oldest cultivated plum species. It originated in China, but was taken to Japan approximately 300 years ago, where horticulturists made improvements to it. It then underwent further development in the nineteenth century after it arrived in California.

Botanical Facts Plum trees share many traits in common. They have dark green, oval-shaped leaves and produce either single or small clusters of white flowers. The European, Japanese, and Canadian plums (*Prunus nigra*) can grow to be 30 ft (9 m) high, while damsons and *P. spinosa* are a bit smaller, about 20 ft (6 m). *P. spinosa*, native to Europe and western Asia, also has thornlike spines on its branches. *P. nigra* has scented white to pale pink flowers and small dark fruits. Plum trees do best in moist, well-drained soil.

The fruits are often what sets these plants apart. Japanese plums are big and round, typically 2–3 in (5–8 cm) across, and are yellow to red in color and quite juicy. European plums are smaller and oval-shaped and have thicker, meatier flesh and a pit that is easily removed. Japanese plums are typically harvested earlier than European plums. Damsons have small, oval, purplish blue fruit and produce a heavy crop in late summer to autumn. Many plums will have what looks like a faint, white coating on their skin that is called bloom and is completely harmless.

Culinary Fare When choosing plums, look for undamaged fruits. They should be firm and plump, and give ever so slightly when gently squeezed. Even very firm plums will ripen in a couple of days on the kitchen counter. Keeping them in the refrigerator will slow the ripening process but not stop it. Both European and Japanese plums can be eaten fresh or used to make preserves, but only European plums can be used to make prunes. Damsons are quite tart—not the best for eating fresh—but make delicious jams and jellies. Plums can be used in cakes, crumbles, and streusels. They can also be candied to make sugar plums. Tart cultivars are good for making savory plum sauces to go with rich meats such as pork or duck. Chunks of them can also be added to stews or braises to provide a tasty counterpoint to the meat.

The Canadian plum's fruits can be eaten raw, though their tartness makes them better suited for use in jams, jellies, chutneys, and other preserves. The fruits of *P. spinosa* are used to make preserves and sloe gin, a liqueur.

LEFT: *The skin of the damson plum is very acidic, which makes it unsuitable for eating out of hand. It is mostly used to make jellies, jams, and preserves.*

Wither the Prune?

The prune, which is an oval black-skinned plum that has been dried, has battled an image problem during the past couple of decades. Prunes became something of a joke, particularly in English-speaking countries, and were associated with the elderly or as a cure for constipation (although the latter is somewhat true; prunes contain sorbitol, which humans cannot digest, thus giving them a laxative effect). In the United States in 2001, the derision was enough to make prune promoters petition the government for approval to change the packaging labels and call them dried plums.

But prunes appear to be making a comeback. In 2006, when British TV chef Nigella Lawson suggested using them in Christmas cake, prune sales at some grocery stores soared by 30 percent. Prunes are commonly used in braised or slow-cooked meat and game dishes as well as in tagines. Prunes can also be used to add moisture to lean ground meat. When soaked in alcohol such as Armagnac or port, prunes can make a very tasty dessert, especially if they are paired with chocolate.

Nectarine and Peach

Prunus persica

RIGHT: *Nectarines are ready to eat when they release a slight aroma and when they give a little to the touch. They are good sources of vitamin C and dietary fiber.*

BELOW: *Peaches with yellow flesh are not as sweet as the white-fleshed variety. All peaches have a thick, velvety skin.*

If any one fruit embodies the essence of summer, it is certainly the peach. Plucking a peach from the tree, standing barefoot in the grass and eating it while its juices run down your chin and arm is one of the most sublime experiences you can have with food. With its juicy and supple flesh, there is little wonder why the peach is often used in erotic art and literature.

Although the nectarine is viewed as a different fruit, it is actually a kind of peach, which is evident from its aroma and flavor.

Historical Origins The peach is native to China where it grows wild, although it has been cultivated for at least 3,000 years. The peach spread westward and was known to the ancient Greeks by 300 BCE according to writings by the philosopher Theophrastus. In the first century CE, the Roman writer Pliny the Elder noted several peach cultivars. Peaches grown in Montreuil, France, are some of the most famous in history. Even Louis XIV was enraptured by their exquisite quality.

The origin of the nectarine is not entirely known. Some research shows nectarines have been around for at least 2,000 years, but it is never clearly referenced in any writing prior to medieval times, when the French word *brugnon* was used, meaning "fuzzless peach." By the early part of the seventeenth century, several cultivars were being cultivated in England, and the term "nectarine" was in use.

Botanical Facts Peach and nectarine trees are deciduous, and their pink or white flowers bloom in the spring before the lance-shaped leaves grow. The trees, which can reach 20 ft (6 m) in height, are quite particular about their environment, and require cool winter temperatures (but not too cold), and warm (but not tropical) summer weather. For optimum flavor, peaches should be allowed to ripen on the tree.

Peaches and nectarines have either white or yellow flesh and are classified as either freestone, meaning the fruit easily separates from the pit, or clingstone, in which the fruit adheres to the stone. Interestingly, nectarines have been known to grow on peach trees and vice versa.

Culinary Fare Although a good, fresh peach is hard to beat, peaches are also delicious baked, poached, or even broiled. They can be used to make pies and jams, and homemade peach ice cream is a delectable treat. Peaches, especially the clingstone type, are also commonly canned.

The most famous peach-based dessert is Escoffier's peach Melba, named for Dame Nellie Melba, an Australian soprano.

Quandong

Santalum acuminatum

For the hearty souls who live in the outback of Australia, the quandong is, with little doubt, a favorite wild fruit. It is not something that city dwellers are likely to see in their neighborhood supermarket, but as Australian bush food becomes more widely known, more people are likely to make the acquaintance of the quandong.

Historical Origins Fossil evidence shows that the quandong was in existence more than 40 million years ago. The fruits were highly valued by Aboriginal Australians, who ate it either fresh or dried, and also used the kernel for medicinal purposes. European settlers quickly learned that quandongs were good to eat and began using them as they would similar fruits in jams, cakes, and pies.

In the 1970s, plant researchers from the CSIRO in South Australia began work to cultivate the quandong and to make it a commercially viable crop. It has proved to be challenging work, but there are now several orchards operating on a small scale. About 27 tons (25 tonnes) of quandongs are sold each year, but about 75 percent of that still comes from wild harvesting.

Botanical Facts The quandong is a parasitic small shrub or tree, which means it needs a host plant growing nearby in order to flourish, and can grow to about 20 ft (6 m) high and wide. The trees have olive-green lance-shaped leaves and small, creamy white flowers. They bear small, shiny, red fruit with a white to brown flesh.

Quandongs should be harvested as soon as their skin changes color, which usually begins to happen in early spring and runs until summer. A mature tree can produce up to 22 lb (10 kg) of cut and dried fruit, which is how it is most commonly marketed.

Culinary Fare Quandongs can be eaten fresh, and have the flavor of tart apricots or rhubarb, although their quality varies from tree to tree and some may taste quite sour. They can also be eaten dried and the dried fruit can be reconstituted in liquid for a variety of recipes. The fruits are a rich source of vitamin C.

Quandongs are used most often in jam making, but they can also be used to make pie fillings, jellies, fruit leather, cordial, chutney, and sauces to serve with meat or to pour over ice cream. When cooking with quandongs, take care not to add too much sugar so as not to overwhelm the fruits' pleasing tartness and flavor.

LEFT: *The quandong fruit ripens from green through yellow-orange to bright red, by which time it measures about 1¼ in (30 mm) across at the widest point.*

Black Chokeberry

Aronia melanocarpa

The chokeberry, which is named for its astringent flesh, grows widely across the United States and is a member of the rose family. The *Aronia* genus contains two species and a naturally occurring hybrid. The black chokeberry is distinguished from the others by the color of its fruit.

Historical Origins The black chokeberry is from the western United States and is less astringent than its eastern counterparts. Native Americans had many traditional uses for the fruit, and both they and early settlers used the berries extensively.

Botanical Facts The black chokeberry is a small, compact deciduous shrub about 3 ft (0.9 m) high, with white or pale pink spring blossoms that are followed by small berrylike fruits that turn shiny black when fully ripe. The foliage colors attractively in autumn in shades of red and crimson.

Culinary Fare The black chokeberry may be eaten directly from the tree, but it is extremely sour and acidic before it becomes fully ripe. It is used to make attractively colored jellies and preserves, or juiced and combined with lemon and cinnamon to make a fruit soup. Native Americans used it in their preserved dried meat, pemmican.

RIGHT: *Black chokeberries form clusters of small black fruits that each contain several small, dark purple seeds. The fruits fall to the ground when they are ripe. They are quite juicy, but shrivel up after ripening.*

Quince

Cydonia oblonga

The most fragrant of pomes, quinces are thought to have been the original golden apples of Cassical mythology. They must be cooked, as they are far too astringent and sour to eat raw.

Historical Origins The quince is a member of the rose family and is thought to have originated in the Caucasus, where wild varieties can still be found. Modern quince preserves can be traced back to classical times when whole quinces were preserved in honey, hence the Roman name *melimelum* ("honey apples").

Botanical Facts The quince is a small deciduous tree some 15–25 ft (4.5–8 m) tall, often bowed by the weight of its branches. The leaves are woolly on the underside and the single flowers range from white to pink. The fruit is shaped like a large, lumpy apple with a downy coating. Quinces are hard, yellow, and highly perfumed when ripe.

Culinary Fare Once cooked, quinces take on a pink tint. In Persian and Moroccan cuisine, quinces are often paired with meat to make perfumed, appetizing stews. Elsewhere, quinces, which are high in pectin, are most commonly used to make preserves such as the French *cotignac* (quince paste) or Spanish *membrillo*. Quinces are often cooked with apples or pears to add color and fragrance.

Persimmons

Diospyros kaki Japanese Persimmon *Diospyros virginiana* American Persimmon

Persimmon trees make a stunning display in autumn as their colorful leaves drop before their fruit ripens, leaving the bright orange globes suspended on bare boughs. Apart from their fruit, the trees are valued for their wood, and as ornamental plants.

Historical Origins The Japanese persimmon has been popular for centuries in the East, and since its introduction to the United States in the mid-nineteenth century, it has overtaken the American persimmon in popularity. The American persimmon has long been eaten by Native Americans and was used by early settlers to make alcoholic beverages such as persimmon beer and wine.

Botanical Facts Botanically, the persimmon is a berry. The fruits of the American persimmon are smaller than those of the Japanese species, but both are round and orange-red in color. The persimmon tree is deciduous with round, glossy, dark green leaves.

Culinary Fare The Japanese persimmon may be eaten fresh while still firm, or it can be cooked, candied, or made into ice cream or preserves. In Asia it is commonly dried and candied. The American persimmon must be allowed to sit until it is extremely soft, like the texture of jam, due to its high astringency. The soft pulp may then be eaten fresh or used in cakes, puddings, and preserves.

ABOVE: *To quickly ripen a persimmon, place it on a plate at room temperature, or in a bowl with a ripe banana or apple.*

Loquat

Eriobotrya japonica

Also known as the Japanese medlar, the loquat was once prized as the first soft summer fruit to mature. Advances in breeding have made this quality redundant, but the loquat remains a popular garden tree.

Historical Origins This subtropical tree is thought to have originated in southeastern China, and was cultivated widely across China, Japan, and India. It was introduced to Europe in the late eighteenth century, mainly for its attractive foliage. The fruits are luscious; the English botanist Sir Joseph Banks believed loquats were as good as mangoes. The loquat is cultivated in parts of Australia, the United States, and southern Europe, where it is often used as a decorative garden tree.

Botanical Facts The loquat is a small evergreen tree that can reach heights of up to 30 ft (9 m), and grows best in warm to hot climates. The foliage is prized for its beauty; the leaves are large, long, dark green and glossy, and the flowers are white and strongly scented. The fruits are small, egg-shaped, and borne in clusters, with large brown seeds and a downy covering.

Culinary Fare Loquats are delicious consumed fresh, but they can also be candied or used in preserves. The flesh is melonlike in texture with hints of the flavors of peaches, apricots, and citrus.

ABOVE: *Loquats can be mixed with other fruits in a fruit salad, or served by themselves, poached in a syrup. They are also used as a filling for pies and tarts.*

Apple

Malus x domestica

The apple is probably the most commonly consumed fruit in the world. It grows readily in temperate climates, and growing it from seed encourages new cultivars, as apples grown from seed can be completely different from the parent tree. Consistent results can be achieved only through grafting.

Historical Origins The apple was one of the first fruits to be cultivated and has come a long way from its beginnings as a wild fruit native to Europe, Asia, and North America. It is now the most important fruit in Europe, North America, and many other temperate regions across the globe.

Apples in Mythology

The apple is significant in the folklore and myths of many cultures. The Apple of Discord, given to Paris by the goddess Eris (Strife) was intended for the most beautiful of the Classical goddesses, and the chaos that ensued upon his presenting it to the goddess of love led to the Trojan War. One of the 12 tasks common to both Greek and Celtic mythology involves the retrieval of sacred apples guarded by maidens and dragons. In Celtic mythology, the apple is a symbol of fruitfulness and sometimes a path to immortality. In Old Norse mythology, the ancient goddess Idun ("the rejuvenating one") guards a casket of magical golden apples intended to rejuvenate the gods should their powers begin to fade. Apples were a symbol of fertility and life in the mythology of northern Europe (as they were in Roman mythology). Apple pips would be thrown into the fire as a means of divination in the nineteenth century.

RIGHT: *Aphrodite, the goddess of love, tries to win the golden apple for herself by offering Paris the beautiful Helen of Troy as a bribe.*

Historically, pears were more prized than apples, as most apple cultivars were small and sour due to the natural tendency of the plant to produce many small, tart fruits rather than fewer, larger, sweeter ones. By the sixteenth century, the reintroduction of grafting techniques first used in Classical times meant that new cultivars of apple were available. Additional cultivars were developed in North America when apple trees were established using seed (rather than using grafted stock) and then crossbred with native crab apples. Cultivation was further increased by the US folk hero Johnny Appleseed (real name John Chapman), who collected seeds from cider mills and planted them across a wide area.

Botanical Facts The cultivated apple is one of the most widely grown of all edible fruits, and many species and cultivars of crab apple are valued as ornamental garden trees. The apples and crab apples comprise a large genus of around 30 species of ornamental and fruiting, small- to medium-sized deciduous trees belonging to the rose (Rosaceae) family. Nearly all have soft green leaves. There are some 7,000–8,000 named cultivars of apple, although many of these are no longer commercially important. The nurseryman and writer Edward Bunyard developed a simple scheme of classification in which apples are divided into eight groups according to color, texture, and flavor. The scheme can be used to identify many cultivars.

Culinary Fare In Britain, apples are divided into two groups—for eating and for cooking. Cooking apples are high in malic acid, which enables them to break down to a puree when cooked. In the United States, apples are more likely to be judged on appearance and color. Red is preferred. The main commercial use for apples is cider, a fermented alcoholic drink (in North America the term refers to unfermented apple juice) that is a major industry in France and Britain. It is also used to make cider vinegar. Other

alcoholic drinks that are made from apples include French calvados and American apple-jack. Apples have long been preserved by drying, usually whole but more recently in slices. Apple butter is a popular preserve made by reducing apple sauce with cider. It was introduced to America by the Dutch and is still extremely popular.

Cooked apples have been served with meat to offset the fattiness since ancient times. The most famous apple dish, however, is undoubtedly apple pie, usually served with cream in Britain and Europe and with ice cream in the United States. The American apple pie has pastry on both top and bottom, whereas the British version is traditionally made with only an upper crust. The French make a simple, open-topped apple tart, and in German-speaking areas both the covered tart and the completely enclosed apple strudel are common. Apple dumplings are traditional in Britain and all over northern Europe and are now most commonly made by peeling and coring the apple, stuffing the cavity with sugar, butter, and spices, then enclosing the whole in pastry, and baking.

Other apple dishes include apple cakes made by mixing raw apples into plain cake mixtures, apple Charlotte—an English pudding using bread and cooked apples—and apple brown Betty—a North American pudding made from alternate layers of spiced fruits and breadcrumbs. Probably the most famous savory apple dish is the Waldorf salad, which dates back to the 1890s and comprises apples, celery, and mayonnaise (with the addition of walnuts in the 1920s).

LEFT: *Green apples have a tart flavor and a bright green skin. They are one of the best apples for use in cooking because of their ability to retain their flavor.*

BELOW: *The 'Fiesta' is one of the newer apple cultivars. Medium in size, the fruit is crisp, sweet, and juicy, and keeps well in storage. The tree also produces a particularly large crop.*

Medlar

Mespilus germanica

This unusual-looking fruit, which ripens in late winter, has a distinctive taste that is not to everyone's liking. It is often served with port wine at the end of a meal.

Historical Origins The medlar is native to Persia, but was cultivated by the Greeks and, from the second century BCE, by the Romans. It later flourished all over Europe. Medlars were a popular dessert in Victorian England and were valued in other parts of Europe for their supposed medicinal qualities.

Botanical Facts The medlar is a hardy, deciduous tree growing to a height of about 25 ft (8 m). Wild trees have thorns, but cultivars do not. It differs from other fruits in the pome family in that the base of the fruit is split open, exposing the 5 seed boxes. The medlar is an attractive garden tree with clusters of large white flowers and elongated leaves that turn deep red in autumn.

Culinary Fare Before eating, the medlar must undergo the process of "bletting," in which the fruit sits (usually in a tray containing sawdust or straw) until it is soft and almost rotten. This internal fermentation produces an acid, aromatic taste. Medlars are best eaten with a spoon and a little cream or sugar, but they can also be used to make jelly, medlar cheese, or medlar butter.

RIGHT: *Medlars are a very hard fruit. They have to be softened until they are almost decayed before they are ripe and ready to eat.*

Sorb Apple

Sorbus domestica

The sorb apple is closely related to the rowan tree. Its tart, acid fruit is now used primarily for flavoring liqueurs.

Historical Origins The sorb apple is native to southern Europe, North Africa, and western Asia. The tree's hard, fine-grained timber was once much prized by wood engravers. In Europe, the sorb apple was valued for its purported medicinal properties, and was also used to make a type of cider. In England, the sorb apple was often cooked in tarts. Commonly known as the "Service Tree" in Europe, in the United States, this name is given to a completely different species.

Botanical Facts A beautiful ornamental garden tree growing to a height of 30–50 ft (9–15 m), the sorb apple is deciduous, with clusters of green to yellow berries that turn bright red in autumn and look like small pears or apples. It has attractive serrated leaves and white flowers.

Culinary Fare Sorb apples are far too astringent to consume raw. They must be left until they are brown and almost rotten—a process known as "bletting." The fruits can then be eaten as is, but they are more commonly used to make liqueur or preserves, especially sharp-flavored jellies to go with meat.

Pears

Pyrus communis Pear *Pyrus pyrifolia* Asian Pear, Nashi Pear

The pear is one of the most long-lived of fruit trees, capable of surviving for more than 250 years. Apart from its delicious fruit, it is prized for its wood and for its beautiful appearance in autumn and spring.

Historical Origins The European or wild pear is native to temperate Europe, including Britain and western Asia where it has been cultivated since ancient times. The Asian pear has been cultivated in China and Japan for more than 4,000 years, but is now also grown in other East Asian countries and in the United States, Australia, and New Zealand. Traditional Asian pears have a gritty flesh and are sometimes known as "sand pears," while modern, high-quality cultivars are known as "nashi" pears. This tendency toward grittiness is due to high concentrations of lignin, a chemical compound present in smaller amounts in all pears. The chemical composition of both the Asian pear and the European pear is virtually identical.

Passion for Pears

At one time, the most important pear-growing areas in Europe were France, Belgium, and western Germany. This was partly due to their climates, but another important factor was that they were home to passionate, amateur pear growers intent on producing high-quality cultivars, such as, in the nineteenth century, the French 'Doyenne du Comice', widely held to be the finest pear of all. The soft, buttery, aromatic pears known as 'Beurre' were developed in Belgium and remain among the best. Of the 2,500 or so cultivars of European pear grown commercially worldwide, only 5 or 6 are consumed today.

Botanical Facts Widely distributed throughout Europe and Asia, the *Pyrus* genus, of about 20 species, is related to the apple and is part of the rose family (Rosaceae). The European pear is a medium-sized tree with rounded or oval, glossy, green leaves. It has thorny branches covered in white blossoms during spring. The Asian pear has oblong, orange and bronze leaves in autumn. Its small white flowers appear just before or with the emerging leaves. If fruits of either species are left to ripen on the tree, they develop a gritty texture.

Culinary Fare Pears are mainly eaten as a dessert fruit, although once they were widely used to make a type of pear cider known as perry. Pears are picked slightly green, and ripen off the tree. However, they deteriorate rapidly and are best eaten at their peak, when the flesh at the stem is slightly soft. They will stay fresh for longer if kept in the refrigerator. Pears are particularly good served with cheese, especially of the strong, sharp variety, but they also go well with chocolate, as in the famous dessert *poires belle Helene*.

ABOVE: *Asian pears are more like an apple in their shape. Not as sweet as European pears, their high water content makes them unsuitable for cooking.*

LEFT: *Mature but unripe European pears will keep well if they are kept cold and ripened later. They are delicious eaten fresh or canned, or used for juice.*

Pineapple

Ananas comosus

RIGHT: *A pineapple nears maturity in a field in Maui, Hawaii. Pineapples were first cultivated in Hawaii in 1901, and today supply most of the US market.*

In 1493, upon setting foot in Guadeloupe, in the West Indies, Columbus and his crew were amazed by a fruit "in the shape of a pine cone, twice as big, which fruit is excellent and it can be cut with a knife, like a turnip and it seems to be wholesome."

Historical Origins This fruit is native to Brazil and Paraguay. The first people to relish pineapples were the Tupi-Guarani Indians, and long before Europeans set foot in the Americas, cultivation of this "excellent fruit," had spread as far as the West Indies.

By the end of the sixteenth century, the Portuguese and Spanish had introduced pineapples to India, China, and Java (in Indonesia), and the west coast of Africa. Until the invention of hothouses in the late seventeenth century, they could be ripened in Europe but not grown. Even then they cost a small fortune, so remained a status symbol until canning made them affordable. Today, some 80 countries grow pineapples (mainly for juice and canning)—the leading producers being Thailand, the Philippines, Brazil, China, and India. The leaf fibers are used in the textile industry to make pina cloth.

Botanical Facts Pineapples are bromeliads, and the plant is a rosette of straplike, gray-green leaves around a terminal bud. The short, stout flower stem bears a globular flowerhead covered in small lavender-blue flowers. The fruit is actually 100–200 "fruitlets" fused in a perfect spiral that conforms to the Fibonacci, "divine proportion" sequence like the snail's shell, pine cone, and sunflower. Flesh ranges from nearly white to yellow.

Culinary Fare Ripe pineapples should feel heavy for their size, have a firm skin, and a tropical aroma. To prepare, cut both ends off, stand upright, and slice away the spiky skin in strips. Then quarter and remove the tough fibrous core from each wedge. Delicious in fruit salads, juices, and desserts, pineapples are also at home in savory dishes like Thai salads or the ever-popular ham and pineapple combination on pizzas. In Asia, half-ripe pineapples are used in Vietnamese and Cambodian sour soups, and in Malaysian and Indonesian curries.

Pineapple Power

The pineapple is the only known natural source of bromelain, an enzyme that helps to break down protein and is thus used to tenderize meat. At the same time it will prevent gelatine from setting and yeast products from rising, unless you cook the pineapple first. In the 1980s pineapples hit the headlines with Judy Mazel's Beverly Hills Diet—she advocated eating pineapple after meals to burn off body fat. According to this diet, papaya softens body fat, pineapple burns it off, and watermelon flushes it out of the body!

Cherimoya

Annona cherimola

Seen as an exotic tropical fruit today, the highly nutritious cherimoya, along with other annonas, has long been cultivated as part of the staple diet of indigenous peoples in South and Central America.

Historical Origins Probably native to the inter-Andean valleys of Ecuador, Colombia, and Bolivia, the cherimoya spread from there to the highlands of Chile and Brazil and throughout Central America. Its native name *quechua* (pronounced "chirimuya") means "cold seeds," as they can germinate at higher altitudes.

Botanical Facts The small spreading tree reaches 17–30 ft (5–9 m) in height. Its deep green, oval to lance-shaped leaves are briefly deciduous just before spring flowering. Fragrant yellowish flowers produce a primitive fruit with spirally arranged carpels, each segment surrounding a single hard, black, beanlike seed that is not for eating. The hermaphroditic flowers do not ripen simultaneously, making pollination a problem. Hand pollinating is often the order of the day where the cherimoya is cultivated.

Culinary Fare Ripe fruit should have a little give but not feel squishy. Best eaten fresh, the fruit is cut in half lengthwise and the sweet creamy flesh spooned out. Dip pieces in lemon or orange juice to prevent darkening before serving. Add cubes to fruit salads or puree for mousse or pie fillings.

ABOVE: *The flesh of the cherimoya is soft and creamy. Its exotic flavor is often described as a combination of banana, pineapple, and strawberry.*

Guanabana, Soursop

Annona muricata

This fruit is so juicy that you might speak of drinking it rather than eating it. It is the largest member of the *Annona* genus, and can weigh up to 6 lb (3 kg).

Historical Origins Ancient Peruvian pottery vessels have been discovered shaped like guanabanas, sometimes topped with a fruit-eating bat as well. Native to the lowlands of Central America and the West Indies, the guanabana was one of the first fruits carried from the New World to the Old World tropics. The fresh fruit is widely available in Malaysia (as "durian belanda") and other parts of Southeast Asia, and with its refreshing sour taste it is the *Annona* most preferred by the Asian palate.

Botanical Facts Normally evergreen, guanabanas reach 20–30 ft (6–9 m) and will flower and fruit more or less continuously in humid tropical areas, although there is usually a principal growing season. The short-stalked, yellow-green, bell-shaped flowers are borne directly on the trunk, branches, or twigs and produce the very large oval or heart-shaped, dark green fruit covered in small, soft, curved spines that break off easily when ripe.

Culinary Fare Guanabanas are found fresh in tropical markets, but the meltingly smooth, somewhat acidic, snow white, fibrous flesh is more commonly processed into ice creams, sorbets, sherberts, jellies, smoothies, and candies. It makes an excellent fruit punch or refreshing daiquiri. In Central America the juice is bottled as a carbonated drink.

ABOVE: *Guanabanas are ready to pick when the skin is deep green in color.*

Apple Annonas

- *Annona reticulata* Bullock's Heart, Custard Apple
- *Annona squamosa* Sugar Apple, Sweetsop

RIGHT: *The conspicuous seeds of* Annona reticulata *are the usual means of propagation, and there can be more than 70 in each fruit.*

Do not be put off by appearances: The lumpy custard apple has a rich creamy texture and a delicate sweet flavor, which is often described as being a cross between a banana and a pineapple.

Historical Origins Numerous annonas, including the hybrid atemoya (*A. cherimola* x *squamosa*—a cross between the sweetsop and cherimoya), are called "custard apples." So, to avoid confusion, opt for botanical names. It is thought that Spaniards carried seeds of *Annona reticulata* from the New World to the Philippines and the Portuguese introduced the sugar apple, along with other annonas, to southern India, Indonesia, and West Africa.

Perhaps not quite the family ugly duckling, the custard apple lacks the commercial clout (and flavor) of the other annonas and the hybrid atemoya that is widely cultivated in Australia and called, of course, custard apple.

Botanical Facts *A. reticulata* has drooping clusters of fragrant, light green flowers that never fully open. The often heart-shaped fruit can have a pink reddish blush when ripe, hence the alternative common name "bullock's heart." Its agreeable flavor lacks the distinctive character of the cherimoya, sugar apple, or atemoya.

A. squamosa has fragrant yellowish flowers borne singly or in clusters of 2–4. Its compound fruit has the typical fleshy center with segments, each enclosing a single black or dark brown seed, which radiate from the central core attached to the stem. There may be a total of 20–38, or perhaps more, seeds in the average fruit although some trees produce seedless fruits.

Culinary Fare Whatever the cultivar, a ripe fruit should give slightly when squeezed. To eat fresh, cut in half lengthwise and twist the halves to separate. Use a small spoon to scoop out the flesh, discarding the seeds and the fibrous center. The fresh flesh goes well with a little cream and sugar. In recipes, add the flesh to fruit salads, make into ice cream, sorbets, drinks (smoothies and cocktails), custards and desserts, fillings for cakes, or try as an accompaniment to spicy dishes such as curry.

Not to Be Taken Lightly

As with many tropical plants, the leaves, bark, and fruit of annonas have a place in traditional natural remedies. A decoction of the leaves alone or with those of other plants is used as a tonic, cold remedy, or digestive, or added to baths to alleviate rheumatic pain. The bark and roots are highly astringent—a bark decoction is given as a tonic and to halt diarrhea. The root, because of its strong purgative action, is administered as a drastic treatment for dysentery. The crushed seeds are toxic and used to treat head lice and parasites, ensuring no contact with the eyes.

Breadfruit

Artocarpus altilis

Famous as the fruit at the center of the mutiny on the *Bounty* in 1789, the breadfruit has been cultivated throughout Southeast Asia and the Pacific for hundreds of years, becoming a dietary staple of the local cuisine wherever grown.

Historical Origins Breadfruit has been a staple crop in the Pacific for over 3,000 years. Migrating Polynesians and Hawaiians spread the breadfruit across the Pacific, which is where European explorers discovered it. Hearing of its nutritional and prolific fruiting virtues, planters in Jamaica petitioned King George III of England for permission to import breadfruit trees to provide a reliable food source for their plantation workers; in 1793 Captain William Bligh delivered 2,126 breadfruit plants to St. Vincent and Jamaica.

Botanical Facts An impressive tree from the mulberry (Moraceae) family growing to 50–85 ft (15–26 m), it has a broad crown of deeply lobed leaves and produces a multitude of tiny flowers; males in small catkins, females in large heads. The large compound fruit that follows is 8 in (20 cm) in diameter, with yellow-green skin and white flesh; they are borne singly or in clusters at the branch tips.

Culinary Fare Used in savory and sweet dishes, breadfruit can be eaten ripe as a fruit or under-ripe as a starchy vegetable. It is usually cooked—boiled, roasted, fried (as chips), or wrapped in leaves and baked in an underground oven. In Hawaii, it can take the place of taro to make *poi*, a sour fermented paste that is a traditional food staple. The seeds can be roasted and eaten like chestnuts.

ABOVE: *Several unripe breadfruit grow on a branch tip. At this stage the fruit is hard and the flesh is white and fibrous.*

Jackfruit

Artocarpus heterophyllus

The jackfruit has no contenders in the heavyweight league. It is massive—weighing in from 10–20 lb (5–9 kg) and occasionally up to 100 lb (45 kg), which explains why the fruit is borne on the main branches, trunk, and even on surface roots.

Historical Origins Found from India to the Malay Peninsula, jackfruit is believed to be indigenous to the rain forests of the Western Ghats of India and has played a significant role in Indian agriculture and culture—archaeological findings reveal that jackfruit was first cultivated in India 3,000–6,000 years ago. It is the national fruit of Bangladesh and Indonesia.

Botanical Facts The jackfruit belongs to the mulberry (Moraceae) family. The handsome tree has shiny, deep green leaves and grows to 30–50 ft (9–15 m). The large compound fruit is formed from an entire flower cluster. Within the fruit there are several hundred yellow to pink, fleshy segments called pericarps, each containing an edible (when boiled or roasted) seed the size of a date. The stringy tissue or "rags" between the segments can be eaten. Unopened ripe fruit has a strong heady aroma.

Culinary Fare Jackfruit, with its pungency and bananalike flavor, may be an acquired taste, but it is a staple, both ripe and immature, in many parts of the world. The segments are used in sweet and savory dishes (soups, fruit salads, curries) and boiled with rice or coconut milk. Before cutting into jackfruit, coat the knife, chopping board, and your hands with oil to protect them from the sticky latex.

ABOVE: *The skin of young jackfruit is light green and deepens to yellow-brown when mature. It is the largest edible tree-grown fruit in the world.*

Bilimbi

Averrhoa bilimbi

ABOVE: *Bilimbi fruit grow in clusters on the trunk and older branches of the tree.*

Rather like a lemon, the sour-tasting, gherkin-shaped bilimbi delivers a tart tang to foods and beverages. But it has another important household role—cleaning and burnishing brass and other metals, as the acid (it contains about 5 percent oxalic acid) dissolves tarnish and rust.

Historical Origins Grown throughout Southeast Asia and much of India, the bilimbi was possibly native to the Maluku Islands (Moluccas). Along with the famous breadfruit, Bligh delivered bilimbi and mango trees (collected from Timor) to St. Vincent and Jamaica in 1793. From there it spread to Cuba, Puerto Rico, and to Central America and northern South America. It was introduced into Queensland, Australia, in 1896.

Botanical Facts This medium-sized tree reaching 50 ft (15 m) from the wood sorrel (Oxalidaceae) family is more often found in a garden than cultivated as a crop. Clusters of small, slightly fragrant, purple to orange-red flowers sprout directly from the branches and produce the slow-ripening, pale yellow-green, watery fruit.

Culinary Fare Bilimbi brings a sharpness to curries, sambals, pickles, and chutneys. It is most often eaten cooked, though in Costa Rica the fresh flesh is made into a relish that is served with rice and beans. The fruit can also be prepared like gherkins or preserved in sugar syrup. Do not use aluminum utensils when preparing this fruit.

Carambola, Star Fruit

Averrhoa carambola

ABOVE: *These yellow carambolas are ready to eat. The cut fruit reveals a star shape.*

With its starlike cross section and adaptable nature, the juicy carambola is used as a garnish, added (skin, seeds, and all) to salads, fruit salads, and mains, and used in preserves and relishes.

Historical Origins Although not found in the wild, it is probably native to the Maluku Islands (Moluccas) and Sri Lanka, and has been cultivated throughout southern China, India, and Southeast Asia for centuries, spreading across the Pacific and introduced to Australia, the Caribbean, and the United States.

Botanical Facts A bushy tree growing to 20–30 ft (6–9 m), its deciduous leaves are sensitive to touch and light, folding at night or if handled. Small clusters of red-stalked, lilac to purple-streaked flowers are borne on the twigs in the leaf axils much of the year, followed by yellow-green to orange 5-pointed fruit. Insects love the ripe fruit—stinging moths can take out a whole crop. Physically attacking the moth (at night) or netting the tree solves the problem for home gardens.

Culinary Fare Juicy, thirst-quenching, and crunchy, the carambola is the closest to a crisp apple that tropical fruit gets, and it can be used as a substitute for apples in recipes. In Asia, the preferred way of eating them is simply dipping half a ripe fruit into salt. Choose those that are deep yellow for eating ripe (just wash and eat, no peeling). For cooking, where they act as a souring agent, buy green fruit.

Peach Palm, Pejibaye

Bactris gasipaes

Little known outside Central and northern South America, the fruit of the peach palm, with its chestnutlike flavor, is one of Costa Rica's unique treats sold from market stalls and stands along the roadside.

Historical Origins The peach palm, from the same palm family as the coconut, is unknown in the wild, but is probably native to southwest Amazonia. The peoples of tropical America cultivated it for centuries for its abundant starchy fruit, which provided an important source of energy (carbohydrates).

Archaeologists have found seeds in Costa Rica dating back some 4,000 years, and accounts from Columbus's fourth voyage (1502) reveal how important it was for the indigenous people at that time. They boiled it, dried it, preserved it, and stored it in trench silos in the ground, then pounded it for making tortillas (as a substitute for maize flour), fermented it to make an alcoholic drink, and used the wood from its thick stems for bows, arrows, and spears. Today the wood is mainly used for quality building materials, such as parquet.

Botanical Facts There are more than 200 palm species in this genus belonging to the Arecaceae family. Primarily from Central America, few are cultivated. One that is cultivated is the peach palm, a clumping, multi-stemmed (usually) palm growing to 65–100 ft (20–30 m). The stems are typically armed with spines in regular bands. It has feathery, dark green leaves. Long sprays of tiny yellowish, mingled male and female flowers are followed by thin-skinned, peach-sized fruit hanging in clusters of 50–500 that weigh up to 25 lb (11 kg). A single stem may bear 5–6 clusters at a time. As the fruit ripens it changes color from yellow or orange to red or purple, although there can be considerable variation in form, size, and color. Some fruits are seedless, but normally there is a white, 3–4 in (8–10 cm) long, conical-shaped kernel.

ABOVE: *Peach palm fruit are commonly sold in local markets. Scarred fruits are considered superior in quality because they are firm, have a low water content, and less fiber in the flesh.*

Culinary Fare Peach palm fruit has to be cooked as it is caustic raw—the flesh contains calcium oxalate crystals. It is usually boiled then peeled, halved, and pitted. The fruit is eaten for breakfast or as a snack with mayonnaise or sauce. Cooked fruits are used in stews, or pureed for making dishes such as Costa Rica's *sopa de pejibaye*, where they are combined with onions, sweet peppers, garlic, and chicken stock. The hard-shelled kernel tastes like coconut and is rich in oil.

Peach palms, like other palms, are also commercially cultivated for their tender growing tip that provides the gourmet vegetable delicacy called heart of palm, also known as palm cabbage or palmito, which is widely used as a salad ingredient.

Akee

Blighia sapida

ABOVE: *The akee is an attractive tree that produces an abundance of fruit. In Jamaica, it is planted in front yards and roadsides and is a naturalized plant.*

The line "Carry me akee go a Linstead Market, not a quattie wud sell," from the popular Jamaican folk song "Linstead Market," features the akee, Jamaica's national fruit. It is the main ingredient in its national dish, akee and saltfish, but the fruit is not actually a "local" at all.

Historical Origins Native to the forests of the Ivory and Gold Coasts of tropical West Africa, akee probably arrived in Jamaica sometime in the 1770s on a slave ship. In 1778, Dr. Thomas Clarke, an early propagator, introduced it to the eastern parishes. The *Blighia* in its botanical name recognizes Captain William Bligh's role in introducing it to the English scientific community. The *sapida* refers to the substances in its seeds, which make water soapy and frothy.

Botanical Facts The large, usually evergreen, tree grows to 50 ft (15 m) and has a dense crown of spreading branches. Long racemes of small greenish flowers are followed by red, yellow, or orange, 3-lobed fruit capsules 2–4 in (5–10 cm) long that split open spontaneously when ripe, revealing the fleshy arils tipped with a shiny black seed. There are two main types identified by the color of the aril—a soft yellow aril is known as "butter"; "cheese" is hard and cream-colored.

Culinary Fare Jamaica is the only place where akee is widely eaten. Only the arils of a mature fresh fruit that has opened naturally can be eaten, as light dispels the toxic hypoglycins (amino acids) found in unripe fruit. The aril must be properly cleaned to remove the pink or purplish membrane under the seeds and the cooking water discarded. Cooked arils look rather like scrambled eggs and are sometimes called "vegetable brains." Canned akee arils are exported for the substantial expatriate Jamaican market.

From Bounty to Blighia

In 1787, Captain William Bligh was charged with transplanting a major food crop (breadfruit) from one part of the world to another. It was a feat of horticultural skill not done before on such a grand scale, and he was accompanied by two gardeners from Kew: David Nelson and William Brown. The trees had to be sourced, seeded, potted, and grown into saplings large enough for transport, then maintained on board ship over a long voyage. The story of the *Bounty*, the mutiny, and Bligh's extraordinary feat navigating some 3,700 nautical miles (6,000 km) across the Pacific to Timor is well known. What is less known is that in 1791, Bligh returned to Tahiti with the *Providence* and the *Assistant* to complete the job and successfully transported breadfruit (and other) trees to Jamaica. He returned to England in 1793 with 465 pots and 2 tubs of plants, including the akee, from the St. Vincent Botanic Gardens, for the Royal Botanic Gardens, Kew.

Peanut Butter Fruit

Bunchosia argentea

Peanut butter may not grow on bushes, but if you smell the fragrant fruit and leaves of *Bunchosia argentea* you could be excused for thinking it does.

Historical Origins Native to South America, this tropical rain forest tree is grown more as an ornamental fruit tree for the front or backyard than as a commercial crop—its ripe fruit needs to be picked every day as it spoils quickly on the tree, and it needs to be handled with care as the skin is very tender.

Botanical Facts The attractive small tree reaches 12 ft (3.5 m) and produces clusters of yellow flowers throughout spring followed by a constant supply of small, dark orange to dark red fruit about the size of a medium fig, with sticky dense flesh that has a texture commonly described as being somewhat like peanut butter. The tree is quick-growing after planting and produces fruit within 2–3 years, which makes it popular for the home gardener.

Culinary Fare The fruit can be eaten fresh, made into milkshakes, added to cakes, or used for preserves, jellies, jams, and dips. Fruit that is just ripening into orange delivers the crunchiest "peanut butter" flavor; fully ripened fruit has a sweeter taste. Fruit can be stored for several days in the refrigerator or separated from the seed and frozen.

RIGHT: *The peanut butter fruit tree requires a hot tropical climate in order to survive. Only in these conditions will it produce a good quantity of fruit.*

Nance

Byrsonima crassifolia

Throughout its natural range, the round fruit of this tree is enjoyed by the locals, the birds, and wild animals, while oil-collecting bees relish the abundance of oil, rather than the sweet nectar its flowers produce.

Historical Origins The indigenous peoples of Central America and northern South America have long made the most of the nourishing fruits produced by a number of species of *Byrsonima*. Of these, the nance is possibly best known as having one of the widest native ranges of all fruits in tropical America—from the Caribbean to Central America and throughout much of South America. The trees are spared when forests are cleared, and maintained in a state of semiwild cultivation. The plant was introduced into the Philippines in 1918.

Botanical Facts This slow-growing tree adapts to a wide range of tropical and subtropical climatic conditions and typically reaches around 33 ft (10 m). Showy yellow-red flowers are followed by plum-sized yellow fruit about 2 in (5 cm) in size, with white, juicy, oily flesh and a single, fairly large stone with 1 or more seeds.

Culinary Fare The fruit is eaten fresh, used in savory and sweet dishes, preserved, made into drinks, used to make the fermented beverage *chicha*, and distilled to produce Costa Rica's *crema de nance*. It is popular in Panama, and is sold along the roadside or in markets.

Papaya

Carica papaya

Whether red, orange, or yellow-fleshed, or labeled "papaya" or "pawpaw" (in Australia), these cultivars are all siblings under their *Carica papaya* skin and can be eaten ripe as a fruit or green as a vegetable. The true papaw (*Asimina triloba*), however, is a completely different fruit related to the cherimoya (*Annona cherimola*).

Historical Origins Exactly where the papaya originated from is unknown, but it is most likely the lowlands of eastern Central America. It had already spread further afield when the Spanish and Portuguese arrived in the New World, and they, in turn, carried it to the Old World and their expanding empires in the East and West Indies. Today it is grown in nearly all tropical and subtropical regions around the world.

Botanical Facts Like bananas, papayas are botanically large herbs growing to 30 ft (9 m) with a hollow green stem and long-stalked leaves that last 4–6 months before falling, leaving behind the familiar leaf scars. Separate male and female flowers are white or cream to green. Once fertilized, the larger female flowers develop the familiar fruit, ranging from a deep orange to a pale green color with soft and juicy, yellow or pinkish flesh that surrounds a cavity of shiny black-gray seeds. Fruit size and shape varies considerably depending on the cultivar.

ABOVE: *The silky smooth flesh of the papaya is juicy with a sweet tart flavor. The black-gray, peppery seeds are edible but are usually discarded.*

Culinary Fare Ripe papayas are most commonly eaten fresh, merely peeled, seeded, cut into wedges, and served with a half or quarter of a lime or lemon, often for breakfast. They are also added to salads and fruit salads, used in salad dressings, pureed for sauces, and made into jams and chutneys. Green papaya can be cooked similarly to zucchini or marrow and served as a vegetable or added to curries. In Thailand, fresh green papaya is finely shredded to make green papaya salad.

The Tender Touch

Papaya fruit (especially green ones) and leaves contain papain, a natural enzyme that aids the digestion of proteins. It is collected by scratching the skin of green fruit and catching the milky latex, which is then used diluted in a marinade or with water to tenderize tough cuts of meat. Cooks have long been well aware of this handy culinary tip—tough meat is often wrapped in papaya leaves, or a little of the "latex" from the fruit, leaves, or stems is drizzled over it before cooking.

An illustration from 1880 detailing the papaya's fruit and foliage.

Natal Plum

Carissa macrocarpa

The Natal plum is a dense, thorny, quick-growing shrub with single, fragrant, pure white, long-tubed, star-shaped flowers, making it popular for home gardens and hedges. It is also suitable for bonsai.

Historical Origins Native to the coastal regions of Natal in South Africa, it was introduced into the United States in 1903 when the US Department of Agriculture distributed a large quantity of seeds (from Durban's Botanical Gardens) to test their growth in different climate zones—these being Florida, the Gulf States, and California.

Botanical Facts The Natal plum is from the dogbane (Apocynaceae) family, and is a cousin of the colorful oleander. It grows to 18 ft (5.5 m) with a similar reach and has glossy, dark green, leathery foliage. It flowers and fruits prolifically, month after month in the right conditions, bearing pinkish red to dark purple-black berries about ½ in (12 mm) long with strawberry flesh and small edible seeds. The tree's stems and unripe fruit both contain a milk sap (latex).

Culinary Fare The ripe berries are eaten whole, including their skin and seeds. They can be used fresh for fruit salads or for topping ice cream, desserts, and cakes; for making jelly; or cooked for pies and tarts. The fruit preserves well for jams, chutneys, or bottling whole fruit.

ABOVE: *The edible fruit of the Natal plum is tender and juicy, and tastes very much like a cranberry.*

Casimiroa, White Sapote

Casimiroa edulis

The word "sapote" from the Nahuatl (Aztec) *tzapotl*, was a term often used to describe soft sweet fruit, hence today's confusion over various "sapotes." They are not all related. Some belong to the actual Sapotaceae family. The casimiroa does not—it is from the rue (Rutaceae) family and is a distant cousin of citrus.

Historical Origins Originating in the Mexican highlands, the casimiroa is found both wild and cultivated in central Mexico, and spread from there to Central America and the West Indies. It was introduced into California by Franciscan monks in about 1810, and is also grown in South Africa, Australia, and New Zealand.

Botanical Facts Growing to 50 ft (15 m), the casimiroa has attractive drooping branches, palmate leaves with 3–5 oval leaflets, and greenish yellow flowers. Trees can be long lived and fruit prolifically in the right conditions, producing up to 1,000 fruits each year. The fruit itself is persimmon- to egg-shaped with apple green to orange-yellow, papery thin skin, and sweet edible flesh.

Culinary Fare It is first and foremost a dessert fruit with a taste similar to a combined pear and banana. The skin is usually thickly peeled as the flesh near it can have a bitter taste. Alternatively, the fruit is halved and the flesh scooped out. It can be served alone or as part of a fruit salad and makes delicious ice creams and smoothies and, with a dash of lemon or lime juice, jellies and sherbets. It can also be used for marmalade, or fruit leather when dried.

Caimito, Star Apple

Chrysophyllum cainito

ABOVE: *The caimito—a green cultivar is shown—has clear white flesh.*

The caimito is not an apple at all, it is a berry that can be as big as a small apple. Its other common name, star apple, describes the fruit's appearance when cut in half.

Historical Origins Native to Central America, the caimito was cultivated long before the Spanish set foot in the New World—conquistador Pedro Cieza de Leon describes it in his 1553 book, *Chronicle of Peru*. It is common throughout the West Indies and a symbol of the Caribbean in *The Star-Apple Kingdom* by Nobel Prize-winning poet, Derek Walcott.

Botanical Facts This shady, ornamental, evergreen tree from the sapodilla (Sapotaceae) family grows to an average 50 ft (15 m) in height and has deep green leaves with gold-bronze felting beneath. The clusters of small, starry, creamy white flowers form in the leaf axils or sprout directly from the branches and are followed by rounded, 4 in (10 cm) wide berries ripening to purple (although there are green cultivars) with edible sweet flesh surrounding seed cells that form a star shape when the fruit is cut across.

Culinary Fare The fruit is mainly eaten fresh, simply cut in half and spooned out. In Jamaica, a traditional "half-sweet, half-sour" dish called matrimony is made by combining the flesh of the caimito (without the seeds) with orange juice, a little sugar, grated nutmeg, and whipped cream, if desired.

Seagrape

Coccoloba uvifera

RIGHT: *The seagrape tree, with its lush foliage and clusters of fruit, happily grows next to the ocean in beach sand. It is salt tolerant but does not like strong winds.*

Seen more as an "exotic" fruit today, once upon a time little of this plant went to waste: the fruit was eaten; the branches used for firewood or furniture; the sap called "kino" for tanning and dyeing; the roots and bark for traditional natural remedies; and the leaves for plates, or pinned together with thorns, as hats.

Historical Origins Widely distributed along the Atlantic and Pacific coasts of tropical and subtropical America, the seagrape is believed to be native to southern Florida, the Bahamas, and the West Indies. The fruit is reddish purple when ripe and hangs in clusters, so the comparison with grapes was quickly made, hence the common name. Today the seagrape is most often used as a landscape tree or ornamental shrub or hedge in coastal areas, and will readily colonize sandy shores and dunes.

Botanical Facts A sprawling bush or small evergreen tree in the knotweed (Polygonaceae) family, the seagrape can reach 20–30 ft (6–9 m) high and 10 ft (3 m) wide. It has large, green, heart-shaped leaves, with reddish veins. The spikes of small, fragrant, greenish white flowers are followed by bunches of fleshy fruit with a pit (technically a small nut).

Culinary Fare The fully mature fruit is tartly sweet in flavor and can be eaten fresh or used for making jams and jellies. It is also possible to make an alcoholic beverage similar to wine, from the grapes.

Elephant Apple

Dillenia indica

There is always room for confusion with common names and "elephant apple" is a good example—it is shared by *Dillenia indica* and *Limonia acidissima* (which is also called the wood apple; see box below). What is also shared is the origin of the name: elephants clearly enjoy both fruits and in the case of *D. indica*, they play an important role in dispersing the seeds.

Historical Origins Native to India and Sri Lanka, *D. indica* is widespread from India through to Java in Indonesia, hence the variety of common names: it is called *chalta* in India and *simpoh* in Malaysia. A handsome, spreading, evergreen tree to 50 ft (15 m) or more with large lustrous leaves, showy flowers, and eye-catching fruit, it is loved as a shady ornamental in parks and gardens. In India, the timber is used for railway sleepers.

Botanical Facts *D. indica* bears large, fragrant, magnolialike flowers that are borne in terminal panicles, in spring and summer. After flowering, the petals drop and the sepals close up to form a thick, fleshy, protective covering surrounding the "true" fruit inside. The juicy acid fruit measures 4 in (10 cm) or more in diameter and looks a little like an artichoke.

Culinary Fare The outer greenish yellow sepals of *D. indica* have a tart taste and are added as a vegetable to curries and used for making jellies, cool drinks, chutneys, and pickles. In the Philippines, the juice serves as vinegar; in Panama, the fruit is mixed with sugar for a refreshing "fresco."

In Sri Lanka, *Limonia acidissima* is mixed with coconut milk for making wood-apple milk, ice cream, and mousse. In Thailand, the young shoots and leaves are eaten with larb and added to sour bamboo salad.

ABOVE: *The fruit of* Dillenia indica *is made up of large fleshy sepals that tightly enclose the "true" fruit inside. Each fruit contains 5 seeds that are embedded in the juicy pulp.*

Apples for Elephants and Fruit for Monkeys

A native of India and Sri Lanka, *Limonia acidissima* (known as the wood apple, but also called the elephant apple) is another fruit that appeals to elephants, as an earlier botanical name, *Feronia elephantum*, attests. Clearly, there was competition for this round gray fruit the size of an apple with a hard shell and sticky brown flesh—the Sanskrit name *kipipriya* means "dear to monkeys."

The small, thorny wood apple tree grows to 30 ft (9 m) and bears large berries with a woody, ¼ in (6 mm) thick rind that is hard to crack. Inside, small white seeds are imbedded in a sticky flesh. The creamy unripe flesh is used to make a sour sambal. The pectin-rich, dark chocolate pulp of fully ripened fruit is used in chutneys and for making jellies and jams.

Longan

Dimocarpus longan

ABOVE: *These fresh longans are ready to peel and eat. Although the flesh is similar to the lychee, many claim the longan to be more fragrant and sweeter in taste.*

The word "longan," which is both the species name and common name in English, comes straight from the Chinese *lung yen* or *long yan*, which means "dragon's eye." Once the outer skin is peeled away, the translucent, jellylike flesh with its huge red-brown to black seed looks very much like a large fleshy eye.

Historical Origins Although it was thought the longan was originally from southern China, experts now extend the native range to Southeast Asia and possibly India and Sri Lanka. Even though it has often taken a backseat to its glamorous cousin, the lychee, the longan has come into its own as an economically important crop in Thailand, China, Taiwan, and Vietnam. In Thailand, longan production is more important than lychee, and in the peak season, from June to August, market tables are piled high with this small, popular brown fruit. It is also grown in Queensland, Australia, and in Florida, California, and Hawaii in the United States.

Botanical Facts The longan, an evergreen tree in the soapberry (Sapindaceae) family, can grow up to 40 ft (12 m). It thrives in tropical areas, and will also grow in more temperate zones that are free from heavy frosts. Clusters of intermixed, creamy white male and female flowers are followed by the prolific panicles of round, leathery, brown-skinned fruit that is smaller than the lychee, with the fleshy edible aril surrounding the seed inside. The skin is smooth by comparison with the lychee's rough checkered skin.

Culinary Fare Like lychees and rambutans, longans are enjoyed fresh. The crisp skin is easy to remove and the white flesh has a texture and fragrant sweetness reminiscent of lychees. The freshest longans will be sold on their twiggy fruit stalk, or at least with a short piece of stem attached. As they are very perishable, they are often sold frozen, canned, or dried. Once thawed, frozen longans can be used like fresh ones. Canned in their own juice (with no added sugar), they retain their flavor and texture well and are served with Asian desserts such as almond jelly, or combined in fruit salads. Dried longans can be boiled in water to make cool refreshing drinks. In Nonya cuisine, dried longans are simmered with black sticky rice to make rice pudding. Dried longan flesh is also an ingredient in traditional natural remedies.

BELOW: *Dried longans in a serving dish. They look like a plump raisin and have a somewhat smoky flavor. They can be eaten as a snack, like other dried fruits.*

Ceylon Gooseberry

Dovyalis hebecarpa

Abundant and prolific describe this vigorous grower that bears fruit in great quantities and can produce multiple crops throughout the year. It is also known as the ketembilla or kitembilla.

Historical Origins Native to Sri Lanka and southern India, the Ceylon gooseberry is grown throughout tropical Asia and was introduced into the United States in the early part of the twentieth century, and flourished in southern Florida. From here, it was distributed to the Hawaiian Islands, the Philippines, and the West Indies. Along with many other tropical and subtropical plants, seeds of this species were exported from Sri Lanka to Israel.

Botanical Facts A spiny, dioecious, evergreen shrub or small tree that grows to 15–20 ft (4.5–6 m), the Ceylon gooseberry has small, inconspicuous, greenish yellow male and female flowers that are borne on separate plants. The fruit with its crimson-red flesh is a berry about 1 in (25 mm) round with a tart taste reminiscent of the unrelated gooseberry (*Ribes uva-crispa*). It has several small seeds, and the velvety skin ripens to a dark purple-red color.

Culinary Fare For most people, the Ceylon gooseberry is an acquired taste for eating out of hand, as it is rather sour and the skin is bitter tasting. The fruit is most often used for making jams, jellies, pickles, preserves, and drinks. In Hawaii, it is combined with other tropical fruits to make ketembilla-papaya jam and ketembilla-guava jelly.

ABOVE: *The skin of Ceylon gooseberry fruit is covered with short, grayish green hairs that are unpleasant in the mouth.*

Durian

Durio zibethinus

From love to loathing, this heavy, spiky, soccerball-sized fruit with its pungent odor produces powerful reactions. Revered in Southeast Asia as the "king of fruits," it remains very much an acquired taste elsewhere in the world.

RIGHT: *The fruit of the durian tree has a thick skin covered in sharp spikes that brown as the fruit ripens. The flesh has a creamy texture and is eaten fresh, or can be frozen or canned.*

Historical Origins Native to the lowland rain forests from Myanmar to Malaysia and Indonesia, durians are currently cultivated throughout Southeast Asia, India, and Sri Lanka and in areas with a similar climate, including Australia's tropical north. The species name derives from the scientific term for the civet cat, *Viverra zibetha*, of Southeast Asia, though it is not clear whether this is because the fruit smells like a civet or is used as bait to trap them.

Botanical Facts The durian is a big tree reaching 80–130 ft (24–40 m) and has simple lance-shaped leaves. Clusters of large, creamy white or pink flowers are borne on older wood. The green to brown oval fruit has 5 segments containing 3–4 seeds surrounded by dense custardlike arils—pale cream to bright yellow in color, depending on the cultivar.

Culinary Fare Ripe durians are eaten fresh but have a short shelf life. In cooking they are used for preserves (jams and pickles), milk-based desserts, and ice creams, cakes, and confectionery. In some parts of Asia unripe durians are used as a vegetable. The seeds can be cooked and eaten, too. In Indonesia, fermented durian, *tempoya*, is a popular side dish.

Tropical Cherries

✳ *Eugenia aggregata* Cherry of the Rio Grande
✳ *Eugenia uniflora* Surinam Cherry, Brazilian Cherry, Cayenne Cherry, Pitanga

ABOVE: *The Surinam cherry tree produces hundreds, even thousands, of fruit each cycle, 2 or 3 times a year.*

Like their temperate namesakes, these tropical cherries are attractive ornamentals in the home garden, especially the Surinam cherry, which with its wine-colored young leaves and dense-growing habit, is ideal for hedges or screening.

Historical Origins The cherry of the Rio Grande is native to Brazil. The native range of the Surinam cherry, the best known of the edible-fruited *Eugenia* species, extends from Central America to southern Brazil and is grown in subtropical and tropical areas around the world today. It is believed the Portuguese took Surinam cherry seeds from Brazil to India, from where it was introduced into Italy and first described botanically from a tree in a garden in Pisa.

Botanical Facts These large shrubs or small evergreen trees from the myrtle (Myrtaceae) family bear fragrant, creamy white flowers in the leaf axils, followed by drupelike red to purple-black berries. The Surinam cherry produces juicy, thin-skinned, 8-ribbed fruit up to 1 in (25 mm) in diameter with 1–2 seeds singly or in clusters falling from slender stems. The cherry of the Rio Grande has egg-shaped fruit about 1 in (25 mm) long with up to 2 seeds.

Culinary Fare The juicy and aromatic fruits, which are very soft once ripe, are eaten fresh; added to fruit salads, fruit pies, and desserts; and used to make juice, jellies, jam, or spicy chutneys. The juice is sometimes fermented into vinegar or wine.

Mangosteens

✳ *Garcinia mangostana* Purple Mangosteen ✳ *Garcinia prainiana* Button Mangosteen, Cherapu

ABOVE: *The thick skin of the purple mangosteen protects the sweet and delicate segments inside.*

The "melt-in-the-mouth" purple mangosteen, with its juicy segments, is often regarded as the "queen of the tropical fruits." Unlike the "king" (the durian), it immediately appeals to most palates.

Historical Origins Native to Malaysia and Indonesia, the purple mangosteen, commonly known as the "mangosteen," has been subsequently introduced throughout the tropics and warmest subtropical regions, as it is very frost sensitive. The purple mangosteen was first introduced into Australia around the 1850s. Long used in traditional natural remedies, the rind of this species has been found to be rich in compounds called xanthones, which have antioxidant properties. Enter marketing and the mangosteen drink, XanGo, craze. Clinical trials are yet to show that these extracts benefit humans.

Botanical Facts The slow-growing (15 years before fruiting) purple mangosteen is a densely foliaged, evergreen tree from the St.-John's-wort (Clusiaceae) family with highly scented male and female flowers on separate plants that open from late in the afternoon. The large mandarin-sized fruit has a thick rind that ripens to a rich purple color. The flesh consists of 4–8 translucent whitish segments, 1 or 2 having soft seeds. Button mangosteens have fragrant reddish pink flowers and produce brilliant orange-colored fruit with sweet-and-sour flesh.

Culinary Fare Fresh is best with purple mangosteens. To open, cut through the rind and lightly pull and twist the fruit apart. They can be used for sorbets and are available canned, but the flavor just isn't the same. Button mangosteens tend to be grown mostly for their fresh fruit.

Dragonfruit, Pitaya

Hylocereus undatus

Short and sweet is the life of the large, fragrant, bell-shaped, white flowers of this sprawling climbing cactus. They open at dusk and last only the one night, earning them the title "queen of the night" or "moonflower."

Historical Origins Of uncertain origin, the dragonfruit is believed to be native to the rain forests of Central and northern South America. Today it is cultivated in dry tropical or subtropical areas in the Americas and Southeast Asia. Vietnam is a major producer of dragonfruit (the French introduced it there 100 years ago), and it is also grown in Israel and northern Australia.

Botanical Facts Dragonfruit is an epiphyte cactus that clings to its support (a trellis, a wall, an old tree stump) and can obtain nutrients from cracks where organic material concentrates. It has fleshy, succulent, 3-angled, jointed stems, with a wavy horny margin and 1–3 short, conical, brown to gray ⅛ in (3 mm) long spines. The 10–12 in (25–30 cm) long flowers rapidly develop into spherical to oval, bright red fruit up to 4 in (10 cm) long, often with large green tentacles. The juicy flesh is scattered with tiny, edible, black seeds.

Culinary Fare Ripe fruit has a light melonlike taste and is best eaten fresh and chilled. A dash of lemon or lime juice enhances the flavor. To eat, cut in half and scoop out the flesh and seeds much like a kiwi fruit. Serve with a fruit platter, in fruit salads, or as a garnish.

ABOVE: *Dragonfruit cultivars have white or red flesh containing tiny black seeds.*

Ice-cream Bean

Inga edulis

A large, quick-growing tree with spreading branches forming a dense canopy, the ice-cream bean is widely used in Central and South America to provide shade for plants (around coffee plantations) and people (in parks and avenues).

Historical Origins Native to the West Indies and Mexico through to subtropical South America, the ice-cream bean has been introduced across most of tropical South America, Panama, and Costa Rica. Like other *Inga* species, it has been associated with cultivating cacao and coffee since pre-Columbian times. Why? It grows fast, provides shade, helps prevent erosion, and, like other members of the legume (Fabaceae) family, maintains soil fertility as a nitrogen-fixing plant.

Botanical Facts Typically reaching 60 ft (18 m) in height, the ice-cream bean has bright green leaves and bears heads of fragrant white flowers with a "pompon" appearance at the tips of the stems, followed by the very long (up to 3 ft/0.9 m) cylindrical, bean-shaped, seed pods. These dangle from the tree and contain a sweet white flesh surrounding several large green seeds.

Culinary Fare The flesh is delicious and tastes very much like vanilla ice cream—monkeys and birds also consider it a delicacy. To eat, split open the pod, remove the seeds, and peel away the juicy cottony pulp. It has traditionally been used to make *cachiri*, a fermented beverage that can be based on maize, manioc, or fruits.

Lychee

Litchi chinensis

ABOVE: *A halved lychee reveals its glossy seed.*

Luscious is often used to describe the lychee (also called litchi), a pinkish red fruit that has inspired countless poems, paintings, and elaborate treatises in China for many centuries.

Historical Origins Native to southern China, the lychee has been cultivated there for more than 2,000 years. The earliest known horticultural monograph (CE 1059) is Ts'ai Hsiang's treatise on lychees called *Li chih pu*, which describes its cultivars, areas of cultivation, methods of preservation, and popularity. The world's major producers today are China, India, Thailand, Taiwan, and Vietnam.

Botanical Facts The slow-growing lychee is a spreading evergreen tree to 40 ft (12 m) with a thick canopy of dark green leaves reaching to the ground. Closely related to the longan (*Dimocarpus longan*), it bears small, greenish white to yellow, insignificant flowers in long panicles at the branch tips, followed by the famously showy fruit. The thin hard skin peels easily away to reveal the pearly, translucent white flesh with a glossy brown seed that is toxic and should not be eaten.

Culinary Fare With their sweet clean flavor, delicate aroma, and slippery texture, juicy lychees are unsurpassed eaten fresh and make a clean-tasting finale to a meal. In cooking, they are best in simple desserts and fruit or savory salads. They can be added to stir-fries and curries—but not until the very end of cooking time. Peeled and seeded lychees are canned in syrup. They are also dried whole and when cracked open reveal a dark, sweet, chewy pulp half its original size.

Acerola

Malpighia glabra

RIGHT: *The fresh ripe fruit from the acerola is high in vitamin C. It is used in various nutritional supplements and its juice is taken as a health tonic in Brazil.*

Acerola is something of a vitamin C superstar being one of the richest known sources, which is why you can find it processed and packaged on the dietary supplement shelf labeled "natural vitamin C."

Historical Origins The native range extends from southern Texas through the Caribbean and Central America to Brazil. There is some confusion over both its common and species names. There were originally three species, all commonly known as acerola or Barbados cherry: *Malpighia emarginata*, *M. glabra*, and *M. punicifolia*. *M. punicifolia* is now considered to be synonymous with *M. glabra*, but *M. emarginata* is still an extant species, and is still called acerola or Barbados cherry—but it is not as widely grown as *M. glabra*.

Botanical Facts These tropical evergreen shrubs growing to 10–20 ft (3–6 m) in favorable conditions, have smooth-edged, glossy leaves and bear pale to deep pink or red flowers—distinctive because of their long-stemmed petals held clear of the central staminal cluster—followed by small, round, red, thin-skinned edible fruit (a drupe).

Culinary Fare The ripe fruits have a pleasant but slightly sour taste, and are more commonly used for making pies, jams, jellies, preserves, juice, ice creams, and flavoring cocktails, rather than being eaten fresh. The fruit flesh or juice is also added to processed foods to increase the vitamin C content.

Mango

Mangifera indica

Chin-dripping, juicy mangoes may be difficult to peel and messy to eat, but few would deny that the effort is worth it. Mango lovers advise eating them in the shower!

Historical Origins Originally from Southeast Asia, especially Myanmar and eastern India, where they have been grown and domesticated since ancient times (perhaps 4,000 years), mangoes were introduced to Africa 1,000 years ago and came to the New World in 1782, when "Captain Marshall of His Majesty's Ship *Flora*, one of Lord Rodney's squadron, captured a French ship bound from Mauritius to Haiti, and on board were found many plants and seeds of economic value, amongst them being the Mango... The ship was sent as a prize to Jamaica, and Captain Marshall, ...deposited the collection of plants and seeds in the garden of Mr. Hinton East."

Today this heavenly fruit is grown in tropical and warm-temperate countries for its handsome foliage and fruit. India is the world's leading producer, and it is Australia's most important tropical fruit crop.

Botanical Facts A cousin of the cashew, the mango grows to about 80 ft (24 m). It has simple, leathery, smooth-edged leaves that are reddish when young, aging to shiny dark green. Dense panicles of small yellowish or reddish bisexual and male flowers are produced on the same plant, followed by the fruit—a large, fleshy, egg-shaped, hanging drupe with a flat fibrous seed. There are countless cultivars, but most have green, red, or yellow inedible skin with yellow-orange, sweet, aromatic flesh.

Culinary Fare There is almost nothing like a fresh mango, especially slightly chilled. Probably the easiest way to eat them fresh (or prepare them for cooking) is to slice off the cheeks on either side of the seed, score the flesh in a criss-cross pattern, flip inside out, and spoon out the dice. Add to fruit and savory salads, desserts, sauces, drinks (including cocktails), jams, and chutneys. Green (unripe) mangoes are used for making chutneys or grating to add piquancy to salads and salsas and are widely used as a souring agent in Indian and Southeast Asian cooking. Tender mango leaves are eaten raw in Thailand.

ABOVE: *Fresh mangoes are often cut to resemble hedgehogs. This is a popular way of presenting the fruit for eating, as the dice come away easily from the skin.*

Tart, Tangy, Tantalizing

Unripe green mangoes are peeled, sliced, and sundried to make *amchur*, a fine, off-white to pale gray powdery spice reminiscent of lemon juice—1 teaspoon of *amchur* has the equivalent acidity to 3 tablespoons of lemon juice. *Am* is Hindi for "mango" and *choor* "powder." A tantilizingly tart spice popular in India, it adds piquancy to curries, soups, dhals, pickles, and chutneys; can be used in marinades for meat and fish; lifts the flavor of stir-fried vegetables, prawns, or chicken; and is added to recipes using beans, chickpeas, and lentils. Mango salsa, combining diced mango with chili, onion, lime juice, salt, and pepper, is a popular accompaniment to barbecued or broiled chicken or fish.

Sapodilla

Manilkara zapota

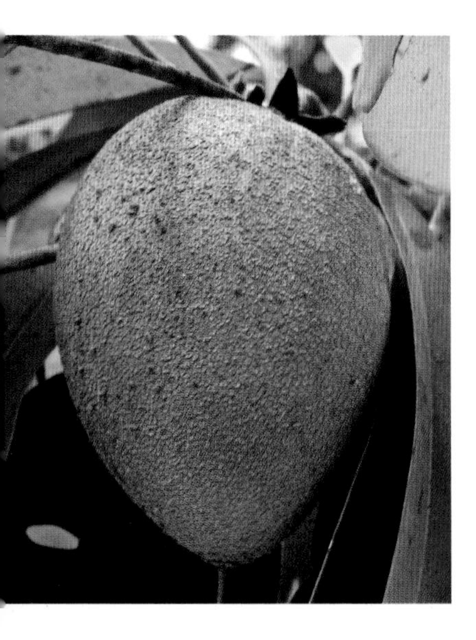

ABOVE: *The skin of the sapodilla fruit is covered with a brown scurf until ripe.*

Chew on this! In 1869, the inventor Thomas Adams popped a piece of "chicle" into his mouth and liked the taste. Chewing away, he had the idea to add flavoring. In February 1871, Adams New York Gum went on sale in drugstores for a penny apiece, and a new industry was born.

Historical Origins Native to Mexico, Guatemala, and Belize, the sapodilla is found throughout Central America where it has been cultivated since ancient times. The tree contains a latex called "chicle" (from the Nahuatl (Aztec) word *tziktli* for "sticky stuff"), which was dried and chewed by the Aztecs, the Maya, and early European settlers. It was introduced into the United States by General Antonio Lopez de Santa Ana, who reputedly gave a piece to Thomas Adams' son to chew.

Botanical Facts A slow-growing, evergreen tree that can reach 100 ft (30 m) in the tropics, the sapodilla has large simple leaves with a thin, papery texture. Its small white flowers are followed by fuzzy-skinned, egg-shaped, golden brown berries to 3 in (8 cm) long with juicy, yellowish brown or pinkish brown flesh and 3–12 hard, inedible, black seeds with a hook at one end that can catch in the throat.

Culinary Fare The caramel-tasting fruit is typically cut in half, the flesh scooped out with a spoon, and eaten fresh; or it is added to fruit salads, or made into purees, iced drinks, ice creams, and desserts. In the West Indies, ripe fruits are crushed and boiled down to make syrup; mashed sapodilla is also added to pancake batter and bread mixes before baking. It is eaten fried in Indonesia.

Brazilian Grape Tree

Myrciaria cauliflora

Little known outside their natural range, Brazilian grape trees are the most popular native fruit-bearing trees in Brazil. An alternative common name, jaboticaba, comes from the Tupi words *jabuti* (tortoise) and *caba* (place), probably meaning "the place where you find tortoises." Brazil is certainly home to many tortoises.

Historical Origins Of the four Brazilian grape tree species, *Myrciaria cauliflora* is the most widespread and has been cultivated in Brazil since pre-Columbian times. Native to an area around Rio de Janeiro and Minas Gerais, the tree features as the "charge" (symbolizing the fruits of the earth) on the coat of arms of Contagem, a city in the center of the region. In Brazil and other parts of South America where it thrives, it is grown in small commercial gardens, and as a backyard tree in other tropical and subtropical zones.

Botanical Facts The Brazilian grape tree is a small to medium evergreen tree to 40 ft (12 m) with peeling bark like a eucalypt. What is most distinctive about the species is that the small white flowers and purplish black, grapelike fruit are borne directly on the trunk, branches, and exposed roots. The fruit has white or pinkish flesh and 4 small seeds. The skin is high in tannins.

Culinary Fare The juicy sweet fruit is eaten fresh soon after picking as it does not keep well, or is used to make jams and jellies, sauces, and fruit salads. Commercially, it is used for jams and jellies, liqueurs, and wines.

Bananas

Musa acuminata ✳ *Musa balbisiana*

ABOVE: Musa acuminata, *shown here, is the more widely grown banana species. There is not much difference in appearance between it and* M. balbisiana.

The banana plant is technically a suckering perennial herb and not a tree. What seems to be the "woody" trunk is actually a "pseudostem" formed from concentrically furling leaf bases, with each true stem flowering and bearing fruit once only.

Historical Origins Edible bananas are native to Southeast Asia, reaching as far south as northern Australia. Archaeological evidence at Kuk Swamp, in Papua New Guinea, suggests they were cultivated there as long ago as 5000 BCE. They spread to Africa, across the Pacific, and to the Mediterranean as far as Egypt 1,000–2,000 years ago. But it was not until the United Fruit Company set up business in the United States in 1899 that bananas became a widely recognized fruit around the world. Today, they are the world's fourth largest fruit crop. As for the name—its origin is West African, passing into English via Spanish or Portuguese.

Botanical Facts The large, mid-gray to green, paddle-shaped leaves are waterproof, and useful for wrapping food for cooking, carrying, and serving in many parts of Asia and the Pacific. The massive yellow, white, or cream flower clusters have female or hermaphrodite flowers near the base and male flowers near the tip. The fruit ripens from deep green to yellow. Most commercial cultivars, usually seedless, belong to the species *Musa acuminata*, native to Southeast Asia and northern Queensland, Australia. *M. balbisiana* of southern Asia and the East Indies, is valued for its disease resistance, playing an important "parent" role in banana breeding.

Culinary Fare Bananas are one of the world's most popular fruits, eaten raw or cooked—whole, sliced, or mashed; or as a snack, smoothie, or part of a dessert, fruit salad or meal. Unlike other fruits, bananas contain both starch and sugars—the starch converting to sugars (sucrose, fructose, and glucose) as they ripen. Commercially, banana puree is used in baby foods, baked goods, ice cream and yogurts, and dairy desserts.

BELOW: *The fruit of* Musa acuminata, *seen here, has the typical banana bend.* M. balbisiana *bananas tend to be straighter.*

Ripeness Is All

Like most fruit, ripening bananas give off ethylene gas. But no other fruit gives off quite as much as bananas do, which is why "cooks' tips" suggest putting unripe fruit in a brown paper bag with a banana, to speed up the ripening process. It is a case of one fruit helping another. The "gassing" principle is not new: the Egyptians, for example, gassed figs to stimulate ripening. In 1901, Dimitry Neljubow, of the Botanical Institute of St. Petersburg, was the first to recognize ethylene as a biologically active plant compound after noticing that the peas in his laboratory near the gas lamps were not growing in the same way as the ones outside, and had abnormal stems and leaves. The reason, he discovered, was that the air in the laboratory had high levels of ethylene, a by-product of the combustion of coal gas used in the lamps.

Rambutan

Nephelium lappaceum

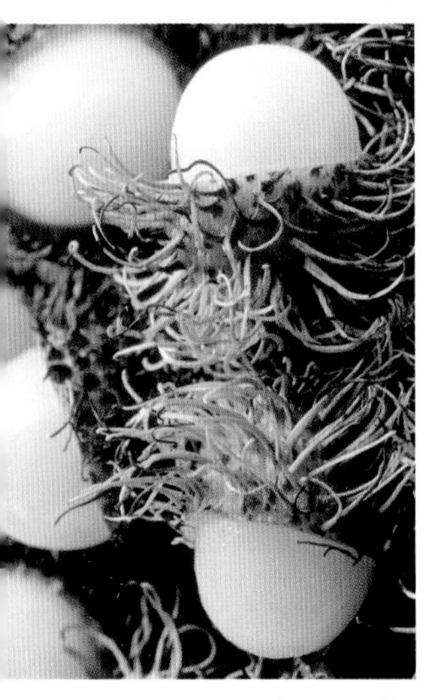

ABOVE: *Rambutans, such as these exotic Thai ones, are succulent and sweet.*

Imagine market stalls piled high with bright crimson or yellow hairy fruit—it is peak rambutan season in Southeast Asia. A close relative of the lychee, the rambutan is just as desirable in this region.

Historical Origins A native of Malaysia and Sumatra, Indonesia, and spreading to other parts of Southeast Asia, the rambutan has been grown for centuries for its fruit and a black dye made from the peel. Commercial cultivation has spread to northern Australia and to Zanzibar and Madagascar. As for the name, it comes from the Malay *rambut*, meaning "hair," describing the long soft spines on the surface of the peel.

Botanical Facts The rambutan is an evergreen tree to 80 ft (24 m) with a dense, usually spreading crown, large leaves, and clusters of small, petal-less, yellowish flowers, followed by—depending on the cultivar—distinctive pinkish red, bright or deep red, orange-red, maroon or dark purple, yellowish red, or all yellow or orange-yellow spiny fruit. Each fruit has a single large seed surrounded by the juicy, pearly white flesh.

Culinary Fare Cultivated primarily for eating fresh, the rambutan is both delicious and nutritious. Commercially, it is canned or cooked for stewed fruit, jams, and jellies. When it comes to eating out of hand, ask an expert: "Make a cut with a sharp paring knife, as if you were going to slice the fruit in half, but only cut through the skin. Then lift off half the skin, leaving the rest as a decorative holder... Hold the rambutan with the fingers... Nibble daintily and detach only the succulent flesh," says Asian food expert, cook, and author Charmaine Solomon.

Otaheite Gooseberry

Phyllanthus acidus

This gooseberry is no gooseberry, despite the name. It merely resembles *Ribes uva-crispa* because of the similarity in the sharp acidic taste of both fruits.

Historical Origins The exact origin of the otaheite gooseberry is not clear. It is thought to be native to Madagascar, spreading from there to Southeast Asia and on to the Indian Ocean islands of Mauritius and Reunion and also across the Pacific. In 1793, it was introduced to Jamaica from Timor by Captain William Bligh, and progressively spread throughout the Caribbean islands, and to Bermuda. It is now naturalized in Central and South America.

LEFT: *The otaheite gooseberry fruits prolifically, forming massive clusters of fruit throughout the tree, creating a spectacular visual display.*

Botanical Facts A fast-growing tree to 30 ft (9 m), the otaheite gooseberry rather resembles the bilimbi. The smooth-edged, ovate leaves are 3 in (8 cm) long, and sit at the branch tips. The small, rosy pink flowers that are borne in clusters on the main trunk and branches can be male, female, or hermaphrodite. They are followed by 6–8 ribbed fruits with crunchy, juicy, acidic-flavored flesh and a hard stone with 4–6 seeds.

Culinary Fare The fruit has a bit too much pucker power to be eaten fresh. It is more commonly made into vinegars, preserves, and pickles. It can also be made into beverages and sauces—the fruits and juice turn bright red when cooked with sugar. In Indonesia, the flesh is added to various dishes as a flavoring. The young leaves are cooked and served as greens in India and Indonesia.

Canistel, Egg Fruit

Pouteria campechiana

Being described as having mealy flesh rather like a hard-boiled egg yolk or a baked sweet potato may not sound like an invitation to tuck in, but history tells that RAF pilots training in the Bahamas during World War II thought the canistel so tasty they bought up Nassau's entire stock.

Historical Origins Native to southern Mexico, Belize, Guatemala, and El Salvador, the canistel is cultivated in Central America, from Mexico to Panama and in Puerto Rico, Jamaica, and Cuba. It was long important to the indigenous peoples for its starchy, nutritious fruit and timber. There are a number of small commercial plantings in many tropical areas, including northern Australia.

Botanical Facts An evergreen typically reaching 60 ft (18 m), the canistel has a milky sap and glossy papery leaves arranged in spirals. Its small, greenish white, tubular flowers are borne along branches followed by the yellow to greenish brown, waxy-skinned fruit with orange-yellow, mealy flesh, which has a sweet flavour, and up to 4 inedible seeds. The fruit varies in shape, and can be nearly round with or without an apex, or oval.

Culinary Fare The ripe fruit is eaten fresh or lightly cooked and flavored with salt, pepper, and lime juice. The pureed flesh is used in custards, beverages (such as smoothies or egg nogs), or added to baking goods such as pies and pancakes.

Tropical Guavas

Psidium guajava Common Guava, Yellow Guava
Psidium littorale Strawberry Guava

The genus name is from the Greek word *psidion* for "pomegranate," prompted perhaps by the hard seeds. But guavas are not related to pomegranates. They are part of the myrtle (Myrtaceae) family and distant cousins of cinnamon and feijoas.

Historical Origins The common guava is indigenous to Central America and an important crop in Central America, India, Asia, Africa, and Australia. The strawberry guava is native to eastern Brazil and is also widely grown in many parts of the tropics—it has even become a serious weed in some areas such as Hawaii and the Caribbean, springing up wherever birds drop seeds.

Botanical Facts Both *Psidium* species are evergreen shrubs or small trees that branch freely, almost to the ground. *P. guajava* grows to 15–30 ft (4.5–9 m), while *P. littorale* reaches 20–25 ft (6–8 m), but is usually smaller. They both bear white, 5-petalled flowers followed by juicy berries with numerous seeds (except in seedless cultivars). The fruit can be rounded or pear-shaped berries almost the size of a baseball, ripening to red or yellow-pink with strong aromatic flesh; strawberry guavas are dark red and walnut sized.

Culinary Fare The whole fruit is edible, although some people cut out the center with its hard seeds. Ripe guavas have a sweet tangy taste plus a distinctive fragrance, partly due to eugenol, an essential oil also found in cloves. They are eaten fresh, added to fruit salads, poached as a dessert, and made into juice. Cooked, they are made into jams, jellies, preserves, and a stiff paste called guava cheese. In Asian countries they are often eaten under-ripe with a salty or spicy dip or sauce. As for canned guavas, delicious though they are, it is like eating a completely different fruit.

ABOVE: *The fruits of the common guava ripen from green to red or yellow-pink. The leaves are leathery, and fragrant when crushed.*

Tamarillo

Solanum betaceum

ABOVE: *Their deep red color shows that these tamarillos are ready to harvest. Simply snap the fruit off the tree, leaving the stem behind.*

Just as they led the charge making the kiwi fruit an international celebrity, New Zealand horticulturists commercialized the tree tomato, coming up with "tamarillo" to give it a more exotic name for overseas markets.

Historical Origins Generally believed to be native to the high altitudes of Peru, Chile, Ecuador, and Bolivia, the tamarillo was once in its own genus, called *Cyphomandra*, but is now known to be from the nightshade (Solanaceae) family (just like the potato, eggplant/aubergine, and garden tomato), and is listed among the lost foods of the Incas. The fast-growing, brittle, shallow-rooted tree was most typically grown in home gardens until the 1930s when breeding trials to develop it as a crop plant began in New Zealand. Production was increased there during World War II when imported fruits such as bananas became hard to get. Today, the world's major producer is New Zealand. Tamarillos are also grown on a commercial scale in Colombia and Ecuador, with smaller plantings in Australia, California, Africa, and Asia.

The New Zealand Tree Tomato Promotions Council wanted a less confusing (and more evocative) name for the fruit they were ready to export. "Tamarillo" was the inspiration of Mr. W. Thompson, who combined the Maori word *tama*, implying leadership, with a Spanish-sounding ending, although the exact inspiration for "rillo" is unknown. On January 31, 1967, the fruit's commercial name was officially changed from the "tree tomato" to "tamarillo," which has been adopted around much of the world.

Botanical Facts The tree is an evergreen reaching 10–18 ft (3–5.5 m) in height with thin, heart-shaped leaves that have soft hairs and conspicuous veins. Loose clusters of fragrant pale-pink or lavender flowers borne near the branch tips are followed by the egg-sized and -shaped fruit with its supersmooth, usually burgundy red skin. The deep orange, orange-red, or orange to yellow or cream-yellow, juicy, tart-sweet flesh surrounds soft edible seeds.

Culinary Fare The fruit is eaten fresh simply by scooping from a halved fruit—children, of course, just bite the top off and squirt the flesh into their mouths. Peeled fresh slices can be added to fruit and savory salads or used as a topping for desserts and cheesecakes. Pureed tamarillo flesh can be used as a sauce for ice cream or as a base for making sorbets or homemade ice cream, or poached to make a compote. It makes tangy chutneys, dips, and sauces for fish, meats, and chicken. In Colombia and Ecuador, tamarillos are blended with water and sugar for juice.

ABOVE: *The flesh of the tamarillo tastes like a cross between a passionfruit and a tomato.*

Miracle Fruit

Richardella dulcifica

This "lemons to lemonade" transformer is rarely found in a recipe ingredient list. It is best known as a "taste-bud tricker." Coat the tongue with a pleasant squirt of its juice and sour foods will seem sweet for up to 2 hours afterward.

Historical Origins The miracle fruit is native to tropical west Africa. Its transforming properties were first documented in 1725 by French explorer Chevalier des Marchais, who noticed the locals eating miracle fruit before meals such as sour-tasting oatmeal, gruels, and palm wine. It will grow in most tropical and subtropical areas but is difficult to market—the perishable fruit turns brown and unpalatable within a day of picking.

Botanical Facts This slow-growing, evergreen shrub or tree reaches 18 ft (5.5 m) in its native habitat, but much less elsewhere. It produces small white flowers in flushes followed by the bright red, pleasant but slightly tart tasting, olive-shaped fruit with a single inedible seed.

Culinary Fare Miracle fruit is more a novelty item for the "sweet lime" party trick than a food crop or even garden plant. Freeze-dried miracle fruit is available, and in Japan it is canned or sold in tablets. It is not a sweetener—the sweet effects depend on what is eaten afterward. It has reputedly been helpful in restoring appetite for some cancer patients whose taste buds have been affected by their treatment.

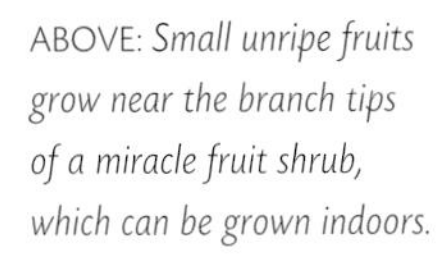

ABOVE: *Small unripe fruits grow near the branch tips of a miracle fruit shrub, which can be grown indoors.*

Mombin

Spondias mombin

The Amazonian Indians called the mombin *tapereba*, meaning "fruit of the tapir"—not only because tapirs are extremely fond of them (and are important seed dispersers), but they used the fruit as bait to hunt them.

Historical Origins This tropical tree is native to the lowland forests of the Amazon in Peru, Brazil, Venezuela, Bolivia, Colombia, and the Guianas, as well as southern Mexico, Belize, Costa Rica, and the West Indies. Fruit is typically harvested from naturally occurring trees rather than grown as a crop. In some areas, cut branches are used to build "living fences" as they take root readily. It is widely cultivated and naturalized in tropical Africa, possibly by natural, long-distance dispersal rather than being introduced.

Botanical Facts A medium-sized, occasionally large, erect, deciduous tree to 65 ft (21 m), the mombin has thick corky bark sometimes with blunt spines on younger trees. Alternate leaves bunched toward the end of branches have a faintly turpentine odor when crushed. Panicles of fragrant whitish flowers are followed by clusters of light golden yellow, plumlike fruit with thin tough skin and juicy, sweet yet tart flesh. It has a single, white, fibrous seed.

Culinary Fare The fruit is eaten fresh or stewed with sugar. A concentrated juice adds a tang to ice cream, drinks, and jellies. The fruit is also made into jams and preserves, or wines, such as Vinho de Tapereba, and liqueurs. In Mexico, green fruit is pickled in vinegar.

Tropical Apples

Syzygium jambos Rose apple *Syzygium malaccense* Malay Apple

ABOVE: *These Malay apples will fall from the tree when ripe.*

Like many other tropical fruits given "apple" as a common name, these *Syzygium* species do not resemble the apple in any way whatsoever, though their texture is crisp.

Historical Origins The fleshy rose apple is native to the Malay Peninsula and Indonesia and is found throughout India and Southeast Asia. Native to Malaysia, the Malay apple (one of several fruits sharing the name "jambu" in Southeast Asia) is cultivated from Java to the Philippines and Vietnam, as well as in Bengal and South India and across the Pacific. The Portuguese carried it from Malacca to Goa and from there it was introduced into east Africa. Bligh introduced it to Jamaica in 1793.

Botanical Facts These tropical apples, from the myrtle (Myrtaceae) family, are distant cousins of cloves. They are large, ornamental, evergreen shrubs or small trees with leathery, dark green leaves that are paler under the surface and shiny pink to wine red in color when new, depending on the species. The rose apple has large, showy, nectar-rich flowers with creamy white stamens and fragrant, pink to yellow edible fruit. The Malay apple has clusters of cream or reddish purple flowers and red, pink, or white fruit with a very shiny skin.

Culinary Fare These juicy, thirst-quenching fruits are mostly eaten fresh by children. They are sometimes stewed with sugar and served as dessert. The fruit can be made into jam or jelly with lemon or lime juice added, or preserved with other fruits with a stronger flavor. In Southeast Asia, the rose apple is dipped in soy or fish sauce.

Tropical Almond

Terminalia catappa

ABOVE: *The fruits of the tropical almond grow within a rosette of leaves.*

Beautiful, shady tropical almond trees with large oblong leaves that turn red before falling are a feature of seashores and coastlines throughout the tropics.

Historical Origins The tropical almond is widespread in tropical Asia, parts of Polynesia, and northern Australia. Like other members of the species, the trees have long been important for communities, providing food, shade, timber, and ingredients for traditional natural remedies. Its vast surface root system binds together both sands and poor soils, making it important for erosion control. It has been introduced into Hawaii, Brazil, the Caribbean, and parts of east Africa.

Botanical Facts It is a medium to large, briefly deciduous tree reaching an average of 90 ft (27 m) in height with a broad spreading crown and tiered horizontal branches. The leaves are often clustered near the shoot tips, hence the scientific name *Terminalia*, which means "end" in Latin. Spikes of small white or creamy flowers, with a slightly unpleasant smell, are followed by the oval drupe fruits (1–4 in/2.5–10 cm long) that ripen to a purplish red and have a fibrous seed containing the edible kernel.

Culinary Fare The delicious milky kernel tastes like an almond, and can be eaten raw or roasted. A cooking oil is made from the sundried kernels. The sweetish tartish flesh of the fruit is edible but fibrous, and sticks to the seed. The foliage is used in traditional medicines, and as food for silkworms.

Babaco

Vasconcellea x heilbornii

With its light texture and tangy strawberry overtones, the babaco is often called the champagne fruit. It is normally seedless and soft, so the entire fruit can be eaten or liquidized, and it is renowned for making refreshing sorbets.

Historical Origins Native to Ecuador, the babaco was classified as *Carica pentagona* in 1921 by plant explorer Otto Heilborn. It is not found in the wild and is now believed to be a hybrid between *Vasconcellea cundinamarcensis* and *V. pubescens*. Today, it is widely cultivated in South America, and as the most cold-tolerant plant in its genus, it thrives as far south as New Zealand, and north to the Channel Islands, off the coast of Normandy.

Botanical Facts The tree grows to 10 ft (3 m) in height. The large leaves of this small tree are shed during the colder months, ringing the trunk with leaf scars. It only has female flowers and can produce from 25–100 5-angled, 12 in (30 cm), "torpedolike" fruits with pale apricot flesh each year. The plant has an average life span of 8 years.

Culinary Fare Although not very well known or widely grown, babacos are versatile and can be eaten fresh with a little lemon or lime. Being seedless, they can easily be made into sauces, desserts, and drinks. Unripe fruit is cooked as a vegetable, used in curries, and for making relishes, chutneys and pickles. Like the papaya, it contains the enzyme papain.

Chinese Jujube

Ziziphus zizyphus

Cultivated in China for over 4,000 years, this highly adaptable plant has been a prodigious traveller for centuries, spreading to southern Europe, Russia, northern Africa, and the Middle East, probably in the traders' and explorers' caravans.

Historical Origins Because the jujube has been cultivated for so many centuries in so many places (there are more than 400 cultivars), it is unclear where it is originally from, but it is believed to be native to southern Asia and southern and central China.

Botanical Facts The jujube is a small deciduous tree growing to 15–35 ft (4.5–10 m). It has spiny branches and shiny green leaves that turn bright yellow in autumn. Clusters of small, yellow-green, rather insignificant flowers are followed by fleshy, cherry- to plum-sized green fruit (a drupe). Each fruit slowly ripens, first to yellow-green and then develops small red spots around the skin that eventually merge to form an all-red fruit. This finally softens and wrinkles like a small date. There is a single, datelike seed in the center.

Culinary Fare The fruit is eaten fresh, dried, and candied. At the yellow-green to fully red stage, fresh fruit is probably at its best, tasting rather like a wonderfully sweet, crispy apple. The fruit has a rather short shelf life (around a week), so is more commonly sold dried as Chinese dates or red dates. The dried fruits are used extensively in Asia for cooking, tea-making, and in traditional natural remedies.

ABOVE: *These Chinese jujubes are at different stages of maturity. Fruits are very high in vitamin C.*

Feijoa

Acca sellowiana

RIGHT: *Whole and halved feijoas. The cut pieces show the seeds within the pulp.*

Also called the pineapple guava, the feijoa has an exotic aroma and bears gorgeous red and white flowers. New Zealand has made the greatest strides in cultivating the South American native.

Historical Origins The feijoa was discovered early in the nineteenth century by German botanist Friedrich Sellow in southern Brazil. The scientific name of the plant is a tribute to him, and the common name commemorates Brazilian botanist, João da Silva Feijó. Although it grew abundantly in Brazil, Argentina, Paraguay, and Uruguay, it appears the plant was used more for medicinal purposes, rather than as a food.

In the late nineteenth century, a French horticulturist introduced the plant to Europe. The Royal Botanic Gardens in Kew, England, listed the feijoa as a newly cultivated species in 1898. Within a few years, it was introduced to California, Australia, and New Zealand, mainly as an ornamental plant because of its attractive flowers. However, the fruit became especially popular in New Zealand where the climate, selective breeding, and lack of pests helped produce larger superior fruit. New Zealand is now the world's leading producer of feijoas, its crop mainly available from spring to early summer. Californian feijoas are available in the autumn months.

BELOW: *The feijoa's spectacular flowers are edible and bloom in spring.*

Botanical Facts Feijoas are small evergreens that can grow to about 10 ft (3 m) high, and sometimes a bit taller. The leaves are oval and glossy gray-green. The single flowers have cupped petals that are white on the outside and purplish red on the inside. Each flower has a cluster of long crimson stamens with yellow tips.

The fruit is oval or pear-shaped, about 1–2 in (2.5–5 cm) long. The skin is dull green and is covered in a whitish bloom. The flesh is white and has a translucent pulp containing several small seeds.

It is best to plant a feijoa in well-drained soil in a sunny location. It is most commonly pruned to form a hedge. Fruit ripens from late summer throughout autumn and falls when it is mature. To ensure an abundance of fruit, two feijoas trees, planted fairly close together, are required.

Culinary Fare Feijoas have a distinct aroma, and the flavor is often described as a cross between a pineapple and a strawberry. The fruit is usually eaten by cutting it in half and scooping out the pulp with a spoon. It can be used to make jams, jellies, and sauces or is used in cakes or muffins, often in place of bananas. They can be added as a flavoring for ice cream, yogurt, or smoothies. Feijoas pair well with ginger. They also freeze well.

Pawpaw

Asimina triloba

The pawpaw—often spelled papaw—should not be confused with the papaya, an unrelated tropical fruit that is called pawpaw in some parts of the world. The pawpaw is somewhat similar to the banana in taste and is even nicknamed the poor man's banana. It is in the same family (Annonaceae) as the custard apple and cherimoya, but is the only member of the family to grow outside the tropics.

Historical Origins Pawpaws are native to eastern North America and are the largest edible fruit indigenous to North America. The pawpaw was first cultivated by Native Americans. A member of Spanish explorer Hernando de Soto's party made note of it during an expedition from Florida to the Mississippi River in the mid-sixteenth century.

The pawpaw was important to the Lewis and Clark expedition in western Missouri, which, toward the end of the trip in 1806, had almost run out of food. Expedition members gathered pawpaws to sustain them in the last days of their journey.

In the early twentieth century, pawpaws were sold in local markets close to growing sites, but the fruit seemed to fall from favor as global trade brought other fruits to the market.

Botanical Facts The pawpaw is a shrub or small tree that grows up to 30 ft (9 m) high. It has large, long, drooping leaves that look almost tropical. It bears purple to brownish red flowers that are about 2 in (5 cm) wide and have an unpleasant odor to attract flies, which help with pollination. The smooth, curved, green fruit is about 4 in (10 cm) long and turns yellowish brown as it ripens. The flesh is creamy white to yellow with 2 rows of large brown seeds.

Trees can be grown from seed and need moist, well-drained soil and partial shade. They should bear fruit within 5–7 years and produce 25–50 lb (11–23 kg) of fruit annually. The fruit drops to the ground when it is fully ripe.

ABOVE: *Pawpaw fruits frequently grow in clusters of up to 9 individual fruits. They are plump and soft with a thin skin when ripe.*

Culinary Fare Pawpaw is rich and creamy with a flavor similar to banana, pear, and mango. It has an intense aroma some may find off-putting. Ripe pawpaws should be used quickly. Fruit that is almost ripe will keep a week or two in the refrigerator. The pulp can also be frozen. Pawpaws can be eaten fresh—the best way is to cut fruit in half and eat with a spoon. They can be used in pies, cakes, and quick breads or for flavoring custards, ice cream, and drinks such as piña coladas. They are even used in beer brewing.

Figs

Ficus carica

Figs have been food for slaves and kings for thousands of years. This sweet nutritious fruit is the most mentioned fruit in the Bible, and is believed to have been a plant in the Garden of Eden. Those who have tasted a truly wonderful fig feel as if they have, indeed, experienced paradise.

Historical Origins The wild fig originated in western Asia and spread to the Mediterranean in prehistoric times. The Ancient Egyptians, or possibly Arabs, began cultivating it as many as 6,000 years ago. Paintings on the wall of an Egyptian tomb from 1900 BCE show people harvesting figs, indicating their cultural importance.

RIGHT: *The ripe pulp of a 'Brown Turkey' fig is sweet and juicy with a creamy texture. It is one of the most popular figs eaten around the world today.*

Fig cultivation eventually moved to Crete and then to Greece by 800 BCE. Homer mentions them several times and both Aristotle and Theophrastus made astute observations about how figs reproduced. The Romans used figs extensively; the finest specimens were eaten by the wealthy, while lesser varieties were food for slaves. Gladiators ate them for quick energy. Dried figs were used as a sweetener, since sugar was still a rare commodity.

Spanish explorers took figs to Haiti and Mexico in the sixteenth century, and by the late eighteenth century figs were being cultivated in both California and Australia.

Botanical Facts Although the fig itself is considered a fruit, what are thought by many to be seeds are actually more than 1,000 tiny fruits encapsulated in the vase-shaped outer flesh, called a syconium. Figs range from green to brown to purplish black outside, and have pale pink to red flesh inside.

Fig trees are deciduous and grow 35 ft (10 m) tall or higher and have a large canopy. They have extensive root systems so should not be planted by walls, houses, or walkways. Trees do best in a climate with hot summers that are relatively dry and prefer low-to-medium fertile soil. They are quite cold hardy and can withstand frosts once established.

Culinary Fare Figs, which, for a fruit, are high in calcium, come in hundreds of cultivars. Some of the main ones are 'Black Genoa', 'Brown Turkey', 'Smyrna', and 'Mission'. Fresh figs are delicate and can command a high price. Look for figs that are soft when pressed and undamaged. Fresh figs go beautifully with prosciutto or a piquant cheese such as goat cheese or gorgonzola. You can bake or broil them and top with mascarpone or ricotta. Use them to make tarts, savory compotes to go with duck or salmon, and fig jam. Dried figs can be used in cakes, puddings, and to make compotes as well.

Strawberry

Fragaria × ananassa

The red ripe strawberry might be the most perfect fruit. But the strawberry we know today is very different to the ones that captured the hearts and imaginations of people long ago.

Historical Origins Strawberries, albeit of different species, are indigenous to both the Old World and New World. European cultivars were first grown in the fourteenth century, but it was not until about 300 years ago that a South American species (*Fragaria chiloensis*) was brought to Europe from Chile by a French spy named Frézier, and hybridized naturally in France with a North American species (*F. virginiana*). The offspring of these two plants was called *F. × ananassa* (the garden strawberry) because it had the aromatic qualities of a pineapple (*Ananas*).

Strawberry cultivation continued with major innovations in nineteenth-century England, particularly with a cultivar called 'Keen's Seedling' that took both Europe and America by storm. Today, California has one of the world's top commercial strawberry breeding programs, with Europe, Asia, Africa, and Australia also producing crops.

Botanical Facts Strawberries are unusual because the "seeds" on the outside are actually the fruit, and the flesh is part of the flower. The low-lying plants spread by runners. Strawberries need rich, well-drained soil and are hardy in all but the coldest of temperate climates. As a rule, berries will not improve after harvest, and must not be picked until fully red and ripe.

Culinary Fare Strawberries do not keep well, even in cold storage, so use them quickly. Wait to wash them until ready to use. They are most commonly eaten fresh and often paired with ice cream, cream, or even sour cream. Some like them with red wine and sugar. Use strawberries to adorn baked goods; strawberry shortcake is a classic dessert. Strawberries are low in natural pectin, so it must be added when making jam.

ABOVE: *An 1824 watercolor of the 'Wilmot's Superb' strawberry. It was one of the first cultivars grown in England.*

What's in a Name?

The origin of the English word "strawberry" is the subject of some debate. Popular belief is that it comes from using straw as mulch to keep the berries moist and off the ground, but the word strawberry was used centuries before they were cultivated. Another theory is that "straw" comes from the root word meaning "to spread or strew" either because the berries spread by runners or because the so-called seeds are scattered about the berry. Still another, quite plausible, explanation is that wild strawberries frequently grew in grassy areas or hay fields and that the berries, which usually ripen when hay is harvested, were frequently found by farmers in the stubble (or straw) underneath the freshly cut hay.

Raisin Tree

Hovenia dulcis

The edible part of this plant is the fleshy stem of the fruit, rather than the fruit itself. It is virtually unknown outside of Asia.

Historical Origins The raisin tree is likely native to China, but it is also called the Japanese raisin tree because it has been cultivated there for so long. The tree has been sold and harvested in China for several millennia but was only introduced to the West in the early nineteenth century where its edible qualities have been largely ignored.

Botanical Facts The raisin tree grows 30 ft (9 m) or more in height, and is a good shade tree. It bears clusters of cream to yellow-green flowers. It is adaptable to many soil conditions. The fruit stalk ripens in autumn, but do not be in a rush to pick it. If it is harvested too early it will taste green and stemlike.

Culinary Fare The Chinese name for the fruit, *chi-chao li*, which translates to "chicken claw pear," describes it quite well. The edible, swollen, gnarled stalks of the raisin tree are not altogether attractive, and taste raisinlike with some flavor qualities of pears. The fruits of the raisin tree can be eaten raw or chopped up and cooked in dishes that call for raisins.

Mulberries

Morus alba White Mulberry *Morus nigra* Black Mulberry

ABOVE: *A cluster of white mulberries ripening on the young shoots of the tree. The fruit is very sweet but lacks the intense tart flavor found in black mulberries.*

These two species have distinctly different uses although both are sources of food. It is usually easy to tell when someone has been eating the black mulberry as it can leave telltale stains on lips and hands.

Historical Origins The white mulberry is native to China and was cultivated more than 5,000 years ago for its leaves to feed silkworms. The black mulberry is native to western Asia and was introduced into southern Europe, presumably from Persia, by the Ancient Greeks and Romans. It is believed to have been introduced to England by the Romans and was first cultivated there early in the sixteenth century.

Botanical Facts The black mulberry tree can grow up to 50 ft (15 m) high, with a short trunk and wide canopy. Its leaves are dark green, oval to heart-shaped, and serrated. Green flowers give way to purplish black fruit. White mulberry trees are smaller in size and have white fruit that colors to red or deep purple as it ripens.

Mulberries can be propagated from cuttings and planted in fertile, well-drained soil. Prune minimally in the winter. Mulberry trees can live for hundreds of years.

Culinary Fare Mulberries are delicate and not grown for commercial purposes, but may be found at farmers markets. Mulberries can be eaten fresh and make a nice addition to fruit salad. They can also be used in pies, preserves, and syrups.

Pomegranate

Punica granatum

With its leathery skin and seedy pulp, the pomegranate poses a bit of a challenge to those who consume it. But this sweet and tangy fruit is replete with history and symbolism that is undoubtedly part of its culinary appeal.

LEFT: *A ripe pomegranate fruit with bright orange-red skin tones. The fruit usually takes 5–7 months to ripen after flowering and should not be picked and eaten until fully mature.*

Historical Origins The pomegranate tree first sprang from the soil in what is now Iran, where it still grows wild. It has been cultivated for at least 5,000 years and spread east into India and China, westward to the Mediterranean, and eventually to the New World. The pomegranate is featured in Greek mythology in the story of Persephone, and in the Bible; Moses told the Israelites, who longed for the fruit while wandering in the Egyptian desert, that they would have pomegranate in the Promised Land.

There have been goblets in the shape of pomegranates since Troy, and European kings and priests had pomegranates embroidered on their finest robes. Catherine of Aragon ate pomegranate seeds—a symbol of fertility—in hopes of bearing Henry VIII a son, but, alas, she failed and he divorced her.

Today, the pomegranate, and especially its juice, is becoming more popular around the globe because of its health benefits. It is high in vitamin C and antioxidants, and studies have shown it might help reduce the risk of heart disease, arthritis, and some cancers.

Botanical Facts The pomegranate is a small tree that grows up to 25 ft (8 m) tall. Its lance-shaped leaves are red-hued in spring, later turning bright green. Flowers have 5–8 scarlet petals, and the fruit is orange-red, about the size of a large orange, and has a seedy pulp, usually red in color. It is a hardy tree that handles both very low and high temperatures but needs hot dry summers to ripen properly.

Culinary Fare Its tough exterior means the pomegranate travels well. Once cut open, the pulp and seeds can be eaten with a spoon, or scoop them out and put them into a nice bowl for an attractive presentation. The seeds are used to decorate a number of dishes and are added to some Middle Eastern stews. In India, the seeds are used to make chutney.

Pomegranate juice is a popular drink, but it can be quite sharp and tannic. If squeezing it, add gelatin or another fining agent to reduce the tannic flavor, although it is probably easier to buy the juice. Juice can be used to make jelly and it is also used to make grenadine, a red syrup used in many cocktails.

RIGHT: *Each tiny pomegranate seed is surrounded by a translucent red pulp that has a sweet tangy flavor.*

Currants

Ribes nigrum Blackcurrant *Ribes odoratum* Buffalo Currant
Ribes silvestre Redcurrant *Ribes uva-crispa* Gooseberry

LEFT: *Blackcurrants are too tart to eat fresh. They are commonly used in preserves, syrups, and liqueurs. Redcurrants are less tart and can be eaten fresh, and white currants are sweet enough to eat straight off the bush.*

Currants, from the genus *Ribes*, are often confused with dried currants, which are the product of small black grapes. However, fresh currants—which also include gooseberries—deserve distinction and appreciation in their own right.

Helping the Medicine Go Down

Because of their distinct taste and odor, blackcurrants were long used for medicinal purposes, but it wasn't until the twentieth century that scientists discovered blackcurrants were incredibly rich in vitamin C. In the 1930s, blackcurrants were made into a cordial called Ribena, which became an important source of vitamin C during World War II when Great Britain was unable to get oranges.

For generations, mothers gave children Ribena because of its health claims, but in 2004, two New Zealand schoolgirls found the vitamin C content in pre-diluted drinks was not consistent with claims on the label. The company that makes Ribena was forced to change its advertising and packaging in New Zealand and Australia.

Most of Britain's blackcurrant crop is now used to make Ribena, and the drink is sold in various forms in more than 20 countries worldwide.

Historical Origins Native to northern temperate regions, currants are relative newcomers to cultivation. One of the first mentions of the redcurrant was in fifteenth-century Germany, and it was domesticated the following century in Denmark and the Netherlands. The blackcurrant was cultivated in Europe in the seventeenth century, but mainly for medicinal purposes.

Gooseberries have been cultivated somewhat longer in Britain, and even inspired clubs in the eighteenth and nineteenth centuries. The buffalo currant is native to central North America and is similar to the American gooseberry (*Ribes hirtellum*).

Botanical Facts Currants grow on deciduous shrubs that range from 3–7 ft (0.9–2 m) in height. The leaves typically have 3–5 lobes and flowers range in color from green to yellow with some red. Fruits can be white, yellow, green, red, and black.

Plant currants in groups to ensure good fruit set after pollination. The plants require winter chilling and shade from the hot summer sun. They do best in well-drained soil. As a rule, currants are rarely grown in North America because they can spread a disease that is deadly to white pines.

Culinary Fare Currants can be eaten fresh, but are more commonly cooked. Redcurrant preserves are quite popular and go well with lamb. They are used for summer pudding in Britain, and redcurrant cake is an Austrian specialty. Gooseberries pair well with oily meats and are used to make tarts, pies, and fools. Currants are frequently used to make juices, and in France, blackcurrants are used mainly for making *crème de cassis*, a sweet liqueur.

Rubus Berries

Rubus caesius Dewberry   *Rubus chamaemorus* Cloudberry
Rubus fruticosus Blackberry   *Rubus ursinus* hybrid Boysenberry

In some parts of the world, the highlight of a hike through a forest or field on a warm summer's day is stumbling upon a patch of any *Rubus* berries. Although some species such as the blackberry and boysenberry, are widely cultivated, the thrill of finding them in the wild seems to enhance their taste.

Historical Origins There are hundreds of cultivars of blackberries, a generic term for this group of *Rubus* berries. They are native to every continent except Antarctica, and typically grow in cooler regions. Blackberries have been harvested since prehistoric times. Europeans used them not only as food and medicine, but also as protection to keep intruders away, thanks to their thick growth of thorny canes.

The boysenberry is a hybrid that was first cultivated around 1920 on a farm in California by a man named Rudolph Boysen. It is believed to be a cross between a loganberry and a dewberry.

Botanical Facts Blackberries are made up of several small fruits, or drupes, that are attached to a central core. When picked, the berry comes away from the plant with the core intact. They grow on shrubs called brambles that range from low growing to upright, most with prickly canes, although there are some thornless cultivars. Plants can be propagated from cuttings or seeds and grow in a variety of habitats. Regular pruning is a must to keep plants under control.

Culinary Fare Look for berries with large drupelets and use them quickly as they do not keep well. They can be enjoyed fresh, but also make scrumptious jellies and jams (although jams can be a bit seedy). Blackberries make delicious pies and cobblers and surprisingly refreshing wines. Boysenberries are preferred over most other hybrids for their rich tart taste, and cloudberries are known for their sweet-and-sour taste and excellent freezing qualities.

ABOVE: *Blackberries growing on a thorny bramble. They are ripe when black or dark purple.*

Cloudberry Wars

Cloudberries are prized not only because they are delicious, but because they are one of the few fruits that grow in the Arctic Circle. They are admired most of all in Scandinavian countries, where there have been known to be cloudberry wars.

There have been disagreements over people picking cloudberries on someone else's land, and neighbors may fight over whose land a cloudberry plant is growing on. The discovery of a good cloudberry patch is kept as a closely guarded secret. Land owners can prohibit cloudberry picking on their land.

To keep disputes to a minimum, some counties in Finland and Norway have strict rules regarding cloudberry picking, and the Swedish Ministry of Foreign Affairs has had a department devoted to berry diplomacy. The cloudberry is so revered in Finland that it is served only on the most special occasions, and a picture of it graces the Finnish Euro coin.

Raspberries

Rubus idaeus Raspberry *Rubus occidentalis* Black Raspberry

ABOVE: *The combination of pureed raspberries and milk chocolate is a classic dessert pairing.*

The raspberry is a lesson in contrasts. It is at once both hearty and delicate; the plant able to withstand punishing winter temperatures, but the fruit perishes easily once picked. It is the red raspberry, rather than the black, that is widely eaten today.

Historical Origins The red raspberry—the name thought to be connected with its rough, slightly hairy countenance—is native to Asia, Europe, and North America and can grow far into the northern latitudes.

The botanical name *idaeus* refers to the wild raspberry's presence on Mt. Ida, according to the Roman historian Pliny (CE 23–79). It was gathered wild in the time of Christ, but attempts at cultivation likely occurred later in the Roman empire, as seeds have been found at old Roman forts in Britain. It was not until the Middle Ages that raspberries were more widely cultivated by the British. Today, about 40 countries produce raspberries worldwide, with Russia growing the most.

BELOW: *Each raspberry is a series of connecting drupelets attached to a single receptacle. The fine hairs help bind the fruit together.*

Black raspberries are native to North America only, mostly in the eastern part of the continent, and were not domesticated until the nineteenth century.

Botanical Facts The raspberry is known as a "bramble," which is a prickly shrub of the genus *Rubus*. The berries grow on canes that are classed as either erect or trailing. Each raspberry is actually a formation of several small fruits attached to a central core that is left behind on the plant when the berry is picked. Raspberries have small fine hairs that help hold the individual small fruits together.

Black raspberries are erect cultivars only and are less cold tolerant, disease resistant, and produce lower yields than red raspberries. The fruit is black in color. They are not sold commercially and are typically a backyard plant.

Rubus idaeus are ripe when they achieve a full red color and can be easily removed from the plant. The berries ripen at different times so plants need to be harvested several times over the course of a few weeks.

Culinary Fare Raspberries are incredibly fragile, which makes fresh ones quite expensive. In fact, only 10 percent are sold fresh; the rest are processed. Use fresh raspberries quickly. Of course they can be eaten plain, but raspberries are used in all sorts of desserts including tarts, crème brûlée, fools, and cobblers. Raspberries and chocolate are a natural pairing.

Both red and black raspberries can be used to make jams, jellies, or sauces to spread on toast, drizzle over ice cream, or to put between layers of cake or trifle. Red raspberry syrup or flavoring is used in many drinks including cordials, liqueurs, and wine.

Loganberry

Rubus × loganobaccus

The loganberry is a hybrid of the raspberry and blackberry. While it is not always popular for its flavor, it has proved to be an invaluable source for creating a number of new hybrid *Rubus* berries.

Historical Origins The loganberry is unusual in that its exact origins are known. It was produced in 1881 in Santa Cruz, California. A judge and amateur horticulturist named James Harvey Logan planted seeds of an American blackberry cultivar called 'Aughinbaugh', which was pollinated by a European 'Red Antwerp' raspberry growing nearby. It was the first known hybrid of its kind. Cultivation in England began in 1897.

It was initially a big success, and was cultivated extensively through the mid-twentieth century. The loganberry's popularity dwindled in part because of low yields, mechanical harvesting difficulties, and, ironically, the success of subsequent hybrids it helped create. Today, only a few acres are still in commercial production in Oregon.

The loganberry has been crossed with other berries to create some popular new hybrids, including the boysenberry (loganberry × dewberry), the olallieberry (black loganberry × youngberry), and its offspring the marionberry. The silvanberry, one of the most popular blackberry hybrids in Australia, is part of the loganberry's lineage.

Botanical Facts Loganberries are aggressive growing shrubs that can reach up to 6 ft (1.8 m) high. The plant can be quite thorny, and the berries are often hidden under the leaves. The fruit is large, often an inch (25 mm) or more long, and somewhat cone-shaped. It is typically reddish purple to reddish black in color. Although it is a blackberry-raspberry cross, it is more like a blackberry when picked because its central core remains intact.

The loganberry propagates well from cane tips and prefers heavy, fertile, well-drained soil and needs a sunny sheltered site. Pruning each year is necessary, and if well taken care of, the plant will produce fruit for up to 15 years. The loganberry is quite robust and is more disease- and frost-resistant than many other berries.

The fruit is typically ready for harvest before blackberries, usually mid-summer to early autumn, and will ripen at different intervals, so it needs to be picked several times. One bush will generally yield 15–17 lb (7–8 kg) of berries.

Culinary Fare When selecting loganberries to eat, choose fruits that are plump, juicy, bright red in color, and similar in size. Avoid moldy or shriveled berries. Do not wash the berries until ready to eat.

The loganberry has a distinct flavor, and is tarter than a blackberry. It can be eaten fresh but can be a bit astringent for some palates. Loganberries are quite soft and do not freeze well, but they are ideal for stewing, canning, and juicing. They can be used in much the same way as blackberries in jams, pies, crumbles, fruit syrup, and wine. They also make fine sorbets and ice cream. Loganberry juice is a popular beverage in southern Ontario, Canada, and western New York, in the United States.

BELOW: *Plump and juicy loganberries are delicious when eaten fresh. They are sweeter in taste when sunripened on the plant.*

Elderberry

Sambucus nigra

This ancient plant, which is found all throughout Europe, western Asia, and North America, has served many functions as food, of course, but it has had medicinal and even musical purposes as well.

Historical Origins Elderberries have been found in prehistoric archaeological sites in Europe. The Greeks used the stems to make musical instruments called *sambuke*, and Native Americans used the stems for whistles and to make arrows. The Hippocratic Corpus (a collection of medical works from Ancient Greece) and Pliny (CE 23–79) both mention medicinal uses for the elderberry including as a laxative and a diuretic.

Botanical Facts The black elder, or European elder, is a deciduous shrub or small tree that can grow up to 30 ft (9 m) high. It has dark green, serrated leaves and large, lacy, white flowers that bloom in late spring to early summer and give way to a generous amount of purplish black berries.

Culinary Fare Elderberries and their flowers should not be eaten raw as they are mildly poisonous, but cooking deactivates the toxins. Elderberries can be used to make jams, jellies, and chutneys and are commonly combined with apples. They are also used to make wine or cordial, a syrupy concentrate, which makes a refreshing cool drink. Elderflowers add flavor to other cooked fruits, jams, and baked goods, and can even be battered and fried.

RIGHT: *Elderberries grow in clusters that often weigh down the branches of the plant. They are used in traditional herbal medicines.*

Chilean Guava

Ugni molinae

This member of the myrtle (Myrtaceae) family is sometimes called *Myrtus ugni* and is also known as the Tazziberry™ in Australia or the New Zealand cranberry in New Zealand.

Historical Origins As the name implies, this plant is native to Chile and also parts of Argentina. Its scientific name honors Chilean botanist Juan Ignacio Molina. The plant was taken to England in the first half of the nineteenth century and is said to have been a favorite fruit of Queen Victoria. Although mainly used as an ornamental shrub, its berries were often used to make jams and jellies.

Commercial growers in Tasmania, Australia, have only recently branded the fruit as a Tazziberry™ in the hopes of marketing it on a wider scale internationally.

Botanical Facts The Chilean guava is a shrub that can grow 6 ft (1.8 m) in height, and has glossy green leaves, and creamy flowers that give way to a small red berry in autumn. The plant propagates from seed or cuttings and needs well-drained soil to thrive. It is particularly drought tolerant.

Culinary Fare The flavor of the Chilean guava, which can be eaten raw, is a combination of strawberries, pineapple, and apple. It can be used for jams, jellies, and sauces for meats. It is also considered to be a good flavoring for yogurt, ice cream, and sorbets, and can be added to fruit salads.

Blueberries

Vaccinium ashei Rabbiteye Blueberry *Vaccinium corymbosum* Highbush Blueberry
Vaccinium lamarckii Lowbush Blueberry

Succulent and subtly sweet, the blueberry is the most recent wild fruit to come under cultivation. Although indigenous to North America, interest in blueberry culture has begun to spread globally.

Historical Origins Blueberries were an important food for Native Americans. They ate them fresh and dried them for use in stews, puddings, and pemmican, a preserved meat mixture. Native American tribes in Alaska preserved the berries in seal oil.

They were exclusively a wild fruit until about 1920, when cultivation began in New Jersey. Since then, about 100 named cultivars have been developed and blueberry production has doubled in recent years. The United States now produces half the world's supply and, with Canada, celebrates National Blueberry Month each July.

Botanical Facts Blueberries come in deciduous or evergreen cultivars. All have similar oval or lance-shaped leaves and white- or cream-colored, urn-shaped flowers. The fruits grow in clusters and are picked from late spring through summer. Cultivated blueberries tend to be larger than the wild cultivars, but the wild ones have a more intense color.

Highbush blueberries are the main cultivated species and grow up to 7 ft (2 m) high. Rabbiteye berries are larger and have more seeds and grow mainly in the southeastern United States. Lowbush blueberries are shrubs, usually no bigger than 12 in (30 cm). The berries are smaller, lighter blue, and almost all are processed after harvest. The lowbush is one of the first plants to grow back after a wildfire.

Culinary Fare Blueberries are sold fresh, frozen, and occasionally canned. They are eaten fresh but are especially popular in baked goods such as muffins, pies, and cakes. Blueberries make lovely sauces and compotes to serve with ice cream or pancakes, and can be used to make jellies and jams. They can be stored in the refrigerator for at least a week.

ABOVE: *The most widely cultivated blueberry in the world is the 'Bluecrop'—a highbush blueberry grown for its firm fruit and flavor.*

Why Blueberries Turn Green

Many cooks have experienced the frustration of a green tinge in blueberry muffins or pancakes when making them. While they do not taste any different, the appearance is a bit strange and it makes one wonder what went wrong.

Blueberries contain red-purple anthocyanins, which are sensitive to changes in pH. The anthocyanins change color in acidic or alkaline conditions. Baking soda, which is alkaline, is usually the culprit that turns blueberries green.

To solve the problem, when a recipe calls for baking soda, mix the dry ingredients well before adding the liquid and berries, and use as little soda as possible. Keep in mind that acid brings the pH back in balance so if a recipe calls for baking soda, then buttermilk or milk with a squeeze of lemon juice will help keep blueberries blue.

Cranberries

Vaccinium macrocarpon American Cranberry *Vaccinium oxycoccus* Northern Cranberry

LEFT: *Cranberries ripening in the wild. They were first discovered by Native Americans, who used them as a food, fabric dye, and healing agent.*

Tart crimson cranberries are delicious but sometimes difficult to find outside of North America where they are a prized element in the most American of meals—Thanksgiving dinner. Their scientific name, *Vaccinium*, comes from the Latin word for cow, *vacca*, because cows are partial to the plants. Bears are also fond of cranberries.

Historical Origins The cranberry is indigenous to North America and was used extensively—both fresh and dried—by Native Americans, especially in pemmican, a preserved meat product made of dried game meat, cranberries or other berries, and fat. Native Americans introduced cranberries to early European settlers who used them in tarts and as a sauce for meats such as venison and wild turkey. Because of its association with the Pilgrims, cranberry sauce has become a mainstay at traditional American Thanksgiving feasts.

Cranberries were gathered wild until cultivation began in the nineteenth century. Today, almost all the world's cranberries (96 percent) are grown in the United States and Canada. The berries have become increasingly popular in recent years because of their medicinal qualities. Doctors often advise women to drink cranberry juice to prevent urinary tract infections.

Botanical Facts Cranberries typically grow in low swampy areas or bogs and need acidic soil to thrive. They are creeping evergreen shrubs that reach about 3 ft (0.9 m) high and have pink or mauve flowers with an extended stamen. The berries are good-sized, about ½ in (12 mm), and are a white creamy color before turning a shiny crimson red in autumn, signaling they are ready for harvest. Cranberries can be harvested dry or by flooding the bogs, but dry-harvested berries tend to keep better.

Northern cranberries are similar to American cranberries but have a grassier, more herbaceous, flavor.

Culinary Fare Less than 10 percent of cranberries are sold fresh. The majority are dried or are used to make juice or pre-made cranberry sauces, but it is worth the effort to start with fresh, or at the very least frozen cranberries. Cranberries are quite tart and acidic and usually require the addition of sugar or other sweeteners to make them palatable. They are also high in pectin, so need little cooking to make them gel.

Cranberries can be used to make sauces, conserves, and preserves to go with meat or poultry. They are also used in baked goods such as muffins, breads, and biscuits. Dried cranberries are much sweeter than fresh ones, and can be used like raisins in baked goods. They also make a nice addition to granola or even tossed green salads.

Bilberry, Whortleberry

Vaccinium myrtillus

People often mistake the bilberry for the blueberry, which is just the beginning of what could be a bilberry identity crisis. It is called the whinberry in England, the blaeberry in Scotland, and the fraughan in Ireland. The term "whortleberry" is a more generic term, often used for other berries in the *Vaccinium* genus.

Historical Origins The bilberry grows in Europe, northern Asia, and in the far north regions of North America. It has long played a role in European folk medicine and is still used for certain ailments. During prehistoric times in County Down, Ireland, a courting ritual involved picking bilberries. Throughout Ireland, Fraughan or Bilberry Sunday, the last Sunday in July, marks the start of harvest and on this day unmarried young women sometimes bake cakes to give to the object of their affection.

Botanical Facts The bilberry grows on a small semideciduous shrub that is about 18 in (45 cm) high, and has small serrated leaves. Flowers bloom green, turning red later. The fruit is distinguishable from the blueberry because it has darker flesh and grows more sparsely, making it more time-consuming to pick.

Culinary Fare Although not typically as good as blueberries, bilberries can be used in place of them. They can be eaten fresh and are used to make jams, sauces, pies, and tarts. In France they are used as a base in liqueurs.

ABOVE: *Bilberries are ready to pick in late summer.*

Lingonberry

Vaccinium vitis-idaea

This small, shiny red fruit, similar to the cranberry, plays an important role in Scandinavian cuisine. Although not previously well known outside Scandinavian countries, the lingonberry has had its profile raised due to its placement in the Swedish food shop at IKEA stores worldwide.

Historical Origins This species is indigenous to Scandinavia; traces of lingonberry wine have been found in Bronze Age Danish graves. The berry is mentioned in some literature during the Middle Ages, but an Italian diplomat provided the first detailed description in the seventeenth century.

The name lingonberry comes from the Swedish word *lingon* meaning "cowberry," another name for the fruit. Wild berry picking is a popular pastime in Scandinavia—particularly Finland—and the lingonberry is the most foraged berry of all.

Botanical Facts The pea-sized lingonberry grows on a small evergreen shrub that is about 6 in (15 cm) high and has small, shiny green leaves. White or pink bell-shaped flowers appear in spring, with berries following in autumn. It grows wild in subarctic and some arctic regions of Europe, Eurasia, and North America. Some efforts have been made toward cultivation.

Culinary Fare Sweeter than the cranberry, lingonberries can be eaten fresh, but are commonly made into jellies and sauces and served with many foods including pancakes, meatballs, and game. In Norway and Sweden, reindeer steak is traditionally eaten with gravy and lingonberry sauce. Lingonberries keep well and can be stored in fresh water.

ABOVE: *The fruits and leaves of a lingonberry growing on a moorland. The plant prefers cool, moist, acidic soil with shelter from the hot summer sun.*

Kiwi Fruits

✳ *Actinidia arguta* Hardy Kiwi ✳ *Actinidia chinensis* Golden Kiwi Fruit ✳ *Actinidia deliciosa* Kiwi Fruit

Long keeping qualities, along with nutritional value and attractiveness, have greatly increased the popularity of kiwi fruits. The 3 species are ideal for export, because they are picked green, like pears, and allowed to ripen slowly. Along with the familiar, green-fleshed kiwi fruit, there are the yellow-fleshed golden kiwi fruit, which was developed in New Zealand, and the smaller hardy kiwi.

Historical Origins The kiwi fruit (*Actinidia deliciosa*) is indigenous to southwestern China, and has been cultivated in European gardens since the early twentieth century. It was first grown commercially in New Zealand after seeds were brought from China in the early 1900s, hence the name "kiwi fruit." Shipments began to arrive in England in the 1950s and kiwi fruit became popular there as an exotic, decorative ingredient or garnish, especially with the advent of nouvelle cuisine.

Botanical Facts Long cultivated in southern China, this vigorous climber has large, furry, heart-shaped leaves up to 8 in (20 cm) long and is prized by gardeners as a decorative and productive climbing plant. It bears scented cream flowers in spring followed in early winter by the ripe fruit, brown and fuzzy with bright green flesh, and black seeds that ripen in early winter. Botanically, the kiwi fruit is a juicy berry containing hundreds of tiny black seeds. Hardy kiwi fruits are hairless, and are smaller, sweeter, and more delicate than regular kiwi fruits, although they look the same when cut open. Golden kiwi fruits are less hairy and less tart, and their flesh has a milder, more tropical flavor.

BELOW: *As well as being rich in vitamin C, kiwi fruits contain as much potassium as bananas, and high levels of beta-carotene and dietary fiber.*

Culinary Fare Kiwi fruits are rich in vitamin C, and one of their advantages is that even after prolonged storage, much of the vitamin is retained. Kiwi fruits contain an enzyme that has a tenderizing effect on meat, and their mildly acidic flavor goes well with all kinds of meat or cheese. The peel can be eaten, but as it is hairy most people prefer to scoop the flesh out with a spoon or peel the fruit before eating. Kiwi fruits are best eaten fresh or lightly warmed. They are too delicate to withstand much cooking. In New Zealand and Australia, kiwi fruits are often served with passionfruit and cream in a soft meringue shell. This dish is known as pavlova. The radial pattern of the black seeds is appealing, and kiwi fruits are often used in fruit salads or sliced and used to fill tart shells. They are also used in ice creams, sorbets, jams, and pickles.

Watermelon

Citrullus lanatus

The wild ancestors of the watermelon originated on the dry, desolate plains of the Kalahari Desert in southern Africa, providing welcome refreshment to people and animals alike. The watermelon is a completely different species to other melons, but is of similar composition, comprising 92 percent water. Watermelon is one of the most refreshing summer fruits and was frequently planted in rows alongside other crops in the south of the United States to refresh workers in the field.

Historical Origins Watermelons are native to Africa and were cultivated in Egypt some 3,000 years ago. Watermelon seeds have been found in tombs, and wall paintings clearly show the large green fruits. By the tenth century CE, watermelons had spread to China, where they became hugely popular, and by the sixteenth century, they could also be found in Southeast Asia and Japan. Slave traders brought the watermelon to the Americas in the early seventeenth century, and farmers developed new breeds there, eliminating bitterness and encouraging the growth of the very large fruits already known in Europe since the sixteenth century.

Botanical Facts The watermelon is an annual climber or trailer of indeterminate size, spreading or climbing throughout the growing season. It has oval, lobed, feathery green leaves with toothed edges and small translucent patches, branched tendrils and pink- to red-fleshed, mottled green fruit measuring up to 20 in (50 cm) long. The fruits of the larger cultivars can weigh as much as 45 lb (20 kg). The majority of cultivars have the characteristic pink flesh, but some are yellow-fleshed. Smaller, seedless watermelons have become popular in recent times. These are generally more round than the larger, cylindrical cultivars. Some of the most popular cultivars of watermelon include 'Sweet Favorite', the smaller 'Sugar Baby', the elongated, striped 'Charleston Gray' and the yellow-fleshed 'Yellow Baby'.

Culinary Fare Watermelon is refreshing chilled and served fresh. The entire fruit can be used, as the rind makes an excellent pickle, and the seeds can be roasted and eaten. In some countries, the seeds are crushed to produce oil for cooking. Like most melons, watermelons spoil easily unless kept cool. The fruit is excellent in salads with onion or cold meats, or as part of a fruit salad. The flesh makes excellent sorbets, ices, and chilled desserts. In China, which produces as much watermelon as the rest of the world combined, the rind is used as a stir-fry vegetable, and in Africa and elsewhere the seeds are processed in various ways for consumption.

ABOVE: *Watermelons are low in calories and very nutritious. They are high in vitamin C, vitamin A, and lycopene, which is a potent antioxidant.*

BELOW: *Watermelons should be picked when they are fully mature. They do not continue to ripen after they have been removed from the vine.*

Melons

Cucumis melo var. *cantalupensis* Cantaloupe *Cucumis melo* var. *inodorus* Honeydew Melon, Winter Melon *Cucumis melo* var. *reticulata* Muskmelon

RIGHT: *Honeydew melons have a light green, thick, juicy flesh. Their skin is a greenish white color when immature, and a creamy yellow color when ripe. Honeydew melons are rich in vitamin C.*

There are many melons, and they all breed easily together, prompting French growers to plant their melons and cucumbers—close relatives of melons—far from each other to prevent crossbreeding. When ripe, melons feel heavy for their size and emit a sweet aroma from their base.

Historical Origins The wild ancestors of *Cucumis melo* were native to a region encompassing Egypt, Iran, and northwest India. Melons have been cultivated since ancient times in the Nile Valley and farther afield, but were not introduced to the continent of Europe until the early fifteenth century, even though adventurer Marco Polo had already discovered their fragrant delights in a thirteenth-century Persian market. In the sixteenth century, melons were introduced to England and were grown under glass or with other protective methods. Melons had by this time reached China and had begun to be used in cooking. Melons were immensely popular with European aristocracy, to the point where it was rumored that people had died from eating an excess of them. The first large-scale crop of melons in Italy was grown from Armenian seeds in the early eighteenth century. This crop was named after its new birthplace—Cantalupa (literally, "wolf cry"). Melons reached the New World in the late fifteenth century when Columbus took melon seeds to Haiti, and the new fruit proved very popular with the native peoples of Central and North America. In the 1920s, Armenian immigrants cultivated the hybridized muskmelon known as "cantaloupe" that is so popular in the United States today.

The most common types of melon in North America are netted (or musk or nutmeg) melons, and of these, the best known are the Persian melon and the "cantaloupe." These melons have highly perfumed orange flesh and weigh about 6–7 lb (2.5–3 kg). The "true" cantaloupe (var. *cantalupensis*) is small, round, and has a rough surface divided into segments. The skin is not netted. Winter melons ripen slowly and do not mature until late autumn. They are more oval than round, with finely ribbed skins. The best known of these is the honeydew melon, which has pale green flesh and a smooth, yellowish skin.

Botanical Facts Melons belong to a genus of about 25 species of climbing or trailing annuals in the pumpkin (Curcubitaceae) family. Originating in warm to tropical areas of Africa and Asia, they are now grown worldwide for their fruit. These dessert

Eastern Exotica

Melons are composed of 95 percent water and contain only 5 percent sugar. Ancient Egyptian travelers are said to have used them as self-contained drinking vessels by cutting a hole in them, resealing them and leaving the whole for a few days to let the flesh liquefy. Numerous modern recipes emulate this technique by scooping out the seeds and adding liquor to the cavity. The liquor can be wine, spirits, liqueurs, or cordials. Other recipes call for the flesh of the melon to be removed, leaving the shell intact. The flesh is then mixed with sugar, other fruits, liqueurs, and wine, returned to the shell, and allowed to sit overnight. Alice B. Toklas has one such recipe entitled "Scheherezade's Melon," evoking the famous collection of folktales, *Tales from the Arabian Nights*, in which melons are mentioned on a number of occasions.

melons are annual vines found in arid regions of Africa, Arabia, southwest Asia, and Australia. Wild types may be bitter, but cultivars produce a range of generally round, sweet fruits with either smooth or rough skins. Dessert melons mostly fall into three groups. In the Cantalupensis group are sweet, fragrant melons, about 4–6 in (10–15 cm) in diameter, with a skin that can be smooth, scaly, grooved, or rough, but not netted. In the Inodorus group are round or oval, sweet melons with green, white, red, pinkish, or yellowish green flesh. In the Reticulatus group are netted melons of assorted shapes and sizes; skins can be ribbed, warty, smooth, or netted, and the seed cavity is large in old cultivars, smaller in newer ones. Some cultivars (muskmelons) have a musky aroma.

ABOVE: *True cantaloupes have a thick, scaly skin. The flesh can be further sweetened with a splash of lemon juice, and their flavor can be enhanced with a sprinkle of ginger or salt.*

Culinary Fare Melons are at their finest picked straight from the vine and slightly warmed by the early morning sun. Chilling them decreases the flavor. A slice of melon is a popular breakfast and is sometimes served sprinkled with powdered ginger (although salt and pepper will do equally well). Melons can be served at the start of a meal as hors d'oeuvres or at the end of a meal as a fruit or dessert. A popular Italian antipasto pairs ripe, juicy melons with thinly sliced, salted, cured meat. Melons can also be paired with smoked meat, fish or fresh cheeses, or used in salads, ice creams, sorbets, and other desserts. They make excellent preserves, but care must be taken, as the fruit has a tendency to disintegrate with prolonged heat. Melon preserves are often spiced with ginger. Commercially, whole melons can be crystallized, but in household kitchens they are more easily candied in slices or cubes. Some melon seeds are edible and are sometimes dried, roasted, and salted and sold commercially as snacks.

LEFT: *Muskmelons, or netted melons, have a soft, juicy flesh that gives off a sweet, musky aroma when they are ripe. Their seeds are contained in a hollow cavity inside the fruit.*

Ceriman

Monstera deliciosa

RIGHT: *The leaves of the ceriman can grow to 3 ft (0.9 m) long, with holes of up to 4 in (10 cm). The white flower spike develops into a cream-colored fruit with a scalelike appearance.*

Once among the most common house plants, and still very widely grown for its highly ornamental foliage, *Monstera deliciosa* is a tropical vine with an unusual and aromatic fruit that has a flavor reminiscent of a combination of tropical fruits, hence the alternative common name "fruit salad plant."

Historical Origins Ceriman is originally a jungle climber native to Guatemala and Mexico, but is now commonly found in gardens in the world's tropical and subtropical regions. *Monstera deliciosa* is the only member of the Araceae family known to produce a delicious fruit. It is cultivated in California and Florida in the United States and also now in Queensland, Australia.

Botanical Facts This is a large, impressive climbing plant, from southern Mexico to Panama. It belongs to a genus of 22 species of climbing evergreen tropical plants from Central and South America; it is a member of the arum (Araceae) family. Ceriman plants are grown as foliage plants indoors or in a greenhouse in all but tropical climates, where they are often used to grow up the trunks of trees. The leaves often have large holes in them, giving them a quite bizarre look and accounting for two more common names, the Swiss cheese plant and windowleaf. A *Monstera deliciosa* will bear white (and sometimes pink to red) flowers of the arum type in suitable conditions, in summer, and fruit after its second or third year from planting, but it rarely fruits in the home, and the large fruits, which resemble corn cobs, take more than a year to ripen on the plant.

Culinary Fare The fruits of the *Monstera*, which are about 6–12 in (15–30 cm) long, are highly aromatic, and the flavor is said to be reminiscent of a combination of banana, mango, and pineapple. It needs to be eaten when fully ripe, as the oxalic acid present in unripe fruit can be an irritant. For this reason it is initially best eaten in small quantities. It is also important to cut the fruit at the right time—when the sections of rind appear to separate a little at the base. The flesh can then be used fresh in fruit salad or with ice cream, or it can be strained and made into a drink or used as a flavoring for ice cream or other desserts. The segments can also be stewed with sugar and lime juice.

Passionfruits

Passiflora edulis Passionfruit *Passiflora laurifolia* Water Lemon *Passiflora ligularis* Sweet Granadilla *Passiflora mollissima* Banana Passionfruit *Passiflora quadrangularis* Granadilla

The edible passionfruits are mostly native to South America but can now be found in warm temperate and subtropical gardens the world over. The strong-growing vines have beautiful flowers and highly scented fruit with sweet, tart, juicy flesh.

Historical Origins Passionfruits belong to the genus *Passiflora*, a group of climbing plants native to tropical America, southeastern Asia, and Australia. In South America, the missionaries used the passionflower in their teachings to represent the passion of Christ. The most important species is the *Passiflora edulis*, the passionfruit (also called the purple granadilla), which is distinguished by its dark purple skin (a yellow-skinned type is also cultivated in Hawaii and Fiji). The water lemon has a yellow shell, and its flesh is similarly highly perfumed and flavored. It is sometimes confused with the sweet granadilla, also known as the water lemon due to its juicy flesh and orange skin. The sweet granadilla is cultivated in Mexico and Hawaii. The banana passionfruit has a host of common names and is particularly popular in Colombia. It is banana-shaped and highly aromatic, with a thin skin that is yellow-orange when ripe. Native to the hotter parts of South America, the granadilla has larger fruits and yellow-green to brown coloring when ripe. It is not as luscious in flavor as the common passionfruit.

Botanical Facts The 500 or so species in the *Passiflora* genus are mainly evergreen tendril-climbing vines from tropical America, though there are a few shrubby species, and the range does extend to Asia and the Pacific islands. Grown mainly for their flowers and fruits, passionfruits, with their great vigor, are also very useful for covering unsightly walls. Fruits of many species are mildly to quite poisonous, though others are edible and delicious. The common passionfruit is an egg-sized berry with a brittle outer skin and juicy, aromatic flesh containing many seeds.

Culinary Fare One of the most delicious and fragrant of tropical fruits, passionfruit has a sweet, acid taste. The fruits are often eaten fresh with a spoon, but commercially most are processed into syrup, which is then used in flavoring ice cream, sorbets, yogurt, confectionary, fruit juices, soft drinks, and cocktails or liqueurs. The pulp is very concentrated and is an excellent addition to fruit salads and desserts. In Australia and New Zealand, passionfruit is called for in traditional recipes for pavlova, a soft meringue shell filled with cream and fruit. Passionfruit is also popular in icings and fillings for cakes.

ABOVE: *In Colombia, the banana passionfruit fruits all year round. The pulp can be eaten fresh, or it can be de-seeded and served with milk and sugar.*

BELOW: *The common passionfruit, or purple granadilla, is grown for its lovely flowers and tasty fruits.*

Grapes

Vitis labrusca Fox Grape *Vitis vinifera* Grape

BELOW: *Grapes grow in clusters of up to 300. Fresh grapes contain high levels of antioxidants that are believed to prevent a build-up of cholesterol, and to help protect against viruses and tumors.*

The European grape (*Vitis vinifera)* is one of the oldest of all cultivated plants, popular for eating fresh and also for producing wine. The most widely grown cultivars belong to this species, but there are dozens of other *Vitis* species, the most important of which is the fox grape (*V. labrusca*) of North America, which has been used to create hybrids with the European grape. There are thousands of grape hybrids and cultivars.

Historical Origins The Caucasus region is thought to be the oldest center of grape cultivation. Ancient Egyptian wall paintings show the growing of grapes, including large-fruited, improved cultivars as well as smaller cultivars, such as those used to produce currants. Just after 1000 BCE, the seafaring Phoenicians brought grapes to Greece, where they flourished in the Mediterranean climate. Grapes grown in Classical times by the Greeks and Romans were used for wine, table grapes, and grape products, such as the range of sweet syrups the Romans made from unfermented grape juice for culinary purposes; such syrups are still used in the Levant. Also in Classical times, the juice of unripe grapes was used to produce verjuice, a souring agent in cooking that is still made in many grape-growing regions today. Grapes were also dried to produce raisins and currants, which in the Middle Ages were a cherished addition to the restricted winter diet.

The Romans introduced vines to much of temperate Europe, including England, but labor shortages, climate change, and imports from France vastly reduced the production of wine in England from about the thirteenth century. Late in the fifteenth century, Columbus brought *V. vinifera* to the New World. During the eighteenth and nineteenth centuries, table grapes were very popular in Britain, grown in the hothouses and conservatories of the great houses of the time. Around the mid-nineteenth century, the North American *Phylloxera* aphid was accidentally introduced into France and devastated European vineyards. However, this aphid does not affect native American vines. European vines were saved by being grafted onto *Phylloxera*-resistant *V. labrusca* rootstocks. Australia proved very

Grape Cultivars

There are approximately 10,000 grape cultivars. Some of the most important for wine making include Chardonnay, which is a small, round, yellow-green grape used to produce white Burgundy, Chablis, and sparkling white wine such as Champagne; Pinot Noir, which is a purplish black grape used for making red Burgundy and also partly for Champagne; Cabernet, which is a blue-black grape used in the production of Bordeaux red wines, sometimes known as claret; and Riesling, which is a collective term for a group of white grapes used to produce Rhine wines.

hospitable to European grapes when settlers introduced them in the late eighteenth and nineteenth centuries, particularly in New South Wales, Victoria, and South Australia.

The fox grape, or slip skin, of North America was crossed with European vines brought by Spanish and Portuguese explorers to the Americas in the early sixteenth century. The black Concord grape is a hybrid of fox grape and is the principal grape gown in northeastern America because it is much more tolerant of cold than *V. vinifera*.

Botanical Facts Grapes belong to the *Vitis* genus of 65 species of woody deciduous vines that are indigenous to the Northern Hemisphere, particularly North America, giving its name to the grape (Vitaceae) family. The vines climb by tendrils; the leaves are mostly simple, toothed, or lobed, and the bark often peels from the stems in strips. The small flowers are sometimes fragrant. There are numerous cultivars, particularly of *V. vinifera*. The fox grape displays a distinctive, musky flavor, as well as astringency and low sugar content. There are also Asian species, but these are primarily grown for their attractive foliage and autumn color.

Grapes can be classified according to color (black or white) or usage: table or dessert grapes are firm and low in acidity; wine grapes have soft flesh and high acidity; and dried grapes have firm flesh, high sugar content, and moderate to low acidity.

Culinary Fare The main use of grapes is in winemaking, which takes 65 percent of grape production. The remaining 35 percent is used for table grapes, dried grapes, and grape juice. Grapes, fresh or dried, go well with all types of cheese, especially soft, white varieties, and also with fish dishes (such as sole Veronique) and certain types of meat, particularly game. They make excellent jelly, preserves, and light, refreshing desserts. Grapes were common in the aspic salads that were popular in the nineteenth and early twentieth centuries.

Sultanas, raisins, and currants make good, portable snacks and are excellent partners for cheese and nuts at the end of a meal. Currants, which have nothing to do with red, black, or white currants (*Rubus* species), have been used by cooks since Classical times, sometimes in savory dishes, but more often in sweet foods and baked goods. Other grape products for culinary uses include grapeseed oil and vine leaves, which are used mainly in Turkish, Greek, and Middle Eastern cuisine.

ABOVE: *Small, dark brown currants are the product of Zante grapes. They are used mostly in cooking. The larger light brown sultanas are made from Sultana grapes. They are used in cooking, as well as being eaten out of hand.*

Vegetables

Introduction to Vegetables

RIGHT: *The snap-fresh crispness of home-grown vegetables cannot be beaten. They typically have more flavor, and the simple joy of going to the garden to select the choices for dinner is second to none.*

Think vibrant color, think vivid flavor, think vegetables. Think also health and well-being. Health professionals around the world unanimously recommend that we all eat more vegetabes, and the greater the variety we tuck into, the better. The evidence from epidemiological studies is clear: "eating your greens" can help ward off heart disease and stroke, prevent some types of cancer, control blood pressure and cholesterol, and guard against cataracts and macular degeneration—two common causes of vision loss.

What actually are these nutritional powerhouses? For the botanist, vegetables are the edible product of herbaceous plants—that is, plants with a soft stem. This can be any part of the plant, from flowerheads, seeds, unripe seed pods, "true fruits," leaves, leaf sheaths, stems of leaves and stems of plants to underground stems or tubers, whole immature plants (sprouts), roots, and bulbs. The vegetables, however, being in the plant kingdom, don't include mushrooms or other fungi. Although often served as vegetables, technically these are not plants, and are therefore outside the scope of this book. They belong to the fungi kingdom, which is an entirely separate biological division.

"Vegetable" is essentially a culinary term. It commonly refers to those herbaceous plants traditionally used for preparing savory dishes, whether raw or cooked—soups, salads, stews, stir-fries, accompaniments—rather than desserts. However, there are many delicious exceptions, such as pumpkin pie and carrot cake.

ABOVE: *Root vegetables such as carrots and turnips typically mature in autumn, to be harvested in winter, making them comforting additions to hearty roast dinners and thick soups.*

What distinguishes vegetables from fruit? It was the taxing matter of tomatoes that took the "is it a fruit or vegetable" question all the way to the US Supreme Court in 1893 (Nix v. Hedden). In his findings, Justice Gray acknowledged that, botanically speaking,

a tomato is a "fruit of the vine," but because tomatoes are usually eaten as a main course instead of a dessert, as are other fruits, they are correctly identified as, and taxed as, vegetables for the purposes of the US *1883 Tariff Act* on imported produce.

WHAT TYPE OF VEGETABLE IS THAT?

The Vegetables chapter includes three sections based on which part of the plant, mainly, is eaten—root vegetables, leaf vegetables, and vegetable fruits and seeds (which includes legumes). Some foods that are often used as vegetables are covered in various other sections of *Edible*. Sweet corn, oddly enough, isn't a vegetable at all, it's a grain; olives are included with olive oil; and the avocado, which can be used in both savory and sweet dishes, is in the stone and drupe section of the Fruits chapter. As for fennel, this "all things to all people" vegetable–herb–spice is included in the Herbs chapter.

Root vegetables include beetroot, carrots, onions, parsnips, radish, and salsify, along with those starchy staples—potatoes, sweet potatoes, yams, and taro. These vegetables are either biennials or perennials, storing food in their swollen roots, which in nature would be used to enhance flowering. Leeks are also included here, although the part eaten is technically a bundle of leaf sheaths; they belong to the genus *Allium*, along with onions, and are used in cooking in a similar way for their pungent flavor.

Leaf vegetables include leafy greens such as beets, spinach, cabbage, lettuce, Asian greens, salad greens, sprouts, and dandelion; the brassicas such as broccoli and cauliflower, where the immature flowerheads are eaten; and celery, which is eaten for its succulent plant stems. Rhubarb, which comes from the knotweed (Polygonaceae) family, is in this section, too, but never eat the leaves—they are poisonous and must be discarded. Only the plant stems are safe to eat, enjoyed for their tart flavor, often sweetened for pies.

Vegetable fruits and seeds include fruits such as aubergines (eggplants), cucumbers, peppers, marrows (such as pumpkin and squash), and tomatoes, along with beans, peas, and other members of the legume (Fabaceae) family. Legumes have the seed enclosed within a pod, and either the seed alone (lima beans) or the seed and pod (French, runner, and butter beans) are eaten. Technically, peanuts are a legume and not a nut; however, because of the ways they are used, they sit more appropriately in the Nuts chapter.

Of all vegetables, legumes come a close second to the cereal grasses in economic importance as a plant food. Dried beans, peas, and lentils have been staple foods throughout the world for thousands of years because they could be stored through the winter to be made into purees and porridges or ground into flour. Not only do they continue to provide nourishing fare for humans, feed for animals, and fiber, along with other products for industry, they are also used in crop rotation to replenish the soil naturally with nitrogen as their root nodules contain bacteria that can convert nitrogen from the air into a form that plants can use.

> *"An onion can make people cry, but there has never been a vegetable invented to make them laugh."*
>
> Will Rogers (1879–1935)

GROWING VEGETABLES

Growing your own vegetables is immensely satisfying and brings with it the opportunity to practice organic gardening techniques that are free from artificial chemicals. More importantly, these vegetables are fresh and tender and have the best flavor and taste.

BELOW: *In California's Half Moon Bay, fields of pumpkins thrive. Here the annual pumpkin festival every autumn celebrates the region's produce. Gourd growers come from far and wide to compete for heavyweight prizes.*

Onions

Allium cepa Onion *Allium fistulosum* Welsh Onion

ABOVE: *Though cultivars, colors, and flavors vary, onions' culinary magic is second to none. They have been written about, prayed over, used as medicine, avoided, and revered to ward off evil spirits.*

Few foods are surrounded with as much mystery and awe as the onion. These pungent bulbs and greens have given food character for thousands of years.

Historical Origins The humble onion is known only in cultivation, and its history is obscure, perhaps originating from a wild species found in central Asia over 5,000 years ago. Selective planting over the years allowed the characteristic large bulb to become dominant and *Allium cepa* was born. The Welsh onion probably originated either in Siberia or central China.

Ancient Egyptians regarded the onion as a symbol of eternal life. They were frequently included in tombs in the belief the odor could bring the dead back to life. No other vegetable was represented more in Egyptian art. Christopher Columbus brought the onion to North America in 1492, where it quickly replaced the smaller wild onions.

Botanical Facts *A. cepa,* a relative of the lily, is characterized by a large bulb, papery skin, and crisp texture. Short-day onions form bulbs when days last 12–14 hours. Long-day onions form bulbs when days last 14–16 hours. The Welsh onion is characterized by slender green leaves and a spicy white or pink base. These onions form little or no bulb and they thrive in moderate environments and will withstand a light frost.

Propagate onions from seeds or bare-root plants in well-drained soil. Pick when the bulbs are large and the tops yellow and wilt. Once harvested, dry and cure to toughen skin and extend storage time. Cutting onions releases a chemical that causes eye irritation.

Culinary Fare The strong taste belies the heavenly sweetness imparted when onions are properly cooked. Exposed to moderate heat, the sharp taste diminishes and glorious subtleties emerge. Onions work well roasted, sautéed, stewed, and fried. They can be eaten alone or as a flavorful addition to vegetables, soups, sauces, or meats. Welsh onions are closely associated with Asian cuisine; they are often served raw or lightly cooked in salads, soups, spring rolls, or stir-fries. The entire onion is edible and has a similar flavor to *A. cepa*. Onions are also associated with countless health benefits. Current studies indicate chemicals found in onions help fight infections, inflammation, and will also improve cholesterol levels.

RIGHT: *In Ancient Rome, gladiators were massaged with onions to promote firm muscles. Onions were associated with strength and wellness—armies were fed onions to produce invincible troops.*

Leek

Allium porrum

Leeks are in the same illustrious family as onions and garlic. Their mild sweet flavor has generated devoted fans wherever they are cultivated. Different types of leeks have found their way into every cuisine around the world. Leeks are also used as a diuretic as they combat fluid retention.

Historical Origins Like onions, modern leeks are only known under cultivation. The first wild version of leeks most likely originated in the Middle East and around the eastern Mediterranean. This theory is further supported by the historic popularity of leeks in those regions.

Leeks were particularly treasured in Ancient Egypt for their superior taste and medicinal properties. Workers on the great pyramids were given a steady supply of leeks to improve their strength and accelerate their efforts. The Roman Emperor Nero was rumored to have eaten leek soup on a daily basis to help keep his singing voice clear. Leeks were so highly valued in Rome that they were also used as a form of currency.

Leeks were later adopted as the national emblem of Wales. Snippets of leeks worn in buttonholes or on hats helped identify friend from foe while in battle.

Botanical Facts *Allium porrum* is an upright plant grown for its long, cylindrical, white bulb. Since the base or bulb is more desirable than the leaves, plants are often light-blanched—covered with soil to block out sunlight. This promotes the formation of an extended white bulb.

Leeks are usually propagated by seed and grow well under a variety of conditions. They are quite hardy and can be left in the ground to overwinter. The following spring, plants will produce dramatic flower spikes with each flowerhead generating hundreds of seeds. Bees are particularly attracted to these flowers. The quality of the bulb decreases with age, and once flower spikes form, the core becomes tough and woody.

Culinary Fare The delightfully fine flavor of leeks makes them a good choice for flavoring soups, stews, stocks, and sauces. The green leaves can be used for flavoring dark stocks. When braised or slow-roasted, the white bulb takes on a buttery texture that complements meat or fowl. Leeks pair particularly well with potatoes and parsnips.

Not only do leeks taste good, they are also good for you. They are an excellent source of vitamin A and are high in antioxidants. In addition, leeks, along with onions, share many medicinal properties, such as improving cholesterol problems.

BELOW: *When harvesting, smaller leeks tend to be more tender and their flavor is more subtle. They will keep for up to a week in the fridge stored inside a sealed plastic bag.*

Garlic

✳ *Allium sativum* Garlic ✳ *Allium scordoprasum* Giant Garlic, Sand Leek ✳ *Allium ursinum* Bear's Garlic, Ramson

Garlic is one of our most odiferous root vegetables and one of the oldest-known cultivated crops. An indispensable flavoring agent for virtually all cuisines, it has been associated with mythology and magic, along with a host of different superstitions.

Historical Origins Central Asia is the home of the wild *Allium sativum* and the only region still supporting abundant wild growth. Earliest references can be found in ancient Sanskrit; later on in Babylonian and Chinese writings. Garlic is believed to have originated in the Mediterranean and Syria.

The first citing of giant garlic (*A. scordoprasum*) did not occur in texts and papers until it was recorded by Gerard, an herbalist from the late 1500s. Bear's garlic (*A. ursinum*) possibly originated in western and central Europe. It was probably first used as livestock fodder in Denmark or Switzerland. Folklore established its name because of the brown bear's habit of consuming massive quantities after waking from hibernation. It has since been associated with renewal and strength.

Images and sculptures of garlic bulbs in tombs show Egyptians revered garlic. Garlic was even used during the mummification process. It is estimated over 1½ million lb (680,000 kg) of garlic was used to feed the workers building the great pyramids at Giza. Garlic shortages due to an unexpected drought caused the workers to strike.

In Ancient Greece and Rome, garlic was thought to offer protection, strength, vitality, and courage. Roman generals were known to plant garlic fields in the countries their armies conquered to transfer courage to the battlefield. Because of the smell and association with magic and superstition, garlic was often shunned by the aristocracy, but it was used

BELOW LEFT: *A nineteenth-century illustration of garlic shows the purple flowers and the bulb. The flowers have a garlicky odor, though cultivars have since been bred to yield sweeter scents.*

BELOW RIGHT: *The strong-flavored hard-necked (purple-striped) and the milder soft-necked garlic (white) are shown, along with a flowerhead and bulbs divided into cloves.*

extensively to feed the lower classes, slaves, and troops. Julius Caesar is reported to have bucked tradition and consumed garlic with abandon.

Garlic eventually became an accepted seasoning and medicine throughout Europe and Asia. It accompanied the Spanish to the Americas during the late 1400s and quickly caught on with indigenous peoples, but immigrants were slow to add this smelly bulb to their larders. It was not until early twentieth century European migrations brought ethnic cuisines to the east coast that the North American love affair with garlic began.

Botanical Facts Garlic is a member of the onion family, grown for its strong-tasting bulb. Each bulb is made up of a number of smaller cloves encased in papery white or pinkish shells. It is propagated by planting individual cloves in well-cultivated sandy soil. Flower stalks should be snapped off to promote bulb formation. Harvest when stalks wither and die. Once removed from the soil, garlic must be dried and cured to improve its keeping quality. Garlic is edible in any stage of growth, but most growers hold out for full bulb development.

Giant garlic is a hardy perennial found mainly in the wild. Although plants are available, its cultivation usually takes place in home gardens in the Mediterranean. It is rarely seen for sale. The small lilac-colored bulbs are often disregarded in lieu of the tiny bulblets formed on the flowerheads. Plants are propagated by planting the bulblets, and grow best in light sandy soil.

Bear's garlic is exclusively grown and harvested in the wild. Wild crops are maintained but areas of growth seldom vary. Plants are propagated by seeds spread mainly by ants in swampy deciduous woodlands. Flowers bloom early and they scent the woods with garlic.

Culinary Fare Although the heat and tang of raw garlic is appreciated by some, cooking tames the volatile compounds and renders garlic more socially acceptable. Garlic can be used raw in dressings, salads, marinades, and sauces. Pickled forms are used as condiments and as an ingredient for other

LEFT: *Bear's garlic grows in forests and flowers in mid-spring. Though it resembles a lily of the valley, its garlic scent reveals its true nature.*

preparations. Cooked, garlic can serve as a flavoring agent for soups, stews, vegetables, meats, and seafood. It can be roasted whole for use as a spread or as an appetizing addition to mashed potatoes.

Although the use of garlic is usually associated with savory applications, aficionados embrace its use in unconventional foods such as ice cream or other sweets. The leaves of bear's garlic can be eaten raw in salads and pureed into a pesto of sorts. Cooked leaves can also be served as a vegetable.

Many early claims to the medicinal properties of garlic still prove true. Among other things, garlic helps ward off infections, regulates blood sugar, and acts as a powerful detoxifier in the digestive tract.

Vampires, Spirits, and Demons!

Garlic has long been associated with mystery, the underworld, and evil. Muhammad wrote of garlic and onions sprouting up where the devil first lay his feet after being cast out of the Garden of Eden. For that reason, it has been seen as a source of protection. In Transylvania, Romania, the legendary home of vampires, garlic is used to supposedly ward off their evil influence. Greek midwives kept evil spirits at bay by preparing birthing rooms with crushed garlic and hanging garlic necklaces around the infant's neck. Early Greek travelers would disguise their route with piles of garlic left at crossroads so demons following them would lose their way.

Celeriac

Apium graveolens var. rapaceum

RIGHT: *The somewhat homely appearance of celeriac cloaks the wonderful flavor it can impart to hearty winter fare.*

Celeriac is the "ugly stepsister" relative to parsnips, carrots, parsley, and fennel.

Historical Origins Celeriac was developed from traditional celery in areas around the Mediterranean sometime in the late 1500s to early 1600s. The flavor of celery was appreciated by the populations of temperate Asia and Europe for thousands of years but the bitter tasting stalks were considered undesirable. The small root of the celery plant was found to be sweet and mild. This discovery prompted botanists to breed celery plants with enlarged roots and fewer stalks. By the 1700s celeriac or celery root became popular across Europe, though it never took hold in North America.

Botanical Facts Celeriac is a long-season plant, taking up to 200 days to fully mature. Its large single root is lumpy, brown, and somewhat hairy, creating a first impression that could easily frighten off all but the most stalwart fan. It is propagated by seed and grows best in loose, well-drained soil.

Culinary Fare Roots need scrubbing and peeling before cooking in soups, stews, or stocks. Celeriac is a delectable addition to roasted vegetables and pairs nicely with potatoes. Young roots can be grated in salads or cut for crudités. However, it oxidizes quickly, so cut roots must be held in acidulated water before preparation.

Horseradish

Armoracia rusticana

BELOW: *After harvesting, horseradish can be stored in the fridge for several weeks if kept dry in a paper bag. It also freezes well.*

Horseradish is a peppery root vegetable that is perhaps most synonymous with German food and British roast beef. Its family tree includes mustard and cabbage.

Historical Origins Horseradish is native to certain areas of western Asia along with parts of southeastern Europe.

It is referenced in Greek mythology as being worth its weight in gold for medicinal and culinary purposes. It was used to relieve muscle pain, as a cough remedy, and an aphrodisiac. It continues to be one of the symbolic bitter herbs on the Passover Seder plate, served during the meal and used to retell the story of the Jews' Exodus from Egypt.

Botanical Facts Horseradish is a tall and somewhat unimpressive perennial plant. Roots are harvested after the first killing frost, taking care to leave some roots behind to produce a crop the following year. The invasive nature of horseradish can wreak havoc in a small garden, so roots are often confined. It is one of the few crops that is routinely planted and harvested by hand.

Culinary Fare Until horseradish is grated, its hot temperament remains cleverly hidden. The heat of the grated root continues to build until the process is neutralized by adding an acid such as vinegar. Horseradish is generally used as a condiment or as an important ingredient in relishes, sauces, marinades, and drinks. Medicinally, horseradish has been found to be helpful as a diuretic and to treat minor infections.

Beetroot

Beta vulgaris

Beetroot belongs to a family of earthy root vegetables grown for food, sugar, and as animal fodder. The typical deep red color of the garden beetroot makes a bold statement on any plate.

Historical Origins The modern beetroot descended from the wild sea beet found in the European and North African coastal regions of the Mediterranean Sea and the Atlantic Ocean. Selective breeding eventually increased its average root size and yield. The first large-rooted, pale red beetroot was referenced around 1550. The dark red beetroot enjoyed today surfaced in the mid-seventeenth century.

Botanical Facts Most beetroots are grown for their large roots but the leaves are also edible. Other cultivars are grown strictly for their attractive leaves. Although red is the dominant color, there are yellow and white cultivars, as well as those sporting rings of alternating colors. Beetroot is propagated by seed and grows best in light-textured sandy soil and cool outdoor temperatures. It easily cross-pollinates, so it is important to keep different strains isolated from one another during the growing season.

Culinary Fare Beetroot is always cooked before eating and usually peeled after cooking to prevent bleeding. It can be roasted, added to soups, and pickled. Pickled beetroot is often served cold in salads, as a side dish, or as a condiment. Slices of pickled beetroot are an essential ingredient in Australian hamburgers. Leafy beetroot greens can be steamed, sautéed, or served raw if they are harvested young, their crimson veins a delightfully colorful addition to green salads.

Kohlrabi

Brassica oleracea Gongylodes Group

Kohlrabi's name originates from the German words *Kohl* or cabbage, and *Rabi* or turnip, indicative of its family heritage.

Historical Origins Northern Europe lays claim as the home of a vegetable thought to be the original version of the kohlrabi. It may have been the "Corinthian turnip" mentioned in *Apicius*, the oldest-known cookbook, in the fourth or fifth century. In the 1600s, kohlrabi found its way to northern India before arriving in China and Africa. North America was not introduced to this unusual crop until the early nineteenth century.

Botanical Facts Kohlrabi is a strange-looking vegetable with a swollen round base and thick-stemmed leaves sprouting off in regular increments around its globe. It is propagated from seed and grows best in light soil with cool outdoor temperatures. It can be harvested at any stage of growth.

LEFT: *The flavor of kohlrabi is somewhat reminiscent of cabbage and turnip, from which it derives its name, but it has a mild quality that makes it unique.*

Culinary Fare If harvested young, kohlrabi can be eaten unpeeled and raw in salads, for which it is usually best grated; or cut it into sticks for dips. Larger specimens benefit from light peeling and cooking. It can be roasted, served as a vegetable, cooked in soups, and braised. It has a sweet turniplike taste without the typical sharpness.

Turnips

✳ *Brassica napus* var. *napobrassica* Swede Turnip ✳ *Brassica napus* var. *rapifera* Turnip

ABOVE: *Turnips were an important food in Roman times when they were a dependable source of nutrition for the masses. Despite since being associated with lower classes as a famine food, there are many references linking their extensive use in all ranks of society.*

Turnips and swede turnips are strong-flavored root vegetables that tend to be most favored as hearty peasant fare.

Historical Origins Turnips most probably originated from wild rape in northern Europe about 4,000 years ago. As they were introduced in other areas of the world, they were readily accepted as animal fodder. Periods of famine caused whole populations to lose their previous reservations about eating turnips and they became more commonly grown.

Swede turnips are said to have originated in central Europe around the seventeenth century. They are thought to be a mutant or a rare hybridization between a turnip and a wild cabbage, based on the differing number of chromosomes found in each type. Swede turnips were first used for animal fodder but eventually people began to appreciate their assertive flavor and generous size. They were an important food of central Europe before the potato was commonly used. They lost favor during World War I after grain and potato crop failures forced many to subsist on a diet of swedes alone.

Botanical Facts Turnips are grown for both their roots and leaves. Some cultivars produce nothing but leaves. They are propagated from seed in light soil and cool outdoor temperatures. Harvesting after a frost ensures the tastiest roots. Turnips can be harvested at any stage of growth.

Swede turnips produce large, yellow-fleshed roots that have the qualities of both turnips and cabbage. They are grown for their root. Propagated by seed, they grow best in well-drained soil and cool outdoor temperatures. While they can be harvested at any stage of growth, doing so after a frost ensures the sweetest roots.

Culinary Fare Young turnips may be scrubbed rather than peeled and steamed whole with their leaves intact. Larger turnips should be peeled and can either be roasted or added to soups, stews, or braised dishes. They can also be pickled.

Swede turnips have a thick skin that should be peeled before cooking They can be served roasted or boiled, or in soups, stews, or braised dishes. They are known as neeps in Scotland and rutabaga elsewhere.

Happy Halloween! Who Wants to Carve a Turnip?

Traditional jack-o'-lanterns originated as hollowed-out turnips with embers or candles inside. They became popular Halloween decorations in Scotland and Ireland several hundred years ago. Folklore claimed they would ward off evil spirits on Halloween. Immigrating Irish families brought the tradition to North America. Turnips were harder to come by there, so they were replaced with plentiful pumpkins. This was fortunate because pumpkins are easier to carve than turnips. It was not long before people began to cut scary faces and other spooky designs into their jack-o'-lanterns.

Rampion

Campanula rapunculus

Rampion, sometimes called little turnip or rampion bellflower, is considered a potherb or a kitchen garden vegetable.

Historical Origins Rampion is native to English field borders and hedgerows. At one time it was commonly cultivated in home gardens but it lost favor because the roots are considered somewhat inferior to other similar edible root species. Although it has slipped into obscurity over the years, champions of the diminutive root still exist in England, France, Germany, and Italy.

Botanical Facts Rampion is a tender biennial often grown as an annual. It is grown for its tender leaves and carrot-shaped white root. The leaves of rampion are borne on long slender stems and form a 12-in (30-cm) rosette at the root crown. Delicate flowers in various shades of blue appear late in the season. If the plant is to be eaten, flowers should be pinched before they open. It is propagated by seed and grows best in moist sandy soil in a partially shaded cool location.

Culinary Fare Leaves may be eaten raw in salads or cooked as a vegetable. Roots may also be eaten raw, boiled as a vegetable, or used in soups or stews. The flavor of the root is said to be similar to that of a radish with a bit more sweetness and nuttiness.

RIGHT: *Although not attractive in its looks, boiled rampion is tasty with butter, oil, and pepper.*

Achira

Canna edulis

Achira is a large-leafed member of the *Canna* lily genus grown for its starch-filled fleshy stems and tender young leaves.

Historical Origins Achira is native to the Andean highlands in South America and is thought to be one of the first cultivated crops of the region. Some cooked remains of achira fleshy stems have been identified in 4,500-year-old coastal tombs in South America. Evidently, the fleshy stems were deemed important enough to be transported long distances. Achira is very cold tolerant, so it was probably used to supplement potato crops in cooler climates.

Achira has been grown successfully on several Caribbean islands and has achieved limited commercial success in other countries, such as Indonesia, the Philippines, and Taiwan. It is currently enjoying a small comeback in some parts of South America as future prospects for cultivation are explored.

Botanical Facts The robust stems of achira produce striking hedgerows for windbreaks and boundary markers. Plants are propagated in all soil types by planting the tips of fleshy stems. They grow rapidly and are ready for early harvest within 6 months of planting. After 8 months, the fleshy stems are full-sized but they can be left in the ground without degrading their quality.

Culinary Fare Achira leaves are cooked and eaten as a green vegetable. The fleshy stems can be eaten raw, but they are usually cooked like potatoes. However, achira is more important as a milled starch crop—its starch content is over three times that of potatoes. The starch granules are the largest ever measured and are especially translucent.

BELOW: *Achira has more to offer than its edible tasty roots and leaves–its pretty flowers make it a very attractive garden plant.*

Taro

Colocasia esculenta

RIGHT: *Because the starch grain in taro is so small, it is highly digestible, making it especially good for infants and people with allergies.*

Often called the "potato of the tropics," taro is the starchy corm of a tall semi-aquatic or dry-cultivated crop. Its starch content is similar to the potato but taro boasts higher protein and nutrient levels.

Historical Origins Taro originated as one of the first cultivated plants in India or the low wetlands of Malaysia. It is argued that taro may have been cultivated before rice and millet. Archaeological evidence of soft plant matter is rarely preserved, making accurate dating very difficult.

Taro found its way to Egypt and was adopted as an important staple crop in both Greece and Rome. The fall of the Roman Empire brought an end to widespread taro use in Europe. It remains popular in Asian countries, along with Africa, Turkey, Cyprus, the US, and some Polynesian islands.

Botanical Facts A taro plant produces large leaves and can grow over 6 ft (1.8 m) tall. Central corms can weigh up to 8 lb (4 kg) with smaller side corms weighing in at 2–4 oz (55–115 g) each. Semiaquatic taro can be grown in paddy fields or as a dry crop. However, dry taro cannot be grown in the same manner as the semiaquatic type. Even dry cultivars require large amounts of water to ensure a good harvest.

Taro is grown for its starchy corms and tender young leaves. It is propagated by planting the top portion of the corms with a tiny piece of the leaf stem still intact. The growing season is about 7 months.

Culinary Fare All parts of the taro plant are toxic and should be handled with care. Gloves are recommended when peeling the roots to prevent skin irritation. Taro roots oxidize quickly, so they must be held in water to avoid discoloration before they are prepared.

Roots and leaves contain calcium oxalate crystals that can become lodged in mucous membranes if the plant is consumed raw. Lengthy cooking times are essential to neutralize these crystals. Roots must be boiled for at least an hour to render them safe to eat. They can also be simmered as an addition to soups or stews. The leaves must be simmered, either alone or in soups or stews, for at least 45 minutes.

Porridgelike Hawaiian *poi* is made from cooked fermented taro paste. Chinese bakeries use taro paste in cakes and buns.

BELOW: *The Hanalei Valley in the Hawaiian island of Kauai is home to wet farms growing the semiaquatic form of taro.*

Carrot

Daucus carota

Few vegetables have as much widespread acceptance as the carrot. The typical bright orange color and sweet delectable crunch were bred into the plant long after its discovery. This ensured the carrot's evolution over the centuries to become one of the most favored of root vegetables.

Historical Origins Carrots were found growing in central Asia over 5,000 years ago. Wild carrots were every color but orange, with purple the dominant hue. Leaves and roots were strong and bitter and were used as an herb for medicinal purposes and flavoring. Cultivation helped roots grow larger and lose some of their bitterness. However, these positive changes did little to promote carrots as a primary food, although their use as medicine and an aphrodisiac continued.

Greeks avoided eating carrots altogether, but the Romans' love and appreciation of strong flavors made them more accepting of this vegetable. Carrots were eaten both raw and cooked. The early Roman cookbook *Apicius* lists a number of recipes for carrots.

In the twelfth century, Moors brought carrots to continental Europe, and they soon spread to England. Carrots arrived in the New World sometime later. Selective breeding brought about the color change to the familiar orange of modern carrots. Once the improvements associated with the new color became widely known, the carrot's status changed from a potherb to a vegetable.

Botanical Facts Carrots belong to the same family as parsnips and skirret, all grown for their fat taproot. Root thickness, length, and color depend on the cultivar. Carrots are propagated from seed and grow best in well-worked sandy soil and cooler temperatures. They are biennial plants, flowering in their second year. Carrots can be harvested as soon as the roots are big enough to eat. Some carrot cultivars produce roots that fatten up extra early so they can be pulled as "baby" carrots.

ABOVE: *"Baby" carrots are an early yielding carrot that is smaller but much sweeter, making it a delicious lunch box snack.*

Culinary Fare Carrots impart a strong sweet flavor when cooked in stocks, soups, stews, and braised dishes. They can be boiled, roasted, or sautéed and served as an accompanying vegetable. As a solo vegetable, they can be served steamed or boiled with a little fresh dill, honey, and butter. When served raw, their color, sweetness, and crunch add interest to vegetable platters and salads.

Purple Carrots?

Dutch botanists are said to be responsible for breeding carrots to form the now-familiar orange color. Growers experimented with mutant yellow carrot seeds from North Africa in order to develop a carrot with less bitterness. Yellow carrots were bred with the common purple and red cultivars to produce the orange carrot. It is rumored to have been adopted as the royal vegetable of the Dutch House of Orange to honor William I during the fight for independence from Spain in the sixteenth century. The orange carrot had a superior taste and texture, and eventually it became the dominant cultivated type.

RIGHT: *Selling carrots or sweet turnips, from a woodcut c. 1515.*

Yams

Dioscorea species

BELOW: *There are several theories as to how yams arrived in the New World. One is that the slave trade introduced yams to the Caribbean islands and they spread from there. Another is that yam plants traveled naturally by water from Africa to French Guiana.*

Starchy wild yam tubers have been a key staple food for thousands of years. Their sturdy nature made them reliable storage crops that fed countless populations during periods of food scarcity.

Historical Origins Evidence of wild yam gathering has been found over a wide area, but North Africa seems to be the point of origin. They were first cultivated in Africa and Asia after 8000 BCE. Population growth and trade took yams to many surrounding regions from their point of origin. The presence of yams in Europe is attributed to the voyages of exploration undertaken by Christopher Columbus.

Yam harvest proved to be quite difficult in some regions where yams were newly introduced. Many yam species produce a huge, long, single root that is very labor-intensive to unearth. Experienced growers knew how to properly trench the rows before planting to allow for easier access to the roots. This knowledge did not necessarily accompany the tubers when they traveled to new places. This, along with the toxicity of some species, kept yams from gaining wide-spread acceptance in Europe and parts of the New World.

Botanical Facts Yams are herbaceous perennial vines grown for their large tubers. Tuber sizes vary, with some plants having the capacity to produce one central tuber over 8 ft (2.4 m) long and a whopping 150 lb (68 kg). Other types have tubers that spread like fingers under the plant and weigh 1–3 lb (0.5–1.4 kg) each. The color of a yam's flesh ranges from white to orange.

Yams are propagated by planting the narrow tip at the end of a tuber in well-drained soil. They respond favorably under adverse conditions and hold fast in the ground for extended periods. Yams require up to a year to fully mature.

Culinary Fare Many cultures, such as those of West Africa and the Pacific Islands, still regard yams as a valuable food resource, despite some species being toxic. These toxic yams require extremely careful handling and cooking to render them safe to eat. All yams have tough skin, which makes them difficult to peel. Heating briefly can help make peeling them easier. Yams may be boiled, roasted, fried, or cooked in soups and stews. Less toxic species such as the Japanese mountain yam can be pickled to remove irritants, then grated and served raw.

Water Chestnut

Eleocharis dulcis

The water chestnut is a semiaquatic sedge requiring similar growing conditions to rice. It produces small crisp corms that are particularly popular in Chinese dishes.

Historical Origins The water chestnut is native to China and is grown extensively in China and the Philippines. Corms have the unique ability to remain crisp regardless of preparation. Their snappy texture and sweet taste make them an accepted vegetable in many countries around the world.

Botanical Facts The water chestnut's white-fleshed corms are borne underground and shaped like a slightly rounded disk. The plant's leafless stems have a tall upright growth habit. The plants are propagated by seed in rich flooded soil. They require very warm growing conditions with a 7-month frost-free season.

Culinary Fare Water chestnuts are often quite muddy when sold fresh, so careful washing is necessary. The surface of the corms is irregular and covered with a papery skin that must be peeled before eating. The small size of the corms makes this a challenge.

The sweet rewards are most definitely worth the effort, especially when fresh water chestnuts are compared to the far inferior canned version. These tasty corms can be eaten raw or pickled as a snack, in salads, or as a condiment. They work well in soups, stir-fries, and as a garnish for vegetable dishes. Corms are also used to produce starch and flour. Store them in the fridge inside a plastic bag and they will keep for up to 2 weeks.

ABOVE: *Belying their name, water chestnuts are not nuts at all but a vegetable, named for their resemblance to chestnuts. They have a lovely crisp texture.*

Jerusalem Artichoke

Helianthus tuberosus

The Jerusalem artichoke is a misnomer as it is unrelated to the artichoke. It grows wild along wood edges in North America.

Historical Origins The Jerusalem artichoke originated in North America and was widely used by many Native American tribes. The striking plants caught the eye of French explorer Samuel de Champlain in 1605, prompting him to take samples back to France. The flavor was reminiscent of artichokes, which gave the tuber that part of its name. By the mid-1600s, Jerusalem artichokes were found throughout Europe. But as the potato grew in popularity, the Jerusalem artichoke lost favor.

Botanical Facts The Jerusalem artichoke is a tall perennial plant grown for its knobby and somewhat unattractive tubers. Tubers can remain in the ground indefinitely, but the quality suffers with age. It produces bright yellow flowers late in the season. Propagate by planting small tubers deep in the ground. Sow in early spring and harvest after the first frost. It can become an invasive weed if all the roots are not removed after harvest.

Culinary Fare The gently delicate, sweet flavor of Jerusalem artichokes can be enjoyed raw in salads, lightly cooked in stir-fries, or as a vegetable. Lengthy cooking breaks down the starch and produces a mushy texture. Tubers must be peeled and held in water before processing to prevent oxidation. All tubers are knobby, but the smoother the tuber, the easier it is to prepare.

BELOW: *The Jerusalem artichoke became an insurance crop that could easily be cultivated during food shortages.*

Sweet Potato

Ipomoea batatas

Sweet potatoes are edible tubers formed from the roots of a perennial vine. Their high sugar content and moist texture make them popular for savory as well as some sweet applications. Also called kumara, sweet potatoes are a highly nutritious food.

Historical Origins Sweet potatoes are native to Peru. Once cultivated, they became a staple food in many New World regions. Spanish and Portuguese explorers probably took them to Asia, Africa, the Polynesian islands, and Europe.

As sweet potatoes thrive in tropical and subtropical climates, they achieved limited success in cooler areas. Their popularity in France waxed and waned, and they were soon replaced by potatoes. Their capacity to produce a large, reliable crop meant that sweet potato cultivation and consumption escalated throughout Asia and Africa, sustaining populations when other crops failed. In the Caribbean, *boniatos*, a dry, pale-fleshed kind of the sweet potato, became popular. Many enjoyed the delicate flavor and fluffy texture. Sweet potato was an important staple in North America until quite recently.

Sweet Potato Superstar

In the early 1900s, Tuskegee University's director of agriculture, George Washington Carver, experimented with this versatile tuber to develop over 100 different products. Notable discoveries included glue for postal stamps, sizing for cotton fabrics, a corn syrup alternative, and its use as a dehydrated food. Carver also revolutionized southern agriculture by showing the benefit of crop rotation on soil regeneration by planting sweet potatoes alternatively with other crops. Years of growing only tobacco and cotton left the soil depleted in the southern regions of the United States at a time when their farming economy was already on a devastating downswing after the Civil War. Carver persuaded farmers to try his crop rotation method to help the region recover.

Botanical Facts Sweet potato tubers form under lush vines capable of shading out even the most noxious of weeds. They grow successfully in poor soil and produce well with a little rain and warm temperatures. Propagate from slips that grow from the roots during storage; the root end of the slip is simply pushed into the ground and watered in. Tubers take anywhere from 90 to 110 days to mature. Like many other root vegetables, sweet potatoes require drying and curing in order to prolong their storage capabilities.

Culinary Fare Sweet potatoes are always served cooked. They can be baked, boiled, or fried and also work well in soups, stews, and braised dishes. Their coral pink color makes them an attractive alternative to potatoes, particularly when served steamed or mashed to accompany white foods such as steamed fish. Medicinally, sweet potatoes have been used to treat diabetes, parasites, and asthma.

RIGHT: *North American interest in the sweet potato decreased over recent years. Now Americans most often use it in holiday meals.*

Cassava

Manihot esculenta

Cassava (manioc) is a tropical shrub grown for its starchy tuberous root and green leaves. Surprisingly, some cassava cultivars are listed among the most toxic staple foods.

Historical Origins The first wild forms of cassava were found and cultivated in Brazil, and it soon became a staple crop throughout the Americas. The spread of cassava cultivation outside the Americas is attributed to Spanish and Portuguese explorers.

Cassava became an essential staple crop in Africa because of its ability to grow under harsh conditions. It can withstand flooding and drought and still produce edible roots. In many Asian countries, however, rice still remained the dominant staple food while cassava was treated as a secondary or insurance crop, for use as animal fodder or as a thickener for desserts.

Botanical Facts Cassava is the only member of the spurge family that produces human food. There are 2 primary types of cassava—sweet and bitter. The sweet types are not nearly as toxic as the bitter ones. The original form of cassava was quite bitter, and the sweet form was developed through selective breeding.

Cassava plants are 3–10 ft (0.9–3 m) tall with roots forming directly under the plant and weighing up to 20 lb (9 kg) apiece. Cassava is propagated by taking stem cuttings at harvest and planting them before the wet season begins. Root maturity takes up to 18 months, depending on conditions.

Culinary Fare Bitter cassava is highly toxic and can only be consumed after it has been thoroughly processed. The toxins found in bitter cassava produce cyanide. The newer short procedure developed by Dr. Howard Bradbury, an Australian plant chemist, is to peel and dry the roots, grind them into flour, mix the flour with water, and spread out the paste in a thin layer to dry in the shade for at least 5 hours. This process reduces the amount of cyanide to a safe level. The toxins in sweet cassava are neutralized during the cooking process, so it is safe to eat boiled, steamed, and fried (after boiling) as a vegetable. It is also used in soups and stews.

Bitter cassava is used in the form of detoxified flour to make dumplings, breads, and porridgelike dishes. The starch is also used to produce tapioca. Leaves of both types have to be thoroughly cooked before eating. They can be served as a vegetable, added to soups or stews, or made into a sort of pesto.

LEFT: *Cassava is grown primarily in Africa, and ranks as the third most important source of carbohydrates for the world's population.*

BELOW: *A mid-nineteenth-century engraving from a painting by Lemaistre shows a group of slaves preparing cassava flour in the French Antilles.*

Lotus

Nelumbo nucifera

Lotus is a flowering aquatic plant grown as an ornamental and as a food crop. It has the unique distinction of having virtually every part of the plant used—either for eating or as a decoration.

Historical Origins Lotus is native to numerous tropical or subtropical regions throughout the Middle East and Asia. It spread into other suitable areas as a food plant and an ornamental. In the late 1700s, lotus was grown in hothouses in western Europe. It is India's national flower and it remains a popular water-garden plant in Asia and Australia. Lotus holds a place of honor in nearly every region where it is grown. It represents purity, perfection, and beauty.

Botanical Facts As a food plant, lotus is grown for its long, unusually shaped, fleshy stems, and its seeds, leaves, and flowers. Fleshy stems root in mud below the water's surface, and long stems anchor the leaves and flowers. As an ornamental, it is a true showstopper, with huge round leaves often measuring 2 ft (0.6 m) in diameter and 7-in (18-cm) flowers in lovely pastel shades.

Lotus is propagated from seed or underground stem division from the remains of the harvested crop. Seeding is generally used when trying to produce new cultivars. Stem division is preferred for plants grown for food. Lotus must be planted completely underwater and requires full sun exposure to thrive. Its fleshy stems mature in about 5–7 months, depending on conditions.

Culinary Fare Externally, whole lotus fleshy stems are shaped something like a link sausage, but they reveal an intricate lacy pattern once peeled and sliced. For this reason, the sliced fleshy stems are often pickled and candied, used in stir-fries and soups, or fried as a garnish or side dish. They are also used to produce starch and flour. Once peeled, the stems must be held in water to prevent darkening before preparation. They have a naturally sweet taste and a delightfully crisp texture.

Lotus leaves can be used as a seasoning, cooked as a vegetable, or used as a wrapper for steaming other food. Lotus seeds are roasted and eaten as a nut or used as a coffee substitute, popped like corn, ground and cooked into sweet fillings for desserts, or ground and used in breads. Lotus flower petals are used as a garnish, and stamens are dried and made into an herbal tea.

All parts of the lotus plant have some kind of medicinal value.

LEFT: *When sliced, the lotus stem reveals its attractive pattern that can dress up any dish it accompanies. Also pictured are seeds, along with leaves in the background.*

Oca

Oxalis tuberosa

Oca is a small perennial plant grown primarily for its edible starchy tubers as well as its young shoots and leaves.

Historical Origins Oca originated in Peru and was introduced to Europe just before the mid-nineteenth century. It failed to become popular because of its specific environmental needs. It became established in New Zealand later with much greater success; a fondness for oca continues there, where it is known as the New Zealand yam.

In the Peruvian highlands, oca continues to be an important staple crop, second only to the potato, largely because of its ability to adapt to poor growing conditions.

Botanical Facts The cloverlike oca plant is compact and bushy. Roots look somewhat like fingerling potatoes and come in a variety of colors, such as red, white, purple, and yellow. They can have a hint of tartness when they first come out of the ground and are often dried in the sun to sweeten up before being stored.

Oca is propagated by shallowly planting tubers in the spring and hilling the plants as they grow to encourage the formation of more tubers. Tubers can be harvested as soon as they are large enough to eat. They mature in approximately 8 months.

Culinary Fare Tubers can be eaten unpeeled, raw, pickled, boiled, fried, or in soups and stews. The leaves can be added raw to salads or cooked as a green vegetable.

ABOVE: *The edible tubers of oca resemble small shriveled carrots and they yield flavors varying from citrus to chestnut.*

Jicama

Pachyrhizus erosus

Jicama is a leguminous tropical vine grown for its edible tuberous root. It is prized for its nutrition, crisp texture, and sweet taste.

Historical Origins Native to Central America and Mexico, jicama was widely cultivated by every key Mesoamerican civilization. Spanish explorers introduced jicama to Asia and the Pacific islands. It was a popular ship staple because it was thirst-quenching, stored well, and could be eaten raw. By the early to mid-1800s, jicama had almost come full circle when it was introduced to French Guiana in South America.

Botanical Facts Jicama is a large legume that produces fleshy tubers off vigorous vines. Tubers are usually quite big, each weighing up to 40 lb (18 kg) though smaller 2-lb (1-kg) tubers are more common for retail sale. The average yield per acre is about 35 tons. It is cultivated by seed and thrives under poor conditions. Because it is a legume, its roots fix nitrogen in the soil, increasing soil fertility. Tubers mature in 3–9 months, depending on the conditions.

Culinary Fare Jicama tubers require no cooking and they retain their raw texture even when they are heated. They can be served raw in salads or with dips, or cooked alone or with other vegetables in stir-fries, as a side dish, or seasoned for a snack.

RIGHT: *The applelike crispness of jicama adds a delicious crunch to salads. Try slices sprinkled with lime juice, salt, and a whisper of dried chili.*

Parsnip

Pastinaca sativa

Parsnips are related to carrots and are grown for their ivory-colored flavorful root. They are considered a true root vegetable because the edible portion is the single swollen taproot of the plant.

Historical Origins Parsnips originated in western Europe and Asia. Roman and Greek literature makes reference to the cultivation of a plant believed to be the parsnip, but the original root vegetable in question may have been a carrot.

Wild parsnips were spindly, fibrous, and had a strong taste that probably encouraged their use for flavoring rather than as a food source. Selective breeding increased the root size and mellowed the taste, making them more acceptable as a vegetable.

By the mid-sixteenth century, parsnips had become one of the staple foods of the lower classes throughout Europe and England. Colonists brought parsnips with them to the New World, and Native Americans quickly adopted this versatile root. Parsnips maintained their status as a primary root vegetable until the potato took their place.

Botanical Facts The parsnip is a hardy perennial with flower spikes and seeds forming in the second year of growth. Parsnip roots form under lacy foliage borne on fleshy stems. Roots vary in size depending on soil type and plant cultivar.

Parsnips are propagated from seed in loose sandy soil and cool temperatures. Seed viability degrades quickly, so freshness is the key for maximum germination. Roots require a long growing season to reach full maturity. Seeds sown in the early spring can be harvested after the first good frost. The flavor is greatly improved when harvested after a frost, due to the roots' ability to convert starch to sugar as a natural defense mechanism to prevent cell damage from the cold. The roots can be held in the soil without incurring damage.

Culinary Fare Much of the flavor of parsnips is located directly under the skin, so many recipes recommend vigorous scrubbing rather than peeling before the roots are prepared. In larger roots, removal of the tough woody core can greatly improve the overall texture.

Parsnips have a somewhat similar taste to carrots, with perhaps a bit more sweetness and complexity. Their fibrous texture is better suited to slow-cooking methods such as simmering, braising, or slow roasting. Parsnips can be cooked in soups and stews and they make a wonderful flavoring agent for all types of stock. They pair well with salty foods such as cod and bacon.

RIGHT: *During the Middle Ages, the sweetness of parsnips was particularly appreciated in light of the scarcity and expense of honey and sugar.*

Hamburg Parsley

Petroselinum crispum var. *tuberosum*

Hamburg parsley is a cultivar of parsley that produces a swollen fleshy taproot in addition to the typically flavored leaves. It provides a two-for-one punch in a small garden.

Historical Origins Hamburg parsley probably developed from wild parsley, which originated in southern Europe and the eastern Mediterranean. It is known only under cultivation. Exactly how it came to be cultivated remains something of a mystery. It is quite possible that Dutch horticulturists are responsible for breeding parsley plants with larger-than-normal root systems. This may have been accidental or could have been purposeful in order to produce a plant with specific characteristics. Eighteenth-century botanist and author Phillip Miller makes mention of this plant in one of his books. By the 1710s, Hamburg parsley was used throughout Europe, although it seemed to be favored in mid-European regions.

Botanical Facts Hamburg parsley would be hard to distinguish from herb parsley if not for its elongated stems and the slight leathery texture of its leaves. Once the root develops, a bit of the crown might peek out of the soil, further giving away its identity. The roots are creamy white and tapered, about 1½–2 in (3.5–5 cm) wide and 8 in (20 cm) long.

Plants are propagated by seed and they grow best in loose rich soil and cool temperatures, with plenty of water. The seeds of this rooted parsley are notoriously hard to germinate. Soaking them in warm water, letting the water cool to lukewarm, and holding the seeds submerged for 12 hours prior to planting is supposed to aid this process. Weeds make fierce competitors for this slow-growing root, so scrupulous weeding is necessary for optimum performance.

Hamburg parsley roots mature about 90 days after the seeds are sown. As with many root crops, harvesting after a hard frost will improve the flavor dramatically. Roots can overwinter in the garden without damage.

ABOVE: *The roots of Hamburg parsley have a flavor a little like a mild parsnip with a texture similar to celeriac; the leaves are reminiscent of parsley but pack a stronger punch.*

Culinary Fare The flavor of Hamburg parsley is described as a combination of celery and parsley with a hint of nuttiness. The leaves can be used as an herb to season and garnish, but the flavor is stronger than that of the average parsley cultivar. A lighter hand is necessary so as not to overpower. A little with mashed potatoes adds a nice touch.

Like parsnips, Hamburg parsley taproots benefit from slow-cooking methods and can be roasted or simmered as a vegetable or added to stocks, soups, stews, and braised dishes. They are, however, somewhat milder than parsnips and make a good alternative when a less powerful flavor is required.

Radish

Raphanus sativus

RIGHT: *Slices of freshly harvested radishes will add crispness, color, and bite to any salad. They are often carved as garnishes.*

The radish is an annual plant that is primarily grown for its peppery edible root. It is in the same family as turnips and mustard.

Historical Origins Radishes are thought to have originated in Europe and West Asia. They were used as food plants and medicine by many early civilizations, including the Greeks and the Romans.

Botanical Facts Radish plants can be quite small or very large, depending on the cultivar. There are many different colors and sizes of radish roots as well. Interior root color is almost always white. Radishes are propagated from seed in well-drained soil in early spring while temperatures are still cool. Seeds germinate very quickly. Smaller radish cultivars can be ready for harvest in as little as 25 days. The larger specimens generally take much longer.

Radishes should be harvested as soon as they reach the suggested optimum size. If the roots are allowed to grow too long past their prime, they will become pithy and soft. Should seedpods be the preferred crop, they need to be collected within a day or two after the flowers drop.

Culinary Fare Radishes are entirely edible, although the root is usually the preferred part. Roots can be eaten raw, pickled, or added to stir-fries. Young greens and immature seedpods can be eaten raw, steamed, or stir-fried.

Salsify

Scorzonera hispanica Black Salsify *Tragopogon porrifolius* Salsify

Salsify and black salsify are members of the sunflower family that are grown for their young shoots and flavorful roots.

Historical Origins Black salsify (*Scorzonera hispanica*) originated in southern Europe and the Middle East. Spanish seeds are believed to be responsible for the plant's spread to other European areas. Until the 1500s, it was considered effective against toxins and the plague. Salsify (*Tragopogon porrifolius*) originated in the Mediterranean and was introduced to many other regions, including England, Africa, and North America. It was being cultivated in Europe by the sixteenth century, but its popularity declined somewhat after the eighteenth century.

LEFT: *Somewhat uninspiring in the looks department, black salsify has a flavor lying between oysters and artichokes.*

Botanical Facts Black salsify has broad upright leaves and sports attractive, bright yellow flowers toward the end of the season. It is propagated from seed in well-worked soil and cool temperatures. The thin black roots can grow to a length of 3 ft (0.9 m) and they require about 120 days to mature.

Propagated in much the same way as black salsify, salsify grows to about 2 ft (0.6 m) tall, producing long fragile roots that are easily damaged during harvest.

Culinary Fare Black salsify roots are very sticky, so peel after cooking to prepare them. They are usually boiled and served along with other vegetables as a side dish. Young salsify roots can be grated and eaten raw. Older roots are usually cooked in soups or stews. They oxidize quickly; hold them in acidulated water until preparation.

Skirret

Sium sisarum

Skirret is a perennial plant in the carrot family that is grown for its edible root. Its flavor is described as a cross between carrot and parsnip with a slight nuance of nuttiness.

Historical Origins Skirret originated in China and was mentioned in first-century CE Roman literature. It has been cultivated in Europe since the mid-fourteenth century and it still remains a popular food plant in China and Japan.

Botanical Facts Skirret plants grow just over 3 ft (0.9 m) tall and produce small white flowers. A hardy species, it resists diseases and pests, and tolerates cold. Firm white roots grow in clusters, each root being 6–8 in (15–20 cm) long. Skirret is propagated from seed or by root division in light rich soil and cool temperatures. Plants prefer full sun but can tolerate periods of shade. Plenty of water is required during the growing season to produce tender roots. The roots mature in 6–8 months.

Culinary Fare Before skirret roots are eaten, the tough fibrous core must be removed. The roots can then be served raw in salads or cooked in soups or stews. They can also be boiled or roasted and served as a vegetable.

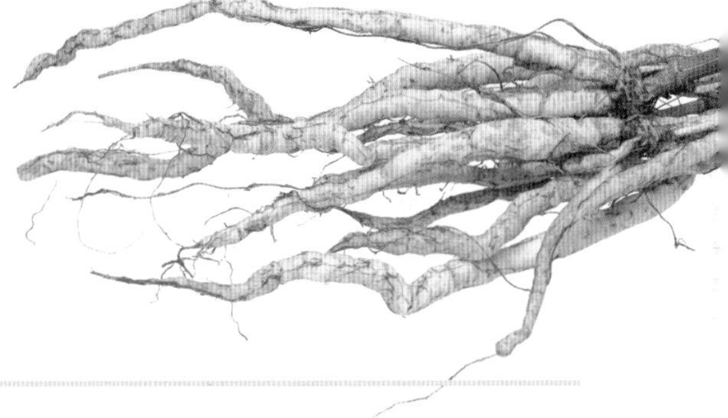

RIGHT: *Like many other root vegetables, skirret can be stored for months in a cool dry place such as a cellar.*

Yacon

Smallanthus sonchifolius (syn. *Polymnia sonchifolia*)

Yacon is tall perennial herb grown for its sweet edible roots and medicinal leaves. Its flavor is said to be a little like a cross between tart apple and celery.

Historical Origins Yacon originated in South America, centering in the lower Andes regions. It was cultivated by early Andean farmers and found in ancient burial sites. Spanish explorers initially ignored it because it was not a high-energy food. The first written record was at the beginning of the seventeenth century in a list of cultivated native Andean crops. It is now exported from parts of South America as a novelty crop.

Botanical Facts Yacon is related to the sunflower and produces fuzzy leaves and inconspicuous yellow flowers. Grown in its native habitat, flesh colors include yellow, pink, red, and orange. Propagate by planting root tops with the growing tip attached. Early planting in well-dug beds will produce a small crop by the first frost. Successive plantings produce a more bountiful harvest. Roots hold in the ground until the first hard freeze; they become sweeter in storage.

Culinary Fare The carbohydrate in yacon is principally inulin, a mainly fructose polymer that offers sweetness without raising blood-sugar levels. Roots can be peeled or scrubbed, then cooked or eaten raw like a fruit, or squeezed for the sweet juice, which tastes a little like a cross between celery and watermelon. Boiling the juice will produce a sweet syrup that has a hint of molasses, but with far fewer calories than sugar. Leaves can be made into a tea that offers antioxidants and helps lower blood sugar. Both leaves and roots have other medicinal uses.

Potato

Solanum tuberosum

BELOW: *The potato was slow to catch on as a staple food around the world, but eventually people fell in love with this unassuming tuber. The potato remains the king of root vegetables.*

The potato is a perennial plant, usually grown like an annual, producing a mild-tasting starchy tuber. Early civilizations cultivated it, worshipped it, and depended on it.

Historical Origins Potatoes originated in South America's Andean region and they have been cultivated for over 7,000 years. Early potatoes looked nothing like modern kinds. Small and knobby, they were most likely dark purple with yellow flesh. Early farmers selected the best for seed crop, and potato quality and size improved.

Cultivation spread throughout South and Middle America, producing our 2 major subspecies: a short-day Andean variety and a long-day Chilean variety. The short-day potatoes were ideal for the Andean environment. In the early nineteenth century, the long-day Chilean potato quickly replaced Andean varieties in Europe because they adapted better to that environment.

Spanish explorers searching for gold came to Peru during the sixteenth century. Though they found no gold, they did return to Spain with the potato, which was at first described as a kind of truffle. Potatoes soon became standard fare aboard ships, helping prevent scurvy on long voyages. By the late sixteenth century, potatoes were present in England, Ireland, and all over Europe. Even though potatoes originated in the New World, they were brought into North America by European settlers.

At first misunderstood, the potato's association with the deadly nightshade family kept this vegetable from gaining widespread acceptance with the general population, and it became a food for the poor and infirm. It was not until the late eighteenth century that the potato gained prominence—largely due to the practical nature of the Irish. Ireland's rich soil and favorable climate made it a perfect location for producing bountiful crops. Once the Irish realized that 1 acre (0.4 ha) of land could easily sustain 10 people if it was planted with potatoes, they readily embraced the tuber as a staple. Unfortunately, their dependence on a single crop as their primary source of sustenance led to a devastating famine.

In 1845, a pathogenic water mold or blight decimated Ireland's entire potato crop. Genetic singularity in the potato cultivar grown in Ireland is suspected to have played a role in this tragedy. With the loss of their major source of food, an estimated million or more people lost their lives as a result of the famine. Social, political, and economic unrest further complicated the distressing state of

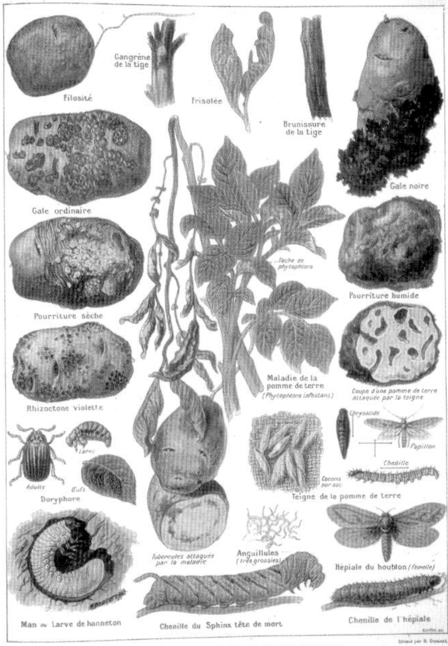

the nation. Within a few years after the famine began, Ireland's population dropped by half, mostly due to migration in an attempt to escape the Great Hunger.

After the traumatic throes of this famine subsided, the potato crept back into the larder of most societies. Though it was still considered food for animals in North America, botanist Luther Burbank introduced the 'Russet Burbank' potato, forever changing the opinion of the humble tuber in the US. Burbank is also attributed with introducing a blight-resistant potato to Ireland.

Botanical Facts Potatoes are propagated by planting cut pieces of tubers in well-worked soil just before the last frost. As plants grow, soil is often mounded around a row to encourage tuber production. Plants can also be propagated from seed in some cultivars, but seeds are usually reserved for producing new cultivars or improving strains.

Most potato cultivars are ready for harvest after plants die back at the end of the season. Tubers must be dried before storage to improve holding properties. The best holding temperature is 45°F (7.2°C) or higher. Cold temperatures prompt potatoes to convert starch to sugar, which changes the taste and texture.

Culinary Fare Russet or baking potatoes are the highest in starch and are therefore the best for roasting, frying, and baking. All-purpose potatoes (the most common and least expensive) are good for use in soups, stews, and for mashing. Red and new potatoes are lower in starch, which makes them the best for boiling, creaming, and for use in cold salads.

Comforting winter warmers are whole potatoes baked in their jackets and teamed with sour cream and chives, or chipped and made into the perennial favorites, french fries or wedges. Potatoes are also used in breads, dumplings, croquettes, and gnocchi.

ABOVE LEFT: *Potato plants are lush with crinkly green leaves borne on sturdy stems. When the flowers appear, they indicate tiny new potatoes are present beneath the soil's surface.*

ABOVE RIGHT: *A poster of potato diseases, c. 1920, shows potato blight in the center. This infection destroyed crops in Ireland, causing a massive famine.*

Green Means Stop!

Green is a color best avoided where the potato is concerned. Potatoes contain high levels of a toxic alkaloid called solanine in their green stems and leaves. Consuming the greens of the potato may not cause a fatal reaction, but it will cause extreme illness. Exposure to sunlight and even strong artificial light can cause "greening" in potato tubers, which also indicates the presence of solanine. Cooking can reduce the effect of the toxin, but peeling the potato completely away is a safer alternative. The level of solanine in a green potato is probably not high enough to do a great deal of damage, but it does impart a bitter taste and could cause digestive upset.

Celery

Apium graveolens

Wild celery was prized for its leaves, which impart quite a strong flavor to soups and stews, and its seeds, which were an important condiment. Cultivated celery is much milder than its original form and its stems can be eaten raw or cooked.

BELOW: *It is not known when celery was taken to the USA, but by the early nineteenth century 4 types of cultivated celery were available and it quickly became popular as a salad vegetable.*

Historical Origins Wild celery can still be found in abundance in Europe and temperate Asia, especially near the sea. It is known as cutting celery or smallage, and has thin stalks and a vigorous leafy top. It is particularly bitter, and was used for flavoring rather than as a vegetable. Wild celery is still used in Southeast Asian, French, and Italian cuisines. Celery was cultivated from this wild herb at some point in the sixteenth or seventeenth century, probably in France or Italy. It is much milder and less bitter in flavor, with thick fleshy stems. This type of celery was also independently cultivated in China, where wild celery has been used since the fifth century. It is stronger, thinner, and juicier than the European kind. The Greeks, Romans, and Egyptians all associated it with funerals, where it was made into garlands, but it was also valued for its medicinal qualities and the seed was used as a condiment. Celery continued to be valued medicinally and as a flavor into the Middle Ages.

Botanical Facts A member of the parsley family, celery originates in Europe and temperate Asia. Modern celery is valued for its thick, fleshy, ribbed stalks, and wild celery for its highly flavored, feathery leaves. Celery is a biennial and reaches 15–24 in (38–60 cm) in length, producing white flowers in its second year. Although some cultivars can be grown without light-blanching, this process reduces the bitterness and also encourages the growth of fine white stems.

Culinary Fare Celery is often eaten raw as a salad vegetable. In Britain, the stalks are eaten as a snack accompanied by strong cheddar cheese, and in France they are served as a first course with sweet butter and salt. Children like to stuff the stalks with peanut butter and sprinkle with sultanas or currants. Celery is an important ingredient in salads (such as Waldorf, Boulestin, and Beaucaire), soups, and stews. It is excellent braised and is frequently stir-fried in Asian dishes. The seeds are utilized in celery salt, a tasty blend of salt and celery seeds.

Asparagus

Asparagus officinalis

Asparagus has always been prized as one of the most delicious of vegetables. It contains asparagine, an essential amino acid and the first to be isolated from its natural source early in the nineteenth century.

Historical Origins Asparagus is a member of the family Asparagaceae. It grows wild in marshy areas of Europe such as Poland and Russia. Selective growing and breeding techniques have resulted in a thick fleshy shoot that has long been esteemed as a delicacy. Asparagus was a favorite of the Romans, who cultivated spears much larger than modern cultivars. They prized the plant for its medicinal virtues, an idea that carried through to the Middle Ages. Recipes for asparagus are included in fourteenth-century Italian and Catalan manuscript collections, and it featured in the menu for a Milanese wedding banquet of 1488. It was a great favorite in Italy, where folklore touted the root as an effective form of birth control. In the US, asparagus was not produced commercially until the latter half of the eighteenth century; it was President Thomas Jefferson's favorite vegetable.

Botanical Facts Asparagus is an herbaceous perennial from Europe, Asia, and North Africa. It has erect, multi-branched stems, and green feathery tips. It bears small, drooping greenish white flowers and bright red berries. Cultivars range in color from dark to light green through to violet and white. Asparagus that is grown for canning is almost always blanched in hot water.

ABOVE: *White asparagus has a milder flavor than the more typical green. It will remain white when, in its growing stages, the stems are heaped with sandy soil to block out sunlight.*

Culinary Fare Asparagus deteriorates rapidly, so it is best eaten as fresh as possible. Although the young stems can be eaten raw, it is normally cooked, preferably in a tall pot known as an asparagus boiler, which boils the stems and steams the delicate tips. Asparagus is usually served lukewarm and eaten teamed with a simple melted butter, vinaigrette, or hollandaise sauce. Cold cooked asparagus is sometimes served as a tasty snack.

In Italy, asparagus is typically sprinkled with breadcrumbs or parmesan cheese, then grilled. Asparagus spears rolled in squares of sliced, buttered bread are baked to make a popular hors d'oeuvre.

RIGHT: *Asparagus grown with the stems exposed to daylight allows chlorophyll to develop. This gives the vegetable its more typical green color.*

Medicinally Speaking

Asparagus has long enjoyed the reputation of being a medicinal plant. Its botanical name *officinalis* means "from the dispensary." Asparagus is a diuretic and a laxative. It was also thought to help with eye problems, toothache, cramps, and sciatica. Asparagus can produce some odd side effects, most notably a distinctive odor in one's urine. This was first recorded in eighteenth-century Britain by Queen Anne's physician, John Arbuthnot. It is thought this unusual property is related to sulfur compounds used as fertilizers and present in the soil.

Orache

Atriplex hortensis

BELOW: *Although orache, often called garden orache, is also known as green or mountain spinach, this leafy vegetable is available in a variety of colors.*

Similar in flavor to spinach, orache was once a popular green vegetable. It is still used in Asian cuisine and is grown as an attractive ornamental plant.

Historical Origins Orache is native to temperate Europe and West Asia, where it is often grown as a potherb. It was popular as a green vegetable in the earliest Mediterranean civilizations and was widely cultivated up until the eighteenth century. Orache was often used in Britain in the sixteenth century but its popularity declined as the use of spinach increased throughout Europe.

Botanical Facts A member of the goosefoot family, orache has heart-shaped to triangular, slightly serrated leaves. They range in color from green through purple, red, and yellow, depending on the cultivar. The flowers are green or red and do not have petals, and the seeds are tiny, brown, and platelike. Due to its attractively colored foliage, orache has long been a popular ornamental plant.

Culinary Fare Originally orache was used in the same way that spinach is now, as the flavor is quite similar. Green orache was used to color pasta in Italy, and was cooked with sorrel to cut its acid flavor. In Asia it is cooked in the same manner as other leafy green vegetables.

Winter Cress

Barbarea vulgaris

Similar in flavor and appearance to rocket and watercress, winter cress is popular in winter when other salad vegetables are more likely to be rare or not in season.

Historical Origins Also known as yellow rocket, bittercress, and wound rocket, winter cress is native to central Europe, but is now found in many other countries. It grows wild across much of North America, where it is considered a weed.

Botanical Facts Winter cress belongs to the mustard family (Crucifereae). It grows well in moist fields, along stream banks, and anywhere the earth has been disturbed. Winter cress can grow to a height of 12–24 in (30–60 cm) and has dark green, lobed leaves. Clusters of bright yellow, 4-petalled flowers appear throughout summer.

Culinary Fare Winter cress is a most welcome addition to salads during the season when not many salad vegetables are available. It resembles watercress in appearance and flavor and can be used in the same ways. The young leaves are sharp and peppery, but the more mature leaves tend to be quite bitter. Winter cress is mostly eaten raw, but it can be substituted for watercress in any cooked recipes, most of which are Asian in origin.

Beets, Chards

Beta vulgaris var. *flavescens* Silver Beet *Beta vulgaris* var. *flavescens* subsp. *cicla* Spinach Beet

The original wild or sea beet is a perennial often found growing on the seashores of Europe, North Africa, and West Asia. Some beets are known as a root vegetable, others for their tasty leaves.

Historical Origins The history of chard (*Beta vulgaris*) has been traced as far back as the hanging gardens of Ancient Babylonia and it was an important vegetable in the Arab world. Evidence indicates that leafy beet forms were being cultivated by the first century CE. The name chard is derived from the Latin word for thistle. By the nineteenth century, "Swiss" was added to the name, probably to distinguish it from cardoons in seed catalogues, though "chard" is still used.

Botanical Facts Beets are divided into 2 groups: the Conditiva Group, comprising root vegetables and fodder crops; and the Cicla Group, the chards. The latter includes spinach beets, closest to wild beet, with slender green leaf stalks and flat blades, and silver beets, with larger leaves puckered to varying degrees, stalks and midribs broad white or colored and the roots not swollen. Beets in this group are valued for their leaves and stalks. Both are often called Swiss chard.

Culinary Fare Silver beet is particularly popular in France, where the stalks are prepared in many different ways, often in a similar manner to celery or cardoons. The leaves are similar in flavor to spinach, but are far more vigorous and resilient. They can be substituted for spinach in any recipe, though they are firmer and tougher than spinach, and will not be as meltingly tender. Spinach beet is often used around the Mediterranean —the word "cicla" derives from "sicula" referring to one of the places where chard first grew, Sicily—and is sometimes cooked in the Arabic style, with pine nuts and raisins.

ABOVE: *The vibrant colored stems of rainbow chard, a type of silver beet, make "eating your greens" far more appealing, and they appear to be a tasty treat, even to fussy children.*

Stems or Leaves?

Swiss chards—silver beet and spinach beet—are sometimes considered inferior to spinach, their leaves being coarser and less delicate. However, they are excellent vegetables in their own right and the enlarged midribs of their leaves are widely held to be a superior vegetable than the leaves, having a more delicate flavor. They can be prepared in the same manner as asparagus or celery. Both the leaves and stems of silver beet can be used; in France, the leaves and stalks are never served together. The leaves are thought to be somewhat coarser than spinach and the stems are often eaten as a separate vegetable. Spinach beet can be eaten whole, including the long green stalk.

Sea Spinach

Beta vulgaris subsp. *maritima*

RIGHT: *In Europe, the tender young leaves of wild sea spinach are harvested and prepared for eating in much the same manner as beet greens or Swiss chard.*

Sea spinach, also sometimes known as sea beet, is an ancient plant from which all currently cultivated beets are derived. In the past it was quite popular as a leafy green vegetable. Other common names for this plant include palangsag and palanki.

Historical Origins The original wild or sea spinach is a perennial growing along the seashores of Europe, North Africa, and western Asia. All cultivated beets are believed to have derived from it, including beetroot, sugar beet, fodder beets (such as mangel-wurzel), silver beet, and spinach beet.

Botanical Facts Sea spinach, as its name suggests, can be found along the high-tide mark on the coasts of Europe and North Africa, particularly the Mediterranean. It has thicker fleshier leaves than cultivated beet, the tuberous root is smaller, and the plant parts are often red. It needs plenty of sun and will not grow in shade.

Culinary Fare Sea spinach is cultivated for its leaves and stems rather than its roots. It is used in the same manner as spinach beet or silver beet and is said to be superior in flavor. It was used medicinally to treat tumors.

Chinese Mustard, Gai Choy

Brassica juncea var. *rugosa*

Often referred to as leaf mustard, many mustard greens are widely used in Asian cuisines, especially in China. Wild mustards have long been a popular potherb in North America, but are now being superseded by oriental leaf mustards.

Historical Origins Wild mustards are native to Europe, and South and East Asia. They are thought to be the ancestors of the modern cultivated leaf mustard, also native to South and East Asia. Chinese mustard has long been cultivated as a leaf vegetable.

LEFT: *Chinese mustard greens make an attractive potherb owing to their vibrant colors, and they are particularly popular as ornamental garden vegetables.*

Botanical Facts Chinese mustard is an annual plant. Its leaves can be green, red, or purple, and may be smooth or puckered, or smooth-edged or lacy, and its small flowers are bright yellow. Like other mustard greens, this is a very hardy plant, making it particularly easy to grow in a home vegetable garden in many environments.

Culinary Fare Chinese mustard greens are used fresh and pickled. They can be cooked in much the same way as spinach, and become very tender. They have a distinctive, slightly peppery flavor of their own. A favorite in Indian cuisine, Chinese mustard greens are also boiled and served with soy sauce in Japan. They can be substituted for any leafy green vegetable in Chinese cooking.

Kale

Brassica oleracea, Acephala Group

Kale is a highly nutritious leafy vegetable that was popular before the advent of spinach and "headed" cabbage.

Historical Origins Before headed cabbages were bred at the end of the Middle Ages, kale was the main green vegetable in the diet of country people across most of Europe. Kale is a Scottish name, and writer J. M. Barrie was a member of the "Kailyard school" of Scottish writers. These writers described rural life in Scotland, where kale fields were common. Kale was so widely used in Scotland that the word became a synonym for dinner. Collard greens, which are also part of the Acephala Group of brassicas, are an important leaf vegetable in southern US cooking.

Botanical Facts Kale is a brassica that doesn't develop a head. A descendent of wild cabbage, it is a hardy plant and can grow in conditions where more delicate cabbages belonging to the same family will fail. Curly leafed kale is less coarse than other cultivars and remains a popular vegetable, while other types are grown for animal feed.

Culinary Fare Kale is often prepared along with collard greens, which are genetically very similar. Both can be prepared in the same way as spinach or other leafy greens—boiled with little to no water added. In southern parts of the US, collard greens and kale are usually simmered with a piece of meat until it is tender.

RIGHT: *A bountiful summer harvest: a variety of different colored kale plants make an excellent addition to a home vegetable garden.*

Chinese Broccoli

Brassica oleracea, Alboglabra Group

Also known as Chinese kale, Chinese broccoli is a strongly flavored leafy vegetable, often appearing at banquets as a tasty dish in its own right.

Historical Origins Like other members of the Alboglabra Group, Chinese broccoli descended from wild cabbage, which has been cultivated for thousands of years and there are a great many cultivars.

Botanical Facts Closely related to European broccoli, members of the Alboglabra Group produce several small succulent heads instead of just one large head. Chinese broccoli is distinguished from European kale and other brassicas by its white flowers and a soft white bloom on the greenish-blue leaves.

Culinary Fare Quite similar in flavor to broccoli, Chinese broccoli is a little stronger and coarser. The entire plant can be eaten, but stems on older plants must be peeled. It is widely used in numerous Asian cuisines, especially in traditional Chinese food, where the cooked stems are considered a delicacy. They are popularly steamed or stir-fried with ginger or laced with oyster sauce. The young plants may be steamed or stir-fried whole without peeling, or used fresh in salads. Chinese broccoli may be substituted in any recipe calling for broccoli or other greens.

Cauliflower

✳ *Brassica oleracea*, Botrytis Group

RIGHT: *An assortment of organic cauliflower shows the remarkable range of colors possible in this usually white vegetable. 'Violet Queen', the purple form, is particularly high in antioxidants; the lime green fractal form is 'Romanesco'.*

Cauliflower is very popular as a winter vegetable. In the past it was valued more highly than its close relative, broccoli, because of its elegant, snowy white curds.

Historical Origins The origins of cauliflower remain uncertain. It is thought that cauliflower was first grown in the Near East and was introduced to Europe by the Arabs after the fall of the Roman Empire. Cauliflower appeared in Italy around the end of the fifteenth century and was being cultivated across Europe by the sixteenth century. In the last 200 years cultivars suitable for growing in hot and tropical climates have been developed in India.

Cauliflower for Lent and Love

After its introduction to central and northern Europe, cauliflower enjoyed a vogue as a Lenten food, as it provided hearty filling dishes that did not break any of the fasting rules. It became popular in France in the seventeenth and eighteenth centuries, and numerous cauliflower dishes were named after Louis XV's mistress, the celebrated Comtesse du Barry (born as the illegitimate seamstress Jean Becu), who was guillotined in 1793. This naming was supposedly in reference to her elaborate powdered wigs, which were composed of curls piled high atop one another like cauliflower curds. Dishes "du Barry" also included non-cauliflower cuisine, but all of them were characterized by the presence of a creamy white sauce.

French women's high fashion encouraged the tallest of powdered wigs.

Botanical Facts Cauliflower is a member of the brassica family, related to cabbage. It belongs to the Botrytis Group, which comprises a range of common cabbages whose flowers have begun to form but have arrested growth at the bud stage. These cultivars are richer in nutrients than other brassicas because the thick stems beneath the buds act as storage organs for vitamins and minerals that would otherwise have gone into the flowers and fruits of the plant. The undeveloped flowers forming the edible part of the plant are known as the "curd." Cauliflower cultivars are not just white—they range in color from yellow-green to orange and purple.

Culinary Fare Cauliflower can be served raw as a salad vegetable or with a vinaigrette, though it is more often steamed or boiled before dressing. Popular cauliflower dishes include cauliflower *au gratin* (popularly known as cauliflower cheese) or cauliflower polonaise. It is sometimes steamed or boiled whole and then masked with sauce for a dramatic dish. Cauliflower is an important ingredient in popular condiments such as mustard pickles or other pickled vegetables.

Cabbage

Brassica oleracea, Capitata Group

Wild cabbage has evolved into a number of cultivars where different parts of the plant are the edible components. The Capitata Group are "heading" cabbages, whose leaves are eaten.

Historical Origins The cabbage and its relatives are in the family Brassicaceae, with the cabbage being the first member to have been cultivated. The wild species are native to southwestern Europe and the Mediterranean region. Cabbage was valued as a food plant by the Ancient Egyptians and the Greeks, as well as having religious significance. These early cabbages were headless. The Greeks and Romans considered cabbage not only to be a nutritious food but also a protection against drunkenness. Although headed cabbages were mentioned as early as the first century CE, they are thought to have developed in Germany around the twelfth century, then spread over Europe replacing the staple, kale. Cabbage became an important staple in Ireland, more so than anywhere else. It was taken to North America in the sixteenth century, but the first written record of its planting does not occur until much later, in the seventeenth century.

Botanical Facts The Capitata Group is composed of herbaceous biennial flowering plants with leaves forming a characteristic compact head. These cabbages have yellow flowers, smooth or crinkled leaves, and range in color from light to dark green through bluish green to purple-red.

Cabbages in the Alba Subgroup are flat with a heavy head and white interior. Cultivars range from 'Dynamo', with a mild flavor to 'Primax', the sweetest type with round light green heads, to the solid red 'Ruby Ball'.

Savoy Subgroup cabbages have crinkled leaves and greenish hearts. They include 'Early Curly', with a small core and sweet flavor; 'Primavoy', with a flat head and dark blue-green leaves; and the subtly flavored 'Savoy King', with a large head and creamy interior.

Culinary Fare Despite the smell it releases on cooking, cabbage is often used in soups and stews. It is also stuffed—a popular French dish with variants. The leaves can be rolled around fillings or added to fillings, as in the Russian *pelmeni* or *kulebiaka*, among myriad eastern European variations. It is a key ingredient in the traditional Russian beetroot soup *borsch*. An Irish cabbage dish is colcannon, a mixture of cabbage, onion, and fried potatoes. A modified version was popular among Britain's upper classes during the eighteenth century. Later that century came a purely British version known as bubble and squeak, now made with fried potatoes and cabbage. In Germany and Austria, where cabbage played an important dietary role, it was used to make sauerkraut, a cabbage pickle flavored with juniper berries.

BELOW: *Sulfur compounds give cabbage its mildly unpleasant smell during cooking. One way to alleviate this is to shred it finely and stir-fry; the hot oil seals the surface and the cabbage retains its flavor as well as a crispy texture.*

Broccoli

Brassica oleracea, Cymosa Group

RIGHT: *While deep green is broccoli's best-known color, it can also be purple, white, or yellowish green. Harvesting the head along with a little of the stem will encourage regrowth.*

There are 3 principal types of broccoli: sprouting broccoli, which is white, green, or purple and harvested in spring; calabrese, an annual broccoli that is green or purple and harvested in summer; and romanesco, a later broccoli with multiple yellow-green heads clustered together.

Historical Origins The true origin of broccoli has puzzled botanists for many years. Like cauliflower, broccoli is a form of cabbage in which the flowers are crowded together to form a dense head. Whereas in cauliflower the buds are so tightly packed they fuse together, in broccoli the heads are more open, with each group of buds at the end of its own stalk.

It is thought that when cabbages were being grown for their shoots (rather than their heads) they began to develop early forms of broccoli that later evolved into cauliflower—more highly prized because of its dense curd and snowy white coloring.

It is generally accepted that broccoli was introduced to Britain (and probably other European countries) from Italy early in the eighteenth century. This vegetable did not actually become popular in North America until the twentieth century, even though it was introduced back in the eighteenth century.

BELOW: *Broccoli is an Italian word meaning "little arms" or "little shoots." It is a highly nutritious cruciferous vegetable with significant levels of beta carotenes and vitamin C.*

Botanical Facts Broccoli is a member of the brassica family, related to cabbage. It belongs to the Cymosa Group, a type of common cabbage whose flowers have begun to form but have stopped growing at the bud stage. Brassicas such as broccoli and cauliflower are particularly high in nutrients because the vitamins and minerals are retained in the stems instead of being used in the development of flowers and fruits.

Broccoli is easily grown. It likes a well-drained, organically enriched soil, but do not add nitrogen as it encourages leaves, not flowerheads. It can be propagated from seed year-round but is best planted in frost-free seasons to avoid bolting to seed.

Culinary Fare Broccoli's rich nutrients are best preserved by briefly cooking with a little water or stir-frying. It tends to lose its vivid color when cooked; brief blanching in hot water and minimal cooking will preserve it. Like other cabbage family members, broccoli contains sulfur compounds that give off a mild odor when cooked. Broccoli can often be substituted in recipes calling for cauliflower. At its simplest, it is boiled or steamed and served with melted butter or hollandaise sauce. In Italy broccoli is improved by the addition of chilli or garlic, and in Asian cuisines is often served with ginger or oyster sauce.

Brussels Sprouts

Brassica oleracea, Gemmifera Group

Sometimes passionately hated, especially by children, when prepared correctly and picked young, brussels sprouts have a delicate flavor and lend themselves well to creamy or buttery sauces.

Historical Origins Brussels sprouts, like all brassicas, are derived from wild cabbage, which is native to southwestern Europe and the Mediterranean region. Brussels sprouts is a many-headed subspecies of cabbage. The miniature head buds grow around the stem, but the main head achieves only limited growth. Brussels sprouts is a relatively recent vegetable, becoming known in French and English gardens at the end of the eighteenth century, and in the US in the early part of the nineteenth century, when Thomas Jefferson planted some in 1812. They were grown in California in the 1900s, and with the introduction of deep-freezing techniques became a popular crop for the frozen-vegetable market.

Botanical Facts Brussels sprouts are a cool-season crop and grow best in cool humid climates. There are 2 main kinds. The tall type reaches 2–4 ft (0.6–1.2 m) in height and the smaller one has a maximum height of around 2 ft (0.6 m). Europeans prefer their sprouts to have a diameter of about ½ in (12 mm), whereas Americans prefer larger ones, with a diameter of 1–2 in (25–50 mm). Young sprouts have a sweet nutty flavor but older sprouts, like other members of the brassica family, can run to bitterness.

Culinary Fare Perhaps a tad unfairly, brussels sprouts are one of the vegetables most intensely disliked by children. The presence of sulfur compounds found in all brassicas causes them to release an unpleasant odor when overcooked but this can be prevented by using only young sprouts and minimal cooking. In older sprouts the sulfur compounds give off a stronger odor that is more pronounced. In Britain, brussels sprouts are commonly served as accompaniments to game or Christmas turkey. In Belgium, they are traditionally cooked with peeled chestnuts. When they are properly cooked, brussels sprouts have a pleasant and delicate flavor. They are often steamed or boiled and served with a buttery sauce, roasted to serve with meat, or stir-fried and accompanied by various nuts.

Brussels sprouts are occasionally served raw as a salad vegetable. The sprout tops that protect the young heads are sometimes sold as greens. They have a mild, cabbage flavor and can be prepared in the same way as many spring or winter greens.

BELOW: *Brussels sprouts have a uniform growth pattern: buds spiraling up the stem. To grow tightly packed firm sprouts—any "blown" buds are tasteless—grow them in firm fertile soil and harvest a few at a time, from the bottom up.*

Asian Greens

* *Brassica rapa* var. *chinensis* Pak Choi * *Brassica rapa* var. *nipposinica* Mizuna
* *Brassica rapa* var. *pekinensis* Chinese Cabbage * *Brassica rapa* var. *rosularis* Chinese Flat Cabbage

RIGHT: *The crispiness of pak choi is best preserved by lightly stir-frying it for no more than 2 minutes. But this versatile vegetable can be used beyond Asian fare—it is delicious steamed and served with a cheese sauce and is an excellent substitute for spinach.*

There are many different kinds of Asian greens and different kinds can be substituted for others in most recipes.

Historical Origins Asian greens belong to the same genus (*Brassica*) as European cabbages. Their development is similar to that of European brassicas. Both developed through the cultivation of wild cabbages and both have numerous cultivars available throughout the year.

Botanical Facts Pak choi, also called bok choi, is available throughout the year. It is an open brassica rather than a hearted one, with smooth, light green leaves on white stalks. Both the young and the more mature plants are used. Mizuna is a popular Japanese salad vegetable with serrated green leaves and a succulent white stalk. It has a mild mustard flavor and is often found in prepared salad mixes. Chinese cabbage is known by many names including napa, Peking or Tientsin cabbage or *wong nga bak*. It is one of the most popular Asian vegetables. It is a very mildly flavored cabbage and comes in a variety of forms—long and narrow or shorter and fatter. These cabbages are the ones most often found in Japan, where they have been cultivated since the nineteenth century. Chinese flat cabbage, also known as rosette cabbage, is plate-shaped, growing up to 2 in (5 cm) high and 15 in (38 cm) wide. The

leaves are round and the stalks green. This type is thought to have a particularly good flavor. Young plants are best.

Culinary Fare All Asian greens can be cut into small pieces for steaming, stir-frying or adding to simmered dishes near the end of cooking. Chinese cabbage is the most delicately flavored of the Asian cabbages and is often shredded raw for salads or added to soups or stir-fries. Pak choi is mild in flavor and should be cooked only briefly.

Favorite Ways to Serve Asian Greens

Chinese cabbage is an important ingredient in the Korean pickle *kimchi* (pictured). Pak choi is deep fried and used to make a dish known as mermaid's tresses. The smaller type of pak choi, known as Shanghai bok choi, is much in demand for banquets, where it is braised whole. Boiled and dressed with oyster sauce, it is a popular dim sum dish. Asian greens of many types are steamed or stir-fried and flavored with aromatics such as ginger, garlic, chili, and spring onions, and pungent liquids such as soy sauce, oyster sauce, sesame oil, sherry, and rice wine.

Gotu Kola

Centella asiatica

Perhaps more familiar by its common English name, pennywort, gotu kola is often sold in nurseries as an "arthritis herb," because eating 2 leaves each day is said to relieve arthritic pain.

Historical Origins Gotu kola is native to northern Australia and many parts of Asia. It has been used medicinally for thousands of years in China, India, and Indonesia and is important in both Ayurvedic and traditional Chinese medicine, especially in its capacity as a wound-healing plant.

Botanical Facts Gotu kola is an aquatic perennial creeper that is native to Asia and the South Pacific. Its leaves are small and fan-shaped, and flowers range in color from white through pink and purple.

Culinary Fare Gotu kola is often used in Vietnam to make a sweetened beverage. The fresh leaves are mixed with water and sugar syrup, then blended with ice cubes to make a refreshing, frothy, bright green drink. It must be served at once before it loses its vivid color. The large-leaf type is a popular Vietnamese herb with a mild bitterness. Gotu kola is a traditional ingredient in the Sri Lankan dish *mallung*, a vegetable coconut curry, and in other vegetarian dishes, and is also eaten as a salad vegetable.

ABOVE: *A creeping wetland plant, gotu kola likes very soggy conditions. It can be grown around aquatic features such as ponds or marshes, providing it has a very sunny position.*

Goosefoot

Chenopodium album Fat Hen, Lamb's Quarters *Chenopodium bonus-henricus* Good King Henry *Chenopodium giganteum* Tree Spinach

Members of the goosefoot family are so named because their leaves resemble the feet of geese. These wild greens grow widely as weeds but are rarely cultivated.

Historical Origins Plants in the goosefoot family are related to spinach and were widely used as green vegetables in Europe before the arrival of spinach in the Middle Ages. They are available year-round, but are now rarely cultivated or used.

Botanical Facts Good King Henry is used mainly as a potherb and grows wild in North America, Europe, and West Asia. The flower clusters can be eaten and the young shoots prepared like asparagus. Fat hen, also known as lamb's quarters, thrives on muck heaps and is commonly referred to in North America as pigweed. The seeds can be ground to produce dark-colored flour. Tree spinach is closely related to lamb's quarters and is sometimes called giant lamb's quarters. It has vivid purplish red coloring and can be used in the same way as lamb's quarters.

Culinary Fare Most goosefoots have a slightly bitter taste that can be alleviated by cooking the leaves in one or two changes of water. They can be cooked in much the same way as spinach and other similar green leafy vegetables—that is, in little or no water.

Endive

Cichorium endivia

Endive is a leafy vegetable prized for its mildly bitter leaves and attractive appearance, particularly the curly cultivars. Flatleaf cultivars are often light-blanched to reduce their bitterness. The main use for endive is as a salad vegetable, although it can also be cooked. Endive is particularly popular in Italy and France.

Historical Origins Endive is thought to have originated in the eastern Mediterranean region. It is descended from wild chicory, which is native to Europe, West Asia, and Africa. Endive is an ancient species and can be traced back to prehistoric times. Wild forms are especially abundant in Italy.

The ancient Egyptians ate endive, and both the Greeks and Romans cultivated it, preferring it to chicory as it is less bitter. By the sixteenth century endive could be found in central Europe. Only flat-leafed types were available; broad-leafed curly endive is not described until the late sixteenth century, and the narrow-leafed curly endive that is now the most popular cultivar did not develop until much later. Endive is now grown all over the world.

Botanical Facts Sometimes known as batavia, escarole, or chicory, endive is a hardy annual or biennial grown for its large, slightly bitter leaves, used in salads or for cooking. Endive can be curly or straight, plain green or green with red midribs. Hearting cultivars have their outer leaves tied up or covered to light-blanch the hearts and reduce bitterness. Curly endive is a large hearting endive that can reach over 15 in (38 cm) across. An easy way to light-blanch it is to place a tile over it, which will result in a mild white heart and bitter green outer leaves.

Culinary Fare Endive leaves are usually used fresh as a salad vegetable, but may be blanched in hot water to reduce bitterness, which is otherwise offset by the use of a sweet or emollient ingredient in the salad. A traditional French dressing of hot bacon fat with small pieces of bacon is sometimes used. Endive needs to be blanched for use in cooked dishes. It can then be grilled or pan-roasted, cooked gently with butter and cream, or used in risotto. In Italy, it is often stuffed or marinated with ingredients such as cooked onions, vinegar, pine nuts, and raisins. The French "endive," also known as witloof, is a different species, *Cichorium intybus*. Hence the popular French dish *endives au gratin* calls for witloof, not curly endive.

BELOW: *When buying curly endive, choose those with tightly packed heads. Endive will keep for a few days if put in the refrigerator crisper, wrapped in a paper towel inside a sealable plastic bag.*

Sea Kale

Crambe maritima

Sea kale is a wild coastal vegetable that never achieved great popularity despite the long-held enthusiasm it has received from many amateur gardeners.

Historical Origins Sea kale grows wild along European coastlines. During the Middle Ages it was grown in Italy and by the seventeenth century it had reached Britain. There it enjoyed a vogue among culinary hobbyists, though it never became very popular. In the US, President Thomas Jefferson is said to have enjoyed it and grew it regularly in the early nineteenth century.

Botanical Facts Sea kale is a perennial found along the coastlines of northern Europe, from the Atlantic to the Baltic Sea and the Black Sea. It has fleshy, blue-green elliptical to rounded leaves with serrated edges and a powdery coating and bears massed white to cream flowers.

Culinary Fare Sea kale is normally bitter and inedible but if light-blanched—by heaping sand around the stems or covering with a pot—the leaf stalk develops into a tender vegetable that can be prepared and eaten in a similar manner to asparagus. The young leaves can be left on or cut off for use in salads; the stems can be boiled and eaten with any of the various sauces that are usually served with asparagus.

ABOVE: *Sea kale growing wild along the beach. Not found in the markets, this hardy salt-tolerant plant grows well in gardens in full sun in an open position.*

Mitsuba

Cryptotaenia japonica

Mitsuba is a salad plant and herb used in Japanese cooking in much the same way that parsley is used in Western cuisine.

Historical Origins Mitsuba is cultivated almost exclusively in Japan. Its other names include Japanese parsley, trefoil, wild chervil, and honeywort. Special care in cultivation and techniques, such as summer and winter blanching, will make the plant more tender.

Botanical Facts This leafy vegetable is a perennial, sometimes evergreen, plant used as an herb and vegetable in Japan. It is a compact plant with feathery leaves on tall narrow stems. There are 2 main types of mitsuba—white and green—but there is also a cultivar that has deep purple-red leaves and stems, and pale pink flowers.

Culinary Fare Mitsuba is often used as a salad vegetable or garnish. As a garnish, the leaves may be used whole or chopped. The stems may be chopped or tied into knots and made into tempura. The vegetable has a subtle flavor that has been described as a combination of coriander, parsley, celery, angelica, and sorrel.

Mitsuba is a popular addition to soups, stewed or simmered dishes, and steamed Japanese custards such as *chawan mushi*. It should be cooked briefly to maintain its fresh perfume, but it can also be blanched in hot water quickly before it is used.

LEFT: *Mitsuba is a very tolerant plant that will grow in most soils. It likes some shade to full shade and does well in containers, indoors or outside.*

Globe Artichoke

Cynara scolymus

ABOVE: *Artichoke can be quite a challenge to eat, for the uninitiated. After cooking, first peel off the hard outer leaves. Take a leaf and scrape off the lower fleshy part with the teeth. Leaves become progressively more tender toward the center.*

The globe artichoke is said to be one of the most delicious vegetables cultivated. The edible part is the immature flower.

Historical Origins The cultivated artichoke is descended from the wild cardoon, which is native to the Mediterranean. The true artichoke is thought to have evolved in North Africa or possibly Sicily. The Greeks and Romans enjoyed it as a food plant. Artichokes were first cultivated in Naples, Italy, in the mid-fifteenth century and gradually spread over other parts of Europe. They took some time to become popular in Britain, but by 1629, apothecary and herbalist John Parkinson stated that even the youngest housewife knew how to prepare them. Despite this, artichokes diminished in popularity in Britain until they were reintroduced as an exotic vegetable in the 1960s. French queen Catherine de Medici had a passion for them and introduced them to France in the 1500s (somewhat tarnishing her reputation, as artichokes were reputed to be an aphrodisiac), and French colonists took them to North America in the 1800s. By the early twentieth century artichokes were being cultivated extensively in California, leading to the "artichoke wars" of the 1920s and the short-lived banning of artichokes in New York.

Botanical Facts The globe artichoke is an herbaceous perennial belonging to the thistle tribe of the daisy family (Asteraceae). They are native to the Mediterranean, northwestern Africa, and the Canary Islands. The plants resemble giant thistles; they have large leaves with pointed lobes, sometimes spiny, and tall heads of thistlelike flowers that, if allowed to attain maturity, can reach up to 7 in (18 cm) across and are a violet-blue in color. The globe artichoke is unrelated to the Jerusalem artichoke.

Culinary Fare The small, immature buds can be eaten whole; however, artichokes are usually harvested at a later stage, by which point the leaves have become tougher and only the fleshy part at their base can be consumed. Once divested of these leaves, the inedible bristly core, called the choke, must be cut out to reveal the tender heart.

Whole artichokes are usually boiled and served warm or hot, with melted butter or vinaigrette in which to dip the bases of the leaves. Small immature artichokes are often braised, fried, or boiled and marinated in vinaigrette and served chilled. Artichokes that have been divested of their leaves and choke can be cooked in a variety of ways. Italians are particularly fond of them, and the bitter Italian aperitif, Cynar, is artichoke-based.

Rocket

Diplotaxis muralis Wild Rocket *Eruca sativa* Rocket *Hesperis matronalis* Sweet Rocket

Rocket refers to several species of leafy green vegetables with a peppery flavor that makes them popular in salads.

Historical Origins All species of rocket grow wild in Asia and the Mediterranean and have been naturalized in temperate regions such as northern Europe and North America. Rocket has been eaten and cultivated around the Mediterranean for centuries and been used as a salad plant since Classical times. The Romans enjoyed rocket in salads and used the seeds as flavoring. In India, Pakistan, and Iran, rocket is grown for oilseeds.

Botanical Facts All 3 species are annual herbaceous plants from the brassica family. Rocket or rocket salad (*Eruca sativa*) has green serrated leaves and violet-veined white flowers. Wild rocket (*Diplotaxis muralis*) has a stronger flavor than *E. sativa*, with smaller, deeply serrated leaves and bright yellow flowers. It is also known as sand rocket or wallrocket. Sweet rocket (*Hesperis matronalis*), also known as dame's or damask violet, is a well-known biennial or short-lived perennial from southern Europe and central Asia. It is the showiest species, with long dark green leaves and clusters of scented blue-violet flowers in late spring to summer. Sweet rocket is sometimes considered an invasive weed but is often grown for its flowers, whose scent becomes stronger in the evening, hence its botanical name. All species are easy to grow but tend to bolt to seed, so for a continuous supply, seeds should be sown at intervals of 2 weeks.

Culinary Fare Rocket's lively taste comes from sulfur compounds. All species have a strong peppery flavor and are mainly eaten raw as a salad vegetable, combining well with other milder leaves. The mature leaves develop a bitter flavor and should be used sparingly. Rocket can be prepared in similar ways to other leafy greens such as spinach.

ABOVE: *Rocket (*Eruca sativa*), also called arugula or roquette, grows rapidly in hot weather, with plants able to double their size in days. All rocket species are rich in vitamin C.*

Popular Rocket Dishes

A popular, near-ubiquitous restaurant salad combines rocket with pears and matured cheese such as parmesan or pecorino. Rocket pairs well with all kinds of cheeses, particularly sharp-flavored varieties. It is sometimes added to or used as a substitute for basil in pesto, or used to stuff pasta shapes and pastries, in which case it is often combined with soft fresh cheeses. In Italy, rocket is often added to pizza or pasta near the end of the cooking process so that it has time to wilt. Rocket is a standard ingredient in prepared salad mixtures.

Lettuce

✻ *Lactuca sativa* Lettuce ✻ *Lactuca sativa* var. *augustana* Celtuce

ABOVE: *There are hundreds of different kinds of lettuce. Collectively, lettuces have so many harvest times that some type of lettuce is available all year round. These green and purple romaine lettuces are providing a bountiful summer crop.*

Lettuce belongs to the family Asteraceae and is the most popular leafy salad vegetable in the world. Celtuce is a type of lettuce grown mainly in China. Only the stem and tender upper leaves are eaten, as the lower leaves are tough and unpalatable.

Historical Origins Wild lettuce is still common in Europe and temperate Asia, but it is tough and bitter. Wild—and to a lesser extent, cultivated—lettuce contains a white sap that is mildly soporific and the plant was probably originally cultivated for this reason.

Ancient Egyptian wall paintings dating back to about 4500 BCE depict lettuce plants, though not the compact hearting cultivars we know today. Instead, they were tall virile plants dedicated to Min, the Egyptian god of fertility and male sexual potency.

The Greeks and Romans ate lettuce and used it medicinally. The Romans would eat lettuce after a meal to induce sleep, or before a meal to stimulate the appetite. Lettuce was being cultivated as a cooking vegetable in China by the fifth century. European settlers took lettuce seeds to the Americas as early as 1494, although bitter wild varieties already existed there.

It was not until the sixteenth century that lettuce began to take on the headed form familiar today, and by the early 1600s, its narcotic effects had been much reduced.

Botanical Facts Lettuce is an annual or biennial plant. If allowed to "bolt"—run to seed—it produces a tall, branched flowering stem with small, pale yellow flowers. Lettuce can be divided roughly into 2 kinds—hearting and non-hearting. There are many cultivars ranging in color from green through to red and bronze. Lettuce grows quickly and seedlings planted 10 days apart will ensure a good continuous crop.

Culinary Fare Lettuce is most commonly used fresh as a salad ingredient in Western countries. In European cooking it is sometimes stewed or braised, and peas are commonly cooked with lettuce leaves. In China, the leaves are shredded and stir-fried. Young tender celtuce leaves can be used in salads, cooked like spinach, or stir-fried. The stem should be peeled before being sliced for use as a raw vegetable, or it can be stir-fried or substituted in recipes calling for celery, celeriac, or cardoons. In this case, cooking time should be slightly reduced.

RIGHT: *Lettuce was sacred to the Egyptian god Min, allegedly enabling him to copulate untiringly. It is not known whether this was due to the plant's upright nature, its milky sap, or its renown as an aphrodisiac.*

Water Spinach

Ipomoea aquatica

Water spinach is a popular Asian vegetable used in salads and stir-fries. Sometimes just the stems are split, soaked, and served as a decorative curly garnish.

Historical Origins Water spinach is a common vegetable in Asian cuisines, especially in southern China and Southeast Asia. It is often planted on the banks of fish farms as food for fish. Although water spinach is commonly eaten in China, the Chinese believe it should not be consumed in vast quantities as doing so will lead to weakness, owing to its hollow stems.

Botanical Facts Also known as swamp cabbage, swamp spinach, and water convolvulus, water spinach is a member of the sweet potato family, Convolvulaceae. It is a perennial creeping plant that grows best in exceptionally damp, swampy areas. Water spinach has pale green tubular stems and long pale to dark green triangular leaves.

Culinary Fare Water spinach is a popular plant in Chinese, Malaysian, Thai, and Indonesian cuisine. It can be eaten fresh in salads or cooked, but needs to be washed thoroughly due to its growing conditions. In Cantonese cooking, water spinach is frequently stir-fried with garlic and fermented bean curd or fish sauce.

Blanching water spinach in hot water before cooking will maintain its attractive bright green color. Cook it just briefly, as prolonged cooking will produce a slightly muddy flavor. The leaves are similar to spinach in texture, taste, and appearance.

RIGHT: *Water spinach is a tropical and subtropical plant that often grows wild in ponds, marshes, and lakes, where it can be invasive.*

Rice Paddy Herb

Limnophila aromatica

A popular Asian herb with a sharp citrus aroma, rice paddy herb is mostly used to flavor soups and curries, or as an aromatic garnish. Vietnamese immigrants introduced it to North America during the 1970s.

Historical Origins Rice paddy herb is grown in rice paddies, ponds, and other warm watery environments. Closely related species are sometimes cultivated as aquarium plants. It is native to Southeast Asia and is a common ingredient in Thai, Cambodian, and Vietnamese cuisines.

Botanical Facts *Limnophila aromatica* belongs to the snapdragon family and has thin spongy stems and small, delicate green leaves that can be used externally to treat insect bites. It bears small, pale lilac flowers. As a garden vegetable, it is very easy to propagate in water and will develop roots within 1–2 weeks, if kept warm, covered, and out of direct sunlight. However, it grows well only in the wet tropics.

Culinary Fare An aromatic plant, rice paddy herb yields a variety of subtle flavors including citrus, sweet basil, and cumin. It is frequently used in the traditional Vietnamese soup, *pho*, and Cambodian sweet and sour fish soup, and is often added to curries or eaten with *nam prik*, a pungent Thai fish and chili sauce.

Alfalfa Sprouts

Medicago sativa

RIGHT: *Alfalfa sprouts are more than an attractive garnish. They are highly nutritious and contain minerals such as potassium, magnesium, phosphorus, zinc, and manganese. It is best to eat them soon, though, as they are prone to food-borne bacteria.*

Alfalfa sprouts are a popular salad vegetable, and in many cases can make a nutritious substitute for lettuce. Alfalfa is often mixed with other sprouts such as radish, mustard, garlic, or broccoli to make a more pungent mix.

Historical Origins Also known as lucerne, alfalfa is native to Iran. Its main use is as green manure or as a forage crop for feeding livestock. Alfalfa is thought to have been introduced to Europe during the Persian invasion of Greece in 491 BCE, and to have been introduced to China in the second century BCE. In Europe it was used for human consumption in times of hardship and shortage, such as during the Spanish Civil War (1936–1939).

Botanical Facts Alfalfa is a perennial forage herb originating from Iran but naturalized worldwide. It grows best in warm temperate and cool subtropical regions, and because of its deep roots is able to withstand dry conditions well. Alfalfa is a member of the clover family and has leaves with 3 oval to narrow leaflets, serrated at the tip, on short stalks. It bears blue to purple pea-flowers in summer through autumn. Because it is a legume and fixes nitrogen in the soil, it is an important agricultural crop for improving soil nutrition.

Alfalfa is easily sprouted at home. The seeds are rinsed first, soaked overnight, then drained. Place them in a thin layer in a container, cover it, and leave them somewhere warm to sprout. The seeds must be rinsed and drained morning and night and will be ready to eat within a few days.

Culinary Fare Mature alfalfa has a grassy flavor. The young leaves are eaten as a vegetable in China. The seeds can be ground to produce meal, but they are more usually used to produce sprouts, which are a very popular salad vegetable and are also highly nutritious. Alfalfa sprouts have a crispy texture and mild nutty flavor. The sprouts are usually purchased growing in their own container and will keep growing in the fridge as long as they are not allowed to freeze or dry out. They make an excellent addition to salads and sandwiches and make a flavorful crispy garnish.

Alfalfa sprouts are considered highly nutritious, being rich in beta-carotene, vitamin K, and minerals. They are used as a diuretic and also in the treatment of urinary tract infections. They have been used in traditional Chinese medicine for digestive ailments and Ayurvedic medicine to treat ulcers, arthritis, and fluid retention. However, all raw legume sprouts contain toxins, so it is advisable to eat them in moderation or cook them.

Miner's Lettuce

Montia perfoliata

Miner's lettuce is an unusual and attractive wild herb that spreads readily and grows wild across much of the United States.

Historical Origins Also known as winter purslane, spring beauty, and by its earlier botanical name, "claytonia," miner's lettuce originated in North America and can be found as far north as Vancouver. It was used as an herb and salad vegetable by both Native Americans and pioneers. Miner's lettuce has an extremely high vitamin C content and was eaten by Californian miners to prevent scurvy, hence its common name.

Botanical Facts Despite its delicate appearance, miner's lettuce is a hardy trailing annual. Early leaves are borne on short stalks and are triangular in shape, while later leaves are rounded and curled around the white- or pink-flowering stalk. This gives the stalk the appearance of having grown through the center of the leaf, but in actual fact it is surrounded by 2 leaves joined together.

Culinary Fare Though related to the nasturtium flower (*Tropaeolum* species), miner's lettuce does not share its pungent characteristics. This mild herb and salad vegetable resembles spinach in flavor. The entire stalk, flower, and leaf can be eaten, all of which make an appealing addition to salads.

ABOVE: *The young leaves of miner's lettuce are the best to use in salads, as older leaves can be bitter. Boiled roots are said to taste like water chestnuts.*

Watercress

Nasturtium officinalis

A sharp-flavored aquatic plant, watercress is popular as a salad vegetable, garnish, or ingredient in soups and stir-fries.

Historical Origins Watercress is found growing wild in Britain, central and southern Europe, and West Asia and also in North America since it was introduced there by European immigrants. Wild watercress has been a popular vegetable since ancient times—the Anglo-Saxons, Romans, and Greeks all valued it medicinally and for its sharp flavor. It is thought cultivation began in Germany in the mid-sixteenth century.

Botanical Facts A hairless aquatic perennial, watercress has white roots and dark green feathery leaves; some cultivars have bronze to purplish brown leaves. It bears small white flowers and small fruits. Plants grow naturally in fresh running streams or springs. As watercress requires clean flowing water of a constant temperature, it is difficult to grow in home gardens.

Culinary Fare Mostly eaten fresh as a salad vegetable, watercress has a pleasant, pungent, peppery flavor. Although cooking diminishes it somewhat, it is used in soups (especially Chinese and French) and as a stir-fried vegetable. It is a popular garnish, especially for game, and makes excellent traditional English-style afternoon tea sandwiches, particularly when teamed with hard-boiled egg and mayonnaise. In western Ireland, watercress has long been cultivated and valued as a highly nutritious food.

ABOVE: *A rich green color and strong stems are the qualities to look for when purchasing watercress. It wilts quickly; if kept cool and dry, it will last for a few days.*

Perilla

Perilla frutescens

Perilla is an attractive leafy herb with many applications in Japanese cuisine. In traditional Chinese medicine it was used to treat respiratory ailments and morning sickness.

Historical Origins Also known as *shiso*, wild sesame, Chinese basil, and beefsteak plant, perilla is native to China, Burma, and the Himalayas. It was used in China over 1,000 years ago, both as a vegetable and for its oil, which is extracted from the seed. It is no longer common in Chinese cuisine, but has become highly popular in Japan.

Botanical Facts A member of the mint family, perilla is an erect, finely hairy, annual herb with attractive serrated leaves. It forms spikes up to 4 in (10 cm) long of small white, pink, or reddish flowers in late summer to autumn. There are green as well as red cultivars.

Culinary Fare Perilla is used extensively in Japanese cuisine. The red cultivar is mainly used as a coloring agent in the production of the delicately pink *umeboshi* (pickled plums), and is also used in the manufacture of confectionery. The green leaves are made into tempura, chopped and added to sushi, or used as a garnish. Perilla is also commonly salted or pickled for use as a condiment. In Vietnamese cooking it is often used raw in salads or wrapped around grilled meat.

Japanese Radish Sprouts

Raphanus sativus

ABOVE: *Under the right conditions, Japanese radish sprouts will begin sprouting within days and are usually eaten later, after the green leafy tops appear.*

In Japan, these spicy peppery sprouts are particularly popular in salads and are also often used as an attractive garnish.

Historical Origins Radish roots have been used as food since prehistoric times. Radishes are thought to have originated in West Asia from a type of mustard green. Oriental radishes are very large in size and come in a variety of colors. The seeds of the *daikon* or giant white radish are used to produce Japanese radish sprouts, also known as *kaiware*.

Botanical Facts Radish sprouts are small, slender, white sprouts with leafy green tops. They can be grown easily in the home. After rinsing the seeds, leave them to soak overnight. In the morning drain and rinse, repeating this daily. In between each rinsing and draining, spread the seeds in a thin layer in a container and cover it. Then leave them somewhere warm to sprout. These sprouts will be ready to eat within a few days.

Culinary Fare Japanese radish sprouts have a hot spicy taste. They are quite commonly used in Japan as a salad vegetable or as a garnish, especially in sushi and sashimi. Like other sprouts, Japanese radish sprouts make a crispy addition to sandwiches and salads, though they have more bite.

Rhubarb

Rheum x cultorum

Horticulturally, rhubarb is a vegetable, though it is usually eaten as a fruit. It has a sharp sour taste that is frequently offset by the addition of sugar.

Historical Origins Wild rhubarb is native to Asia. It was known in China by 206 BCE and was valued medicinally, developing into an article of trade from China to West Asia in the tenth century. The Greeks and Romans knew it as an imported dried root with medicinal qualities. Rhubarb became known in a medicinal context in England during the sixteenth century, but it was not until the early 1800s that rhubarb recipes began to appear in cookbooks.

Botanical Facts Rhubarb is a perennial with large, wavy-edged leaves on stout red or green stalks. Only the stalks are eaten. The leaves have been associated with cases of poisoning due to their high concentrations of oxalic acid and anthraquinone, and must not be eaten.

Culinary Fare Rhubarb is mostly used for pies and similar dishes, hence its common name "pieplant" in the US. Ginger, orange, and angelica are the traditional flavorings, especially in jam-making. Rhubarb and custard, and rhubarb crumble, are traditional English dishes.

Rhubarb is sometimes cooked as a vegetable, notably in Polish, Iranian, and Afghani cuisines, and is used to make an Italian aperitif, Rabarbaro Zucca Bitters.

RIGHT: *When preparing rhubarb for cooking, remove all traces of leaves, wash, then trim the base. Chop the stems into short lengths.*

Marsh Samphire

Salicornia europaea

Marsh samphire is a seaside plant that is found along the coasts of Europe.

Historical Origins Also known as glasswort, due to its use in the manufacture of soda glass, marsh samphire is abundant in soda. It was harvested, dried, and burned, after which the ashes were used in the production of glass. Marsh samphire was traditionally pickled, along with its cousin rock samphire (*Crithmum maritimum*)—with which it is often confused—as early as the beginning of the seventeenth century.

Botanical Facts Marsh samphire is found near the sea, especially around estuaries where high levels of minerals and trace elements are washed down from the highlands. This vegetable is a member of the beet family, and the plant grows as a small shrub with long, thin, fleshy leaves.

Culinary Fare Slightly saltier than rock samphire, marsh samphire has a less powerful scent. In Britain it is boiled and eaten as a vegetable, and is often served in a similar manner to asparagus, with melted butter or hollandaise sauce. This vegetable is regarded as a delicacy. It is at times misleadingly sold by fishmongers as a type of seaweed.

BELOW: *The leaves of samphire are best eaten before the plant flowers and while the leaves are still a fresh vivid green.*

White Mustard Sprouts

Sinapis alba

White mustard sprouts are one half of the ingredients in traditional English "mustard and cress." Children often grew them, sowing the cress a few days later.

Historical Origins Mustard is a member of the cabbage family and its seeds are used to produce the condiment mustard, as well as mustard oil. White mustard sprouts have a sharp, tangy, hot flavor that is reminiscent of horseradish. This is due to the presence of sulfur compounds known as glycosides and the enzyme myrosinase.

Botanical Facts White mustard sprouts are small and fine, with bright green leafy tops. They can be grown very easily in the home. First rinse the seeds, soak them overnight, then drain. Place them in a thin layer in a container, cover, and leave somewhere warm to sprout. The seeds must be rinsed and drained morning and night. They will be ready to eat within a few days.

Culinary Fare In England, white mustard sprouts are traditionally combined with cress and sold as "mustard and cress," also known simply as "mustard cress." This combination makes a particularly good sandwich and is often served plain on buttered bread, or with cucumber or avocado. It is also a popular spicy salad ingredient or garnish.

RIGHT: *A vibrant green carpet of white mustard is an attractive groundcover. Best suited to a temperate climate, it grows quickly. Freshly sprouted leaves are best in salads while older leaves are an ideal potherb.*

Alexanders, Black Lovage

Smyrnium olusatrum

A member of the Umbelliferae family, in medieval times alexanders was used in the same way as celery or as an alternative to it, which back then was quite a bitter herb.

Historical Origins Alexanders is originally native to Macedonia, but has been naturalized in Britain for 2,000 years. It was used by the Greeks and Romans as a potherb, and is often found growing wild in Europe and Britain, especially around the sites of medieval monastery gardens.

Botanical Facts Alexanders has feathery dark green leaves and greenish yellow flowers. The fruits are small and black, hence its other common name, black lovage. It has a bitter aftertaste that can be alleviated by earthing up to light-blanch the young shoots. Alexanders is a biennial, growing to medium height, and can be found wild in meadows and wastelands near the sea, particularly around the Mediterranean, though it is widely naturalized elsewhere. It is also sometimes called horse parsely.

LEFT: *The use of alexanders diminished with the increase in the availability of mild sweet cultivars of celery that emerged during the eighteenth century.*

Culinary Fare Alexanders is most similar in flavor to celery and parsley. The entire plant can be eaten. The seeds make a particularly nice spicy substitute for pepper, the flowers and leaves can be used in salads, and the stems and roots give a flavor to soups and stews that is similar to celery.

Spinach

Spinacia oleracea

Spinach seems to have achieved rapid success as a vegetable, probably because it was superior in flavor and texture to other, coarser greens commonly used in medieval times.

Historical Origins Spinach originated in Persia, where it was cultivated by the fourth century CE, possibly earlier. It had spread to China via Nepal by the seventh century, and to Japan and Korea in the fourteenth to seventeenth centuries. The Arabs introduced spinach to Spain in the eleventh century. It was introduced to Britain in the mid-1500s; in his *Herball* of 1568, William Turner notes: "Spinage or spinech is an herbe lately found and not long in use." The Spanish planted spinach in the Americas; widespread cultivation began in earnest in Europe in the eighteenth century.

Botanical Facts Spinach is an annual with large bright green, oval to triangular, smooth or wrinkled leaves. It is a member of the goosefoot family, growing to about 1 ft (0.3 m) in height, with small, inconspicuous yellow-green flowers.

Sweet Spinach

During medieval times, spinach was quite a popular ingredient in sweet dishes in Asian and European cuisines. This habit seems to have diminished during the eighteenth century. Back then, spinach was often combined with ingredients such as eggs, honey, nuts, and spices to make a sweet dessert tart, which would have been a refreshing juicy change from the dried fruits available in winter. It is thought the common practice of adding nutmeg to cooked spinach derives from such dishes. In Provence, France, a sweet spinach tart is still traditionally eaten on Christmas Eve.

To grow, spinach requires deep rich soil to accommodate its lengthy taproot. A top-dressing of nitrogen will encourage rapid growth. Water well throughout the growing period. This leafy vegetable is best suited to cool season growing in spring and autumn. It is liable to bolt to seed during hot weather. Spinach is susceptible to downy mildew and leaf spot. It is propagated from seed.

Culinary Fare Spinach is often boiled in little or no water (the water left clinging to the leaves after rinsing is usually sufficient), and it shrinks considerably in bulk during cooking. A pound (0.5 kg) of leaves can reduce down to a single cup, so use large quantities. Spinach is used in soups, tarts, pies, soufflés, and cheese and egg dishes. In Italy, spinach is used to color pasta. The French culinary term "à la Florentine" usually denotes a dish that includes spinach; eggs Florentine is poached eggs on a bed of spinach with hollandaise sauce. Spinach is also available frozen or canned.

ABOVE: *Young spinach leaves are excellent as a salad vegetable. Its mildly acidic taste is due to its low concentrations of oxalic acid. It quickly became popular in medieval Europe, as it was a superior green to those usually used, such as orache, sorrels, and other goosefoot family plants.*

Dandelion

Taraxacum officinale

RIGHT: *Usually best known as an annoying lawn weed, dandelion has a long history of use as a food plant. The young leaves, high in vitamins A and C, are best eaten raw. Older leaves can be used as a potherb.*

Dandelion is one of the most widespread weeds in the world, growing in numerous temperate regions. It is related to chicory and shares that herb's bitter characteristics.

Historical Origins Dandelion is native to Europe and Asia, but different species of dandelion are common in temperate regions the world over. The common name is a corruption of the French *dent de lion*, meaning lion's tooth, a reference to the plant's serrated leaves. Cultivation of dandelion began in France and Britain in the mid-nineteenth century. Settlers took the dandelion to North America, where it thrived. Native Americans used it medicinally and as a food.

Botanical Facts Dandelion is a particularly common weed originating from the Northern Hemisphere. Leaves vary from broad to narrow, and from deeply cut, even-fringed edges to nearly smooth. Yellow ray flowers produce a ball of tufted seeds dispersed by wind in spring to summer.

Culinary Fare All parts of the dandelion are edible, and the leaves are particularly high in vitamin C. Young leaves may be eaten raw in salads and older leaves can be boiled or wilted. A tasty traditional French salad dressing for dandelion is made using hot bacon fat with tiny pieces of bacon. The roots of dandelion are used to make a caffeine-free beverage that tastes like coffee.

New Zealand Spinach

Tetragonia tetragonoides

ABOVE: *New Zealand spinach is a salt- and heat-tolerant plant resembling spinach. It is easily grown in sunny frost-free gardens with light dry soil.*

This creeping groundcover was quite popular as a summer spinach in Britain and North America during the early nineteenth century.

Historical Origins Despite its name, this plant is unrelated to spinach. In New Zealand it was a coastal plant eaten by Maori. In 1770, while exploring New Zealand, Captain James Cook recognized the importance of this green vegetable and had it served to his crew in salads and pottages to prevent scurvy, as it is high in vitamin C. His botanist, Sir Joseph Banks, took the plant back to England where it was grown in Kew Gardens, and by the nineteenth century, "Botany Bay greens" became a particularly popular summer spinach in both the US and England.

Botanical Facts Commonly known as warrigal greens or *tétragone* in France, New Zealand spinach is a short-lived creeping perennial native to New Zealand and Australia, but naturalized elsewhere. It has dark green, oblong to triangular, pointed, fleshy leaves with shiny undersides. Its succulent stems bear small, greenish yellow, daisylike flowers, usually in spring.

Culinary Fare As its name implies, New Zealand spinach resembles spinach in flavor as well as appearance, and can be cooked in a similar manner. It has a tendency toward bitterness, but this can be alleviated by changing the water during cooking.

Corn Salad, Lamb's Lettuce

Valerianella locusta

Corn salad is a wild green with a mild nutty flavor similar to spinach. It is sometimes known as mache, or lamb's tongues, due to the shape and soft velvety feel of the leaves, along with the fact that lambs are said to enjoy eating it.

Historical Origins Corn salad is a common wild plant that is native to Europe, West Asia, and North Africa. It has been naturalized in North America and cultivated in Europe since the sixteenth century, though it was eaten as a wild salad herb long before its cultivation. Corn salad's name arose from its common appearance among corn crops, but it is also known by many other names. In Europe it is often called field lettuce and rampion, and in Germany it is sometimes called rapunzel, which was the plant that was so strongly desired by the pregnant mother in Grimm's fairytale she traded her unborn child for it. Corn salad grows in winter when other greens are rarely available, which may explain her somewhat insatiable craving.

Corn salad is still most popular in France and Germany, and its appeal has increased in North America since its introduction there, but its use in Britain has declined since the eighteenth century.

Botanical Facts Corn salad forms large rosettes of slightly succulent, smooth-edged or slightly serrated, spoon-shaped to round, velvety, dark green leaves. It has a branched flower stalk with smaller stem leaves tipped with rounded clusters of small, bluish, mauve, or white flowers in spring.

Corn salad is a very hardy plant and is particularly appreciated because it grows in winter and early spring when mild-flavored salad leaves are at a premium.

Culinary Fare Corn salad is high in vitamin C and is a nutritious addition to the winter diet. It is mainly used as a salad vegetable, where its delicate, faintly nutty flavor is complemented well by mild dressings based on olive, hazelnut, or walnut oil. Corn salad does not keep well and should be used within a few days of purchase; vibrant green leaves indicate freshness and any yellow leaves should be discarded. In mixed salads it is often combined with sliced beetroot and potato, or with apple or walnuts.

Corn salad can be cooked and eaten in the same way as spinach and related greens, though its color fades in cooking; avoid this is by blanching it briefly in hot water before use. Lightly dressed leaves can also accompany cheeses.

BELOW: *Picked fresh from the garden, the crisp young leaves of corn salad make a delicious change from lettuce. An adaptable plant, it grows well in most soils and positions but dislikes lengthy dry periods.*

Standard Okra

Abelmoschus esculentus

Also known as lady's fingers, okra is a delicate elegant vegetable with an unusual texture that is either loved or hated.

Historical Origins Okra is a member of the mallow (Malvaceae) family and is a relative of cotton. It is thought to be native to Africa and may have been cultivated in West Africa or Ethiopia, but it is not known when it spread to North Africa, the Middle East, India, or the eastern Mediterranean. Although it was recorded as growing by the Nile as early as the thirteenth century BCE, there is no evidence of it in Egyptian tombs. It is thought that okra traveled to the New World via the slave trade in Africa. By the mid-seventeenth century it had reached Brazil, and by the late seventeenth century it had reached Suriname. During the seventeenth century it may also have arrived in America's south, and by the eighteenth century okra was being grown as far north as Philadelphia and Virginia. From India, okra traveled slowly to Southeast Asia and was found there in the nineteenth century, and in China shortly afterward.

Botanical Facts Okra is a widely naturalized tropical and subtropical annual grown extensively as a vegetable. It has long and broad serrated leaves with 5–7 lobes and single white or yellow flowers marked with red or purple toward the base. The podlike fruit is fleshy with green or red skin, a downy exterior, and ridges that run from stalk to tip. In order to produce fruit, okra must have a long, frost-free season.

Culinary Fare The main characteristic of okra is its glutinous quality. In the United States it is one of the staple vegetables in Creole and Cajun cooking. It is famously used in gumbo, a spicy soupy stew with many ingredients, including meat, seafood, and vegetables. Okra was used to give the dish its characteristic glutinous texture, but this is no longer considered as desirable as it was in the past. The term "gumbo" comes from the West Indian name for okra. Okra is also used extensively as a vegetable in Indian and Middle Eastern cuisines. In order to avoid the glutinous texture, care must be taken not to break the pods and to cook them gently. The texture can also be minimized by cooking with cornmeal, steeping the okra in acidulated water, frying before cooking, or choosing very young pods. Okra can be dried and powdered for use as a thickening agent.

LEFT: *The 'Louisiana Green Velvet' cultivar is a long, spineless, dark green okra that remains tender when large.*

Pigeon Pea

Cajanus cajan

Also known as the Congo pea or red gram, the pigeon pea is one of the most popular pulses in India.

Historical Origins The pigeon pea belongs to the pea-flower subfamily of the legumes (Fabaceae) family. It is native to the Old World tropics and is naturalized in other warmer regions. It is thought to be one of the earliest cultivated food crops. Grown mainly for its edible pulses, the pigeon pea is also used as a fodder crop, and the dried stems are used for fuel and for basketry.

Botanical Facts The pigeon pea is a shrubby perennial with downy stems and trifoliate leaves that are green above and grayish green beneath. It has red and yellow flowers that are followed by yellow pods up to 4 in (10 cm) long.

Culinary Fare Pigeon peas are a favorite pulse in India, where they are used to make *dhal* and the spice blends known as *sambhars*. Pigeon peas and their pods can be served fresh as a vegetable, and the seeds can be sprouted. When dried they are a source of flour or can be split and used in soups and stews. In some places the pigeon pea is grown for canning or freezing.

ABOVE: *The seeds of the pigeon pea are high in fiber and are a rich source of vitamin C.*

Capers

Capparis spinosa

Capers are the pickled or salted, green, unopened flower buds of a Mediterranean shrub that are widely used as a condiment. They differ from caper berries, the fruit of the same plant, which set after the flowers have opened and fallen off.

Historical Origins Capers have been known for thousands of years and are widely cultivated in many Mediterranean countries, including France, Spain, Italy, Greece, and Cyprus. They were introduced to the French around 600 BCE by the Greeks and were a favorite condiment of the Romans.

Botanical Facts From southern Europe, Africa, and Asia, the caper bush is a salt-tolerant scrambling shrub with semiprostrate branches. The leaves are very broad, rounded, and arranged in 2 rows, and the flowers—grown on slender stalks and appearing in summer through to autumn—are white with pale purple stamens. The small fruits are elongated and ribbed.

Culinary Fare Capers must be pickled in vinegar in order to develop their slightly bitter flavor, which is due to the formation of capric acid during the pickling process. The most prized capers are the smallest ones, known in France as *nonpareilles*, followed in increasing order of size by *surfines*, *capucines*, *fines*, and *capotes*. They go well with seafood and lamb, and are used in condiments and sauces.

ABOVE: *Capers are tangy and vinegary. They are an essential component, along with parsley, chives, and shallot, of tartare sauce.*

Chilies and Peppers

Capsicum annuum Bell Pepper, Chili Pepper
Capsicum chinense Habanero *Capsicum frutescens* Tabasco

Capsicum and chili peppers are prized in many cultures for their tangy flavor. This is due to the presence of capsaicin, a powerful alkaloid that is mainly found, in varying degrees, in the inner membranes of the fruits. Individual fruits from the same bush can vary greatly in their capsaicin content. Capsaicin is odorless, and will irritate the skin and any delicate areas. Synthetic capsaicin is used to make pepper spray.

Historical Origins Capsicums are a genus of the nightshade (Solanaceae) family and are related to the New World tomatoes and potatoes, and to the Old World aubergine and deadly nightshade, the latter being highly toxic. All capsicums are native to the Americas, and in Central America the word "chili" refers to all of them. Wild chilies were being eaten in Mexico around 7000 BCE, and cultivated before 3500 BCE. Columbus probably brought plants back to Europe in the late fifteenth century, and the Spanish and Portuguese took them to India and Southeast Asia shortly after. Chilies spread quickly to the Middle East, the Balkans, and Europe. They reached Italy by 1526, Germany by 1543, and Hungary by 1569. Unlike other members of the nightshade family, which were regarded with suspicion, chilies were welcomed in all areas whose inhabitants were used to eating food with hot spices.

BELOW: *Bell peppers are sweet and mild, and different colors have different uses. They are roasted or stewed for the dishes* ratatouille *and* pepperonata.

Botanical Facts The *Capsicum* genus has 10 species, 5 of which are commercially important. Most capsicums are perennials but are treated as annuals and are grown for their vibrant fruit. There are 2 main types of capsicums: hot, or chili, peppers, which are hot to taste; and sweet, or bell, peppers, which are mild. Some types are also grown for ornamental purposes. Capsicums have small, white, cream, or purple flowers followed by hollow fruits that are full of seeds. They grow best in tropical, subtropical, and warm to hot climates, but can be grown in cooler climates if they are protected.

All modern, sweet, cultivated chilies and most of the hot cultivars belong to the *Capsicum annuum* species, including sweet (bell) peppers, the Jalapeño, and Cayenne. Despite the name, they are perennial plants in their native habitat. *C. frutescens* is a perennial usually grown as an annual with flowers borne in groups rather than singly. The fruits are usually hotter than those of *C. annuum*, but they are used in the same way and are widely cultivated in many tropical countries. To this group belong the Tabasco, African Birdseye, and Thai Hot cultivars. Tabasco is also the trade name of a bottled, hot, fermented chili sauce made since the 1870s in Louisiana. *C. chinense* is a third species and is widespread in the northeast of South America. The species includes Habanero and Scotch Bonnet peppers, so-called because their shape is reminiscent of a bonnet or lantern. Habanero chilies are extremely hot. They are short, wide, orange-colored pods with a tropical fruit scent and are thought to have originated in Cuba (as the name, meaning "from Havana," suggests). Scotch Bonnet chilies are also thought to have originated in the Caribbean and range in color from light green through to golden yellow and bright red.

Culinary Fare Chilies are used by cooks for flavor, as well as for heat. Caution should be used when preparing them. Cucumber, rice, and dairy products such as yogurt are recommended for relieving the burn of overly hot dishes. The flavor of chilies is subtle and, in small doses, its mild warmth can stimulate the appetite and digestion. In larger amounts it inflames the stomach and burns the mouth. Red and green chilies are both commonly used. Chilies are eaten whole and fresh in many countries throughout the world, including Africa, China, India, Mexico, and Thailand. They are chopped and used as a garnish, or ground and mixed with other ingredients in cooked dishes. They are also dried and roasted before use, which gives them a more concentrated flavor.

RIGHT: *Chili sauces are an important condiment in many cuisines. Green or red chilies can be used to make chili sauce, with the addition of salt, oil, vinegar, and spices.*

Hot Stuff

The term "chili" is derived from the Nahuatl (Aztec) language. Chilies or chili peppers vary greatly in size, shape, color, and taste. They range from mild to pungent and extremely hot. In the United States, the heat of chilies is determined by a system of measurements devised in 1912, called Scoville Heat Units. This scale refers to the number of times dissolved chili extracts can be diluted with sugar water before the capsaicin can no longer be tasted. Mild sweet (bell) peppers score zero, the Jalapeño is rated at 2,500–5,000, and Tabasco and Cayenne peppers rate at 30,000–50,000. Habaneros are at the top of the range, scoring 100,000–300,000 Scoville Units. In Asia there is no such system for measuring the heat of chilies, but the general rule of thumb is the smaller the chilies, the hotter they are.

Tabasco chilies have a dry smoky flavor combined with a fiery pungency.

Chickpea

Cicer arietinum

ABOVE: *Cooked chickpeas have a firm texture and a mild nutty taste. They make a wonderful addition to salads, stews, and soups.*

Also known as garbanzos, chickpeas are one of the three most important pulses in the world, and are used extensively in the cuisines of many countries.

Historical Origins Chickpeas are thought to have evolved from a wild plant and were first domesticated in the Fertile Crescent of the Middle East. The earliest record of chickpeas found at Hacilar, Turkey, is thought to be about 7,500 years old. Chickpeas reached the Mediterranean region by 4000 BCE and India by 2000 BCE. They were taken to the New World in the sixteenth century by Spanish and Portuguese travelers, but never became as important as the native haricot bean. Chickpeas were grown in the Hanging Gardens of Babylon and vast fields of them were common in Ancient Egypt. The Romans regarded chickpeas as food for poor people and peasants, but this did not stop all classes from enjoying them, often with pasta. In India during the second millennium BCE, chickpeas were a major foodstuff known as *channa*. Chickpeas are now India's most important pulse.

Botanical Facts Belonging to the pea-flower subfamily of the legume (Fabaceae) family, the chickpea is an annual plant with sprawling to erect, downy stems, and leaves with up to 17 pairs of small leaflets. Flowers are white to violet and appear in spring to early summer, followed by hairy pods up to 1½ in (35 mm) long, containing globular white, brown, or blackish seeds. The seeds are round and curled at the sides, hence the botanical name referring to the ram's skull. Chickpeas are grown mainly in the Middle East, North Africa, and India. Compared to other pulses, they have a lower protein and higher fat content.

Culinary Fare Chickpeas can be eaten fresh but are more often dried and sold whole or split. They are pale brown in color. Like most pulses, chickpeas must undergo prolonged soaking before cooking. In India, whole chickpeas are known as Bengal gram and chickpea flour is known as besan flour; it is used extensively in sweets and in savory dishes such as pakoras. Probably the most famous chickpea dishes are both Middle Eastern: hummus, which is the Arabic word for chickpea and describes a paste of chickpeas and sesame paste with lemon, salt, and garlic; and felafel, which are small, deep-fried balls of ground chickpeas and herbs. Chickpeas are also used in North African couscous and Spanish and Italian soups and stews. In Italy, a dish of chickpeas and pasta is known as *tuoni e lampo* (thunder and lightning), a reference, perhaps, to the differing textures offered by the firm chickpeas and soft pasta.

Cucumbers

Cucumis metuliferus Horned Cucumber *Cucumis sativa* Cucumber

Cucumbers are an ancient, highly refreshing fruit normally eaten as a vegetable. They contain up to 96 percent water and are popular for tempering the effects of hot and spicy foods.

Historical Origins Cucumbers are one of the oldest cultivated vegetables. They are thought to have originated from a wild species in the foothills of the Himalayas. The cucumber was highly prized in Mesopotamia where it was grown 4,000 years ago, and it is thought to have been cultivated in India about 3,000 years ago. The Roman Emperor Tiberius is said to have been particularly fond of cucumbers. Cucumbers were also known in both Ancient Egypt and Greece and were first cultivated in China in the sixth century. Cucumbers are thought to have appeared in France in the ninth century, in England during the fourteenth century, and in Germany during the sixteenth century. They were introduced to Haiti by Christopher Columbus in the late fifteenth century, and from there rapidly spread all over North America.

Botanical Facts The cucumber is a member of the pumpkin (Cucurbitaceae) family and is a hairy, climbing or trailing plant with both male and female yellow flowers born in clusters or pairs. It climbs via tendrils and has large toothed leaves that shade the delicate fruits. Cucumbers are long and cylindrical and can reach up to 20 in (50 cm) in length. They are normally eaten before they mature because they become bitter and unpalatable as they ripen further. About two-thirds of cultivated cucumbers are eaten fresh; the remaining third are used for pickling and are usually picked when smaller. Cucumbers grow best in temperate climates and are available throughout the year. The horned cucumber is native to Africa and is grown in many temperate countries. It is oval, bright yellow-orange in color, and has many prominent spines, and a green watery interior with many seeds.

Culinary Fare Young cucumbers are no longer as bitter as they were in the past, due to the bitterness having been bred out of them. They are combined with yogurt to make a cooling refreshing condiment (such as Indian *raita* or Greek *tzatziki*) or soup. Examples of chilled cucumber soups include Turkish *cacik*, Spanish *gazpacho*, and Russian sorrel and cucumber soup. Cucumbers are a conventional garnish for cold salmon and are peeled or semipeeled to make English-style tea sandwiches. Cucumbers are often added to salads and are sometimes served cooked.

ABOVE: *'Telegraph' cucumbers (also known as burpless cucumbers), have a thin dark skin and white flesh with smaller and fewer seeds than other cultivars.*

LEFT: *Chopped cucumber is combined with yogurt and mint to make a refreshing dip called* raita. *It is often served with hot and spicy dishes to help soothe the heat.*

Pumpkins, Winter Squash

Cucurbita maxima *Cucurbita moschata* *Cucurbita pepo*

ABOVE: *Three kinds of widely eaten pumpkins are 'Fairytale' (the largest), 'Ambercup' (the small orange ones), and butternut (one cut in half).*

Three different species in the *Cucurbita* genus are known as pumpkins (also called winter squash), which are grown and used in many countries. They range widely in size, shape, and color, but all have thick skins and are harvested when fully mature. Pumpkins are used in both sweet and savory dishes and the seeds are eaten as a snack food.

Historical Origins *Cucurbita maxima*, the principal parent of pumpkins, is native to South America, and seeds found in Peru have been dated to CE 1200, although no remains have been found at archaeological sites in Mexico or Central America. Pumpkins are thought to have been first domesticated in South America and, today, a wide range of cultivars are grown worldwide, the largest producers being China, India, Ukraine, and the United States. The name "pumpkin" is thought to have come from the old French word *pompon*, which itself came from the Ancient Greek word *pepon*, which was also applied to melons.

Botanical Facts *Cucurbita* is a genus of about 27 trailing and climbing species of annuals and perennials from North, Central, and South America in the pumpkin (Cucurbitaceae) family. It is divided into 2 groups: gourds and pumpkins. *Cucurbita* species are grown for both their edible and their ornamental fruit. Their leaves can be quite large, and are sometimes spotted or lobed and usually rough or prickly, as are their stems. They produce short-lived yellow or orange male and female flowers that often require hand pollination to fertilize the stigma of the female flower. *Cucurbita* are generally classified according to when they are harvested (summer or winter) or by their shape. "Pumpkins" refers to those kinds of *Cucurbita* that are allowed to remain on the vine until they mature, and are then stored for use in winter. The skin is hard and inedible, while the flesh becomes firm and is frequently superior in flavor to most other types of summer squash, which are harvested when immature. Zucchini are derived from *C. pepo*.

Culinary Fare Pumpkins are eaten when fully ripe and cope well with prolonged storage. They often grow to a very large size and the flesh is fibrous and has a nutty earthy taste. In France, pumpkin is used almost exclusively for making soups, however in the south of France it is also used to make a pumpkin-flavored bread and the pumpkin pie known as *citronillat*. In Argentina, large pumpkins are hollowed out and a meat stew is cooked inside. Many recipes using pumpkin start with the cooked puree, which is made by boiling pumpkin in a little water until soft. Pumpkin soups are also very popular in Australia, and as pumpkins are widely and easily grown, and readily take on other flavors, they are often used as an economy vegetable in dishes such as pumpkin scones or in jams and preserves. Pumpkin is a popular roast vegetable and is sometimes mixed with potato and made into mash. Whole pumpkins can be baked and the flesh served with cream, salt, pepper, and a little nutmeg.

When pumpkin is used for sweet dishes it is common practice to add spices such as ginger, cinnamon, or nutmeg. In the United States, pumpkin pie is a traditional Thanksgiving dish. It is thought that this dish originated from old English recipes for sweet pies using "tartstuff," a thick pulp of spiced boiled fruit. In Cyprus, pastries stuffed with crushed wheat and pumpkin are a popular breakfast in winter. They are flavored with a mixture of cloves, cinnamon, and sultana raisins. In Majorca, Spain, an unusual sweet known as Angel's Hair is made from cooked pumpkin that has been cooked again with sugar to produce a stringy jam, which is used in a variety of ways such as in the filling for its famous sweet yeast buns, *ensaimadas*. The filling is readily available canned. In Spain, Mexico, and other countries, pumpkin seeds are eaten roasted, fried, or salted as a snack food. In Italy, the unshelled seeds are dried, salted, and eaten as a popular snack known as *passatempo* (pass the time).

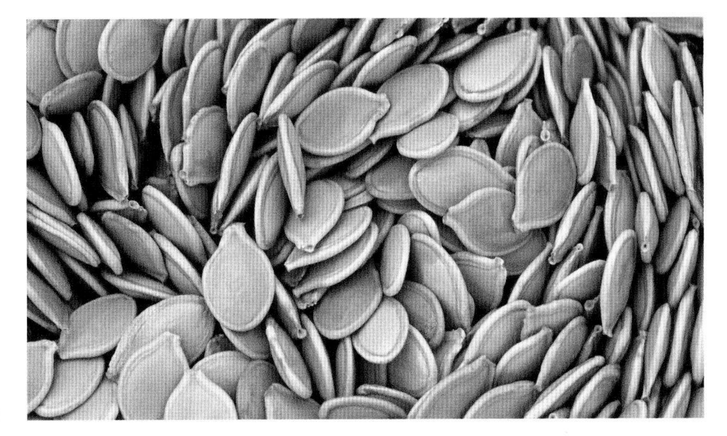

ABOVE: *Pumpkin seeds are a delicious snack and are easy to prepare at home. Thoroughly wash the seeds, removing any flesh clinging to them, and dry in a single layer in a slow oven.*

Pumpkin Cultivars

'Australian Blue': Also known as 'Queensland Blue,' this large pumpkin has blue-gray skin with deep ridges, bright orange flesh, and a sweet flavor. A similar cultivar, the 'Jarrahdale,' is gray and less deeply ridged.
Butternut: This is a popular small, creamy-skinned, bottleneck-shaped pumpkin with bright orange, sweet, nutty flesh.
'Delicata': Also known as the 'Bohemian' or sweet potato squash, this is a large oblong pumpkin with bright yellow skin streaked with green or orange, and creamy flesh that tastes like sweet potato.
'Golden Nugget': This is a small, bright orange-colored pumpkin with a rather bland flavor, mainly used for its attractive appearance.
Kabocha: Also known as the Japanese pumpkin, there are numerous cultivars, the best known probably being 'Kent.' It has gray-green, mottled skin and sweet, yellowy orange flesh that is somewhat drier than other cultivars.

Ripe 'Naguri' from the kabocha type of pumpkin.

Summer Squash, Zucchini

Cucurbita pepo

BELOW: *The 'Tigress' cultivar of summer squash produces papery yellow flowers and smooth, cylindrical, medium green fruits that retain a fine texture.*

Zucchini is the Italian name for the marrowlike forms of summer squash, which are also known by their French name, courgette. The flowers and young fruits are very popular in Italy but less so in other countries, probably because their delicate flavor and texture are easily marred by improper preparation, or because the use of such young fruits was uneconomical.

Historical Origins The origins of summer squash are difficult to pinpoint, but they appear to have originated in Mexico from the giant pumpkin. Squash have been cultivated in South America for more than 5,000 years and are an integral part of its cuisine. It was not widely eaten in Europe until early in the twentieth century, and was developed there from squash introduced from the Americas by Christopher Columbus. It was the Italians who introduced zucchini to the French sometime during the nineteenth century, and in the 1920s they were still being referred to as *courgettes d'Italie*. They are found in only a few English cookbooks of the 1930s and did not become popular until British food writer Elizabeth David wrote about them in the 1950s and 1960s.

Botanical Facts *Cucurbita* is a genus of about 27 trailing or climbing species of annuals and perennials from North, Central, and South America in the pumpkin (Cucurbitaceae) family. *Cucurbita* species are grown for their edible and ornamental fruit. They are generally classified according to when they are harvested (summer or winter), or by their shape. *Cucurbita pepo* is a trailing or bushy species with many cultivars. The leaves are lobed, triangular, and prickly and care should be taken when collecting the fruits. Botanically, summer squash are immature fruits. They are very easy to grow in the home garden but are extremely persistent, vigorous, and abundant. Most cultivars are best eaten immediately after harvest. The young leaves and delicate flowers can also be eaten.

Culinary Fare Summer squash tend to be watery, so choose young firm fruit and avoid overcooking. The fruit can be boiled or steamed or, for better flavor, fried, roasted, or broiled. They can also be stuffed, and combine well with tomatoes and aubergines (eggplant). They can also be used in breads, pasta dishes, and soups. Their mild flavor can be enhanced by adding herbs such as parsley, chives, and basil. The male and female flowers can be eaten and are often stuffed and then crumbed or battered before being fried, often with a miniature fruit still attached.

Hyacinth Bean

Dolichos lablab

The hyacinth bean is a tropical climbing plant grown as an ornamental and for food. The seeds and pods can be eaten.

Historical Origins The hyacinth bean is an ancient legume originating in Africa or Asia that has been cultivated in India since earliest times. It is widely grown in Southeast Asia, Egypt, and the Sudan and cultivated in many tropical and subtropical regions where it is used for food and animal fodder.

Botanical Facts The hyacinth bean is a decorative, fast-growing, tropical, twining plant that grows best in warm climates. The purple cultivar is a popular ornamental garden plant with beautiful foliage and striking purple pods and bracts of violet to purple-red flowers. Other cultivars are green with white flowers.

Culinary Fare Young pods and young or mature seeds can be eaten. Fresh mature seeds have a very hard coating and require prolonged cooking in order to become edible. They can be toxic if consumed in large quantities, and when cooking it is advisable to change the cooking water several times. Dried seeds can be cooked and eaten as is, processed into bean cakes, or fermented to produce a pressed, fermented bean cake similar to tempeh. In Myanmar, the seeds are used in curries.

ABOVE: *The seeds of the hyacinth bean have a white scar along one side.*

Calabash, Cucuzzi

Lagenaria siceraria

The calabash is frequently grown for its hard shell, which is dried and used as a container. The young fruits are also eaten.

Historical Origins Also called the bottle gourd or white-flowered gourd, calabash is usually regarded as native to Africa, though it has been cultivated in the Americas and Asia since prehistoric times. Archaeological evidence places it in the Peruvian Highlands as early as 13,000–11,000 BCE and along the Peruvian coast and in Mexico around 7000–4000 BCE. It is uncertain how the calabash traveled to the Americas.

Botanical Facts Belonging to the pumpkin (Cucurbitaceae) family, the calabash is a widely grown, tendril-climbing annual with dull green, hairy, oval to heart-shaped, serrated leaves. In summer it bears solitary, white, trumpet-shaped flowers. Various cultures use the hard shell of the calabash to make utensils (such as bowls and ladles), musical instruments, and containers.

Culinary Fare In its wild form, calabash is bitter and inedible, but cultivation has resulted in many forms that are sweet and highly valued. Only the tender young fruit is used in cooking. It can be boiled, steamed, fried, pickled, or added to soups, curries, or fritters. Calabash have a delicate flavor and high water content, so are used in soups or stews.

ABOVE: *These rattle type calabash are commonly made into instruments.*

Lentils

Lens culinaris

Along with peas and chickpeas, lentils are one of the oldest crops cultivated. They are high in protein and very versatile. The main producer of lentils is India, but they are also cultivated in most tropical and warm temperate countries.

Historical Origins Lentils originated in the Middle East and have been cultivated since ancient times in Egypt. Remains of lentils have been found in many prehistoric sites throughout Europe, particularly in Greece, Syria, and Turkey. Lentils were first domesticated around 8000 BCE in the Fertile Crescent of the Middle East, and spread into Europe, India, China, and Egypt. The red pottage referred to in the Old Testament was made of lentils, and Esau hungered for them so much that he exchanged his birthright for them.

Botanical Facts The *Lens* genus comprises 4 species of annuals closely related to peas and vetches, in the pea-flower subfamily of the legume (Fabaceae) family, native to the Mediterranean region, western Asia, and Africa. The various cultivated races of lentils were once grouped under the names *Lens ervoides* and *L. culinaris* but are now all treated as a single species, probably originating in the wild in Turkey. It is an erect to sprawling annual with feathery leaves that usually have 6 pairs of narrow leaflets about ¾ in (18 mm) long. It bears stalked clusters of 1–3 small flowers in spring, and pods under ¾ in (18 mm) long.

The fruit is a small flattened pod containing flattened disc-shaped seeds, convex on both surfaces. Lentils are best adapted to regions with hot dry summers and good winter or spring rainfall, in order to produce high yields.

BELOW: *Cultivated lentil plants in bloom during the mid-growth phase. Each flower produces a short, flattened pod containing 1–2 lens-shaped seeds.*

ABOVE: *Even though they have a different texture, red and green lentils have a somewhat nutty flavor when cooked.*

Culinary Fare Unlike other pulses, lentils do not require soaking before cooking. Lentils are high in protein, and for this reason they are valued in Asia and are a favorite food in Roman Catholic countries during Lent. Lentils come in a variety of sizes and colors. The main types are those that are large and light-colored or yellow, and those that are smaller and colored red, brown, or gray. In Europe, the variety *Verte du Puy*, which is small and green, is considered a delicacy. Lentil seeds, either whole or split, are used in soups, stews, and casseroles, and are sometimes fried, seasoned, and eaten as a snack food. They are also used in the popular Indian dish, *dhal*. Flour made from the ground seed can be mixed with cereals in cakes, as well as added to children's food and food for the elderly. In India, the young pods can be eaten as a vegetable.

Chinese Okra, Luffa

Luffa acutangula

Chinese okra is not related to standard okra but is similar in appearance and texture, having a bland taste and distinctive ridges. It is used in many different Asian cuisines and is especially popular in India.

Historical Origins Also known as the ridged gourd or angled luffa, Chinese okra is native to India. Today, it is cultivated in India, Southeast Asia, China, and Japan. Like its close relative, *Luffa cylindrica*, which is often cultivated for use as a textile or sponge, Chinese okra has a fibrous spongy interior.

Botanical Facts Chinese okra is a member of the pumpkin (Cucurbitaceae) family. It is a vigorous, tropical running vine with ribbed, cylindrical, elongated, light to dark green fruits that can reach up to 12–24 in (30–60 cm) in length. Like okra, Chinese okra is distinguished by ridges that run from stem to end and give its cross-sections an appealing star shape. Chinese okra has rounded hairy leaves and bears male and female yellow flowers that open late in the afternoon and remain open during the night. They are receptive to pollen for only one day, and must be pollinated by hand if grown in an area deficient in insects. They do best in warm temperate climates and are often grown as ornamental plants.

Culinary Fare The flesh of Chinese okra has a bland and slightly bitter taste. It is best eaten young, when it is less than 4 in (10 cm) long and tender throughout. It is usually steamed, sautéed, or braised and can be cooked in the same manner as similar vegetables, such as zucchini and squash. Unless very young, the ribs must be trimmed with a vegetable peeler before cooking to reduce the bitterness. In China it is often stir-fried and in India it is used in curries or eaten deep-fried. In Indonesia it is steamed and used as a side dish, or sliced and cooked with coconut milk and other ingredients. In Hong Kong the young fruits are cooked briefly and served as an accompaniment to stewed mutton. In Jamaica they are peeled, boiled, and seasoned with butter, pepper, and salt before being added to curries. In Japan the fruits are sometimes sliced and dried before being cooked.

LEFT: *The Chinese okra is a vigorous vine that produces cylindrical fruits and blue-green foliage. Young fruits have tender skin and need to be handled gently.*

Tomato

Lycopersicon esculentum

ABOVE: *Ripe, firm hothouse tomatoes are popular due to their uniform red color and smooth skin.*

Although botanically a fruit, the tomato is used as a vegetable. It is one of the world's most popular salad vegetables and a staple of many of the world's cuisines.

Historical Origins Tomatoes are thought to have evolved from the cherry tomato, which grows wild in Peru, Ecuador, and other parts of tropical America. First domesticated in Mexico and cultivated by the Aztecs, tomatoes were taken to Europe early in the sixteenth century by the Spanish and very gradually were accepted into Mediterranean cuisines. Toward the end of the eighteenth century they were introduced from Europe to the United States. By the end of the nineteenth century, large-scale cultivation and canning had begun in Italy and from the mid-nineteenth century the United States was also canning tomatoes on a large scale. By the 1830s production of tomato ketchup, the US "national condiment," was well under way.

Botanical Facts Botanically, the tomato is a fleshy berry. It is a member of the nightshade (Solanaceae) family and is native to western South America. It is grown widely as an annual food crop. It has an erect or scrambling hairy stem and deeply cut, mid-green leaves with up to 9 lobes. It bears up to 12 yellow, open, star-shaped flowers per stalk. Tomatoes can be red, yellow, orange, or blackish in color. Some cultivars display distinctive stripy patterns.

Culinary Fare Tomatoes are one of the most widely used and versatile vegetable or salad plants in the world. They can be consumed raw, cooked, or made into preserves. They are canned, dried, and made into various different sauces (such as tomato paste or concentrate). It is difficult to imagine Italian cuisine without them. Tomatoes are famously partnered by basil in many dishes. They are commonly used in soups, stews, with pasta or as a pizza topping, in salads, sandwiches, and many other dishes.

Climate of Fear

Tomatoes were regarded with a great deal of suspicion when they were first introduced to Europe. This was in part due to the fact that they belong to the same family as a number of highly poisonous plants such as the deadly nightshade, black henbane, and tobacco. Later, in the sixteenth century, tomatoes were also associated with the herb mandrake, and as a result they came to share its aphrodisiac qualities, leading to the common name of "love apples." Tomatoes were still regarded with suspicion in England well into the nineteenth century, even though they had been widely grown in greenhouses during this time.

Tepary Bean

Phaseolus acutifolius

Also known as the tepari, yori mui, pavi, and Texas bean, tepary beans are related to haricot beans and are of ancient origin.

Historical Origins The tepary bean is thought to be native to Central America and has a long history in Mexico, where it was first cultivated 5,000 years ago. It has also been grown in Arizona for thousands of years. The name is thought to have derived from the Papago Indian name for "bean." It has since been introduced to parts of East and West Africa, where it is suited to the dry hot conditions.

Botanical Facts The tepary bean is part of the pea-flower subfamily of the legume (Fabaceae) family. It is highly resistant to drought and disease, and has a high nutritional value. It is a bushy or twining plant with small, slightly hairy, green pods. The beans vary a great deal in color and range from reddish or purplish browns to greenish yellow and white. They are about ¼ in (6 mm) in size.

Culinary Fare Tepary beans have a rich nutty flavor. They must be shelled and dried before use, and can then be cooked in stews or casseroles, ground and used in *pinole* (a flour made from various different seeds used in Mexico and southwestern America), or prepared in similar ways to other legumes. The Hopi Indians use the bean to break a fast by parching them in hot sand and mixing them with saltwater before eating.

ABOVE: *A watercolor shows the heart-shaped leaflets, pods, and reddish brown seeds of the tepary bean.*

Adzuki Bean

Phaseolus angularis

Adzuki beans are an especially popular legume in Japan and China because their red color represents good fortune.

Historical Origins Also known as the azuki or red bean, this legume is native to Japan, Korea, India, China, and neighboring areas of southern Asia. It is believed to have originated in China and spread to Japan sometime between the third and twelfth centuries. They have been introduced to Hawaii, the southern part of the United States, South America, New Zealand, and various African countries.

Botanical Facts Adzuki beans are an annual, tropical climbing plant growing to about 12–24 in (30–60 cm) in height. The pale yellow flowers are followed by smooth cylindrical pods containing the seeds or beans. The beans come in a range of colors from red and white to black and gray, but they are most often dark red.

Culinary Fare Adzuki beans are tender with a mild sweet flavor. Along with the soybean, they are one of the most widely used legumes in Japan. Adzuki beans are a staple in the preparation of sweets; the fresh, sweet bean paste known as *an* is the basis of many Japanese confections. In Japan they can also be ground into flour, used in soups, or boiled with rice to make "red rice." Adzuki beans are an ingredient in shaved ice desserts such as *ice kacang* in Singapore and Malaysia, and *halo-halo* in the Philippines.

ABOVE: *The round seeds of the adzuki bean have a distinctive white ridge along one side. The bean is widely grown in South Asia and is not found in the wild.*

Mung Bean

Phaseolus aureus

Mung beans are a versatile popular legume widely used in many Asian countries in both sweet and savory dishes. They are sold sprouted in Western countries.

Historical Facts Mung beans are native to India and have long been cultivated there. They are used as a food crop and also as a forage crop and as green manure. Mung beans have now spread to China, Southeast Asia, the United States, the Caribbean, and East and central Africa. China and India are the main producers today.

Botanical Facts The mung bean is a small bean with a green seed coat and a light yellow interior. It is known as green or golden gram in India, where there are approximately 2,000 cultivars grown. Mung beans are a popular pulse as they are easily digestible, do not require soaking, cook quickly, and have a good flavor. Although mung beans are tropical and subtropical in origin, they also grow well in warm temperate areas. The plant produces clusters of thin, hairy black pods, each containing as many as 15 small green seeds.

Culinary Fare Mung beans are available whole, split with the skin on, or split with the skin removed. They do not need to be soaked prior to cooking. They are used in soups or porridge, or as a savory accompaniment to rice. Mung bean starch is used to make bean starch sheets and bean thread vermicelli, also known as jelly noodles, cellophane noodles, bean starch threads, and spring rain noodles. Along with soybeans, they are the bean most commonly used for sprouts and can easily be sprouted at home, although they will be curly instead of straight. Their tails should be pinched off before using either fresh or briefly cooked. In traditional Chinese medicine the sprouts are considered to be a cooling (*yin*) food. Mung beans also provide a basis for confections. In Malaysia they are served as a sweetened porridge with palm sugar and coconut milk. In Sri Lanka they are roasted and ground before being sweetened and used in a variety of confections. In Thailand the husked split beans are cooked and mixed with sugar and coconut cream to form a pliable paste that is then shaped into tiny fruits. In China the beans are used to make sweet bean paste for filling steamed buns and other sweets. In Vietnam they are used in sweet drinks as well as in rice flour cakes.

RIGHT: *In Western countries, mung beans are mostly eaten sprouted. They should be eaten fresh or lightly cooked, as prolonged cooking destroys the vitamins and flavor.*

Runner Bean

Phaseolus coccineus

Runner beans are commonly grown as an ornamental garden plant and for their tender green pods.

Historical Origins Runner beans, also known as scarlet runner beans because of their bright red flowers, are thought to have originated in Central America, most likely in Mexico. They are native to the uplands of Mexico, Guatemala, and possibly some other countries in this region, and were domesticated here. Evidence places them in Mexico's Tehuacán valley some 2,000 years ago. They were introduced to Europe in the sixteenth century and were initially grown only for their showy flowers.

Botanical Facts Runner beans were originally cultivated as a perennial grown for their pods, fresh and dried seeds, and starchy tuberous roots. They are popular in Britain for their tender green pods and for their flowers, which indicate the color of the seeds, ranging from white through to pink, deep purple, and jet black.

Culinary Fare Due to the presence of toxins, the dried beans and roots of runner beans must be boiled in water before eating. The fresh pods can be steamed, boiled, stir-fried, or prepared in the same way as other beans. In the United States the mature beans are sold as snail beans, butter beans, or pea beans. They can be used as a substitute for lima beans in recipes.

ABOVE: *As runner beans can grow to 6 ft (1.8 m) tall, they need support.*

Lima Bean

Phaseolus lunatus

Lima beans are a large, white to pale yellow bean particularly popular in the United States.

Historical Origins Also known as the butter bean or Madagascar bean, lima beans are of Peruvian origin. Archaeological evidence dates them back to 7000 BCE. They were domesticated in both Central and South America around 5000 BCE, and European slave traders took the bean to Africa, from where it traveled to Asia. Lima beans are now also grown in many tropical, subtropical, and warm temperate regions of North America, Africa, and Asia.

Botanical Facts Depending on the cultivar, lima beans can be a small annual bush or a large climber that is cultivated as an annual or perennial. The mature seeds are white, 1 in (25 mm) long, and flat or kidney-shaped.

Culinary Fare In the United States lima beans are a popular legume sold fresh or frozen. In the Philippines they are ground into a flour that is used to make noodles and bread, and in Japan they are used to make a bean paste. They are a staple food in parts of tropical Africa and Asia, including Myanmar. Due to the presence of toxins in the mature bean, lima beans should not be eaten raw, but soaked and boiled in several changes of water.

ABOVE: *Once cooked, these raw lima beans provide a high amount of fiber and protein, and help lower cholesterol in the body.*

Haricot Bean, Common Bean

Phaseolus vulgaris

ABOVE: *These fresh pods of the haricot, or common, bean are eaten in various cuisines throughout the world and are popularly grown in the home garden.*

The haricot, or common, bean is known by a variety of names. It is the best known and most widely cultivated bean in the world. Both the fresh pods and the dried beans can be eaten.

Historical Origins Haricot beans are native to Central America, where they grow wild in the mountains and were domesticated over 5,000 years ago. Archaeological remains of the bean in Peru and in Mexico's Tehuacán valley have been dated to around 5000 BCE. When Europeans arrived in the Americas they found that numerous cultivars had been developed in both South and North America. The first beans introduced to Europe by the Spanish and Portuguese in the sixteenth century were dark red and kidney-shaped, hence the common name of kidney bean. They quickly became popular, as they had more flavor and were easier to digest than the broad beans commonly used in Europe. The Spanish and Portuguese also introduced beans to Africa and other parts of the Old World. The haricot bean is now the main pulse grown in tropical America.

Botanical Facts The haricot bean is an erect or twining annual from tropical regions of the Americas. It is widely grown in cooler areas but requires some protection. The plant has green compound leaves with rounded or oval leaflets up to 4 in (10 cm) long, and stalks bearing up to 6 white, purple, or pink flowers, up to ¾ in (18 mm) across, in spring. Fruits are flat or nearly cylindrical pods up to 20 in (50 cm) long, containing elongated or spherical red, brown, black, white, or mottled seeds that are harvested in summer. Yellow-pod forms are known as wax or butter beans.

Culinary Fare Fresh young beans in their pods can be boiled and served with butter, cream, and herbs or with olive oil and garlic or lemon juice. In Greece and the Middle East they are stewed with olive oil, onion, tomato, and spices (dried beans are often treated in the same way). In France they are usually served with toasted almonds. Fresh beans can be lightly cooked and used in salads. Dried beans must be soaked before cooking.

Beans form the basis of many celebrated dishes such as the French *cassoulet*, American baked beans, chile con carne (the official dish of Texas), and Mexican *frijoles refritos* (refried beans). In Asian countries they are fermented and used as the basis of many sauces, added to stir-fries, or fried quickly until they blister then dressed with garlic, ginger, and chili, or other sauces. Beans lend themselves well to canning and fresh beans are often sold packaged and frozen.

Peas

✳ *Pisum sativum* Peas ✳ *Pisum sativum* var. *macrocarpon* Snap Peas, Snow Peas

Peas have been a staple food since ancient times. Snap and snow (or sugar) peas differ from other peas because the entire pod is tender and edible.

Historical Origins Relics of the garden pea dated to about 3000 BCE have been found in Bronze Age settlements in Switzerland. Peas were grown in the Classical era by the Greeks and Romans and spread quickly through India, where they are still popular, reaching China in the seventh century. Dried peas were a staple food for poorer people throughout Europe in the Middle Ages. In the sixteenth century the garden pea was introduced to England. Around the end of the seventeenth century there came a sudden craze for eating immature peas fresh, especially among the aristocratic ladies of the French court. By the end of the nineteenth century peas were a popular canning vegetable, but with the advent of frozen vegetables in the 1920s and 1930s, peas came into their own, being particularly suited to that method of preservation.

Botanical Facts *Pisum* is a genus of bushy or climbing annual herbs from the pea-flower subfamily of the legume (Fabaceae) family. They are native to the Mediterranean region. Divided leaves have 4–6 opposite leaflets with rounded leaves and the main stem ending as a tendril. They bear single or groups of 2–3 showy butterflylike flowers. The fruit is a flattened oblong pod containing few to several seeds.

Culinary Fare Fresh peas deteriorate quickly and should be eaten as fresh as possible. In France fresh peas are gently stewed with lettuce leaves. In England they are boiled and served with butter and mint. Traditional English dried pea dishes include pea and ham soup and pease pudding, as well as peasecod soup, made from the tough pods.

ABOVE: *The large pods of the* 'Alderman' *cultivar are easy to pick, and contain plump tender peas with an excellent flavor.*

Peas and Presidents

Peas are not usually thought of as a glamorous vegetable. Nevertheless, they have aroused fervent passion in various circles. Thomas Jefferson, who had a passion for gardening, adored peas. Among his neighborhood friends there was an annual competition to see who could grow them first. The winner would invite the others to a triumphant dinner showcasing the first peas of the season. Invariably, Mr. George Divers was the winner, but one year, Jefferson's peas were ready first. When reminded by his family of the winner's privilege, Jefferson nobly declined to take advantage of it, on the grounds that Divers would be happier thinking that his peas were always the first.

US President Thomas Jefferson was an avid gardener who enjoyed growing vegetables.

Tomatillo

Physalis ixocarpa

RIGHT: *The husk of this tomatillo has split open to reveal the fruit inside. Purple tomatillos are much sweeter than the green cultivars, and can be eaten straight off the plant.*

Tomatillos are indispensable in Mexican cuisine and have a delicate, slightly acid flavor. They can be used raw or cooked.

Historical Origins A native of Mexico, the tomatillo was cultivated there before the tomato and dates back to 800 BCE. It was first domesticated by the Aztecs. Tomatillos were introduced into India during the 1950s. They are also cultivated today in Australia and South Africa.

Botanical Facts The tomatillo plant is a member of the nightshade (Solanaceae) family and is a perennial often grown as an annual. Tomatillos grow inside a thin papery husk that splits open as the fruit ripens. They have a thin skin that ranges in color from green through to yellow and purple. The flesh is yellow, crisp, acid, or sweet and contains many small seeds.

Culinary Fare Tomatillos are very important in Mexican and Guatemalan cooking, in that they provide a distinctive flavor to its "green" sauces. In India they are used to make chutney. Ideally, tomatillos should be cooked for a few minutes, in order to release their full flavor, which is tart and lemony. They are often combined with green chilies to alleviate the chilies' heat. When cooked, they release a pectinlike substance that thickens the sauce as it chills.

Chayote, Choko

Sechium edule

As it is bland, the chayote lends itself well to other flavors and is commonly used as the main ingredient in preserves, along with other flavorings.

Historical Origins The chayote was cultivated as a vegetable by the Aztecs in Central America, where it originated, and after the conquest of Mexico it spread to other tropical and subtropical regions. It runs wild in Australia, where it is a common garden vine and sometimes a pest.

Botanical Facts The *Sechium* genus belongs to the pumpkin (Cucurbitaceae) family and comprises some 6–8 species of climbing plants from the cooler mountain regions of tropical America. The chayote is a perennial with large, green, ridged, oval fruits that are fleshy and have hairy, spiny, or smooth skin. They contain a single large seed that can germinate within the fruit. The seed, fruit, young shoots and leaves, and large, starchy tubers can all be eaten.

Culinary Fare Chayote has a subtle taste and readily takes on the flavors of other ingredients. Some people think it is insipid and watery, while others think it subtly sweet. Chayote grows vigorously and abundantly, and for this reason it is often used as an economy vegetable in chutneys, preserves, pies, desserts, and stews. It is usually served cooked.

Aubergine, Eggplant

Solanum melongena

Botanically a berry, aubergine is used as a vegetable and is popular in Italian, Greek, Middle Eastern, and Asian cuisines. Due to its popularity, aubergine is now cultivated throughout tropical, subtropical, and warm temperate countries as a field crop, and in cool countries in greenhouses.

Historical Origins The aubergine is native to tropical Asia. It was first cultivated in India, where wild forms occur, and became popular in the Near East and Asia, as its mild flavor and spongy texture combine well with other ingredients. The Persians introduced it to Africa, and the Arabs introduced it to Spain in the eighth century, and to Italy in the thirteenth century, but it did not become popular until the fifteenth century. In the 1500s Spanish and Portuguese colonists took it to the New World.

Botanical Facts The aubergine is a member of the nightshade (Solanaceae) family. It is a bushy annual and is grown as a vegetable for its egg-shaped fruit. The stems are sometimes spiny, and the leaves are soft, green, downy, oval, and shallowly lobed. Aubergines bear attractive, star-shaped violet to light blue flowers with drooping petals. The fruits are white to purple-black in color and up to 8 in (20 cm) wide. Some cultivars produce small white, mauve, or orange-red fruits.

Culinary Fare It is no longer necessary to salt aubergine before cooking unless the fruit is large and old. However, the technique of salting can be used to diminish aubergine's tendency to soak up large amounts of oil or in preparation for pickling. In Asian cuisine the aubergine is stir-fried, stewed, or used in hotpots and simmered dishes. Aubergines are commonly used in India, Afghanistan, and Iran to make a hot pickle. In Italy it is also pickled for use in antipasto dishes. Favored in many Mediterranean countries, aubergines feature in dishes such as the Greek *moussaka*, French *ratatouille*, Italian *melanzane parmigiana*, and Sicilian *caponata*.

ABOVE: *These ripening purple aubergines are the most widely recognized and eaten aubergine cultivar in the Western world.*

Aubergine Exotica

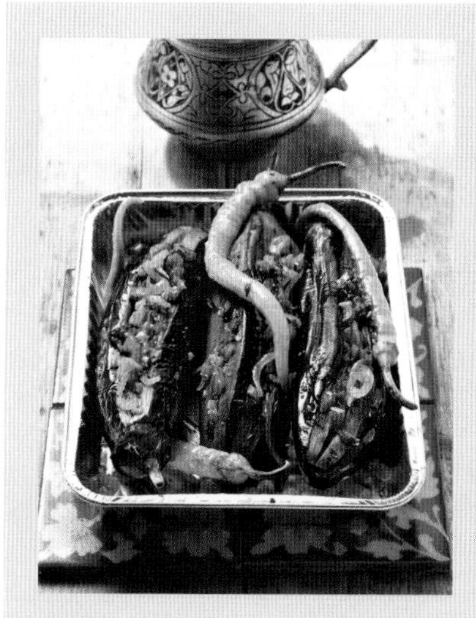

The most popular dishes using aubergine are of Middle Eastern origin and their names reflect the Arabic love for this vegetable. Sultan's Delight is a rich smoky puree of grilled aubergine seasoned with cheese, while *baba ghanoush* is made with grilled aubergines, sesame paste (tahini), lemons, salt, and garlic. Aubergines are especially popular stuffed. *Imam bayildi* ("the Imam fainted") is a Turkish specialty of aubergine with a tomato-based stuffing. Sources differ on the Imam's reaction to first tasting this dish. He is said to have fainted, either from pleasure, or from shock, after he realized the cost of this new dish.

A specialty Turkish dish, Imam bayildi, *ready to serve.*

Asparagus Pea

Tetragonolobus purpureus

Asparagus pea is a common spring wildflower found in the Mediterranean. It is prized for its flowers and its ability to grow well in poor soil.

Historical Origins The asparagus, or winged, pea is a legume native to the Mediterranean and the Middle East. It has an unusual appearance, with 4 wavy ridges extending along the length of the pods, and is sometimes confused with the unrelated winged bean (*Psophocarpus tetragonolobus*).

Botanical Facts The asparagus pea is an attractive, hairy, trailing annual with broad, oval, gray-green leaves and bright red flowers up to 2 in (5 cm) long, which are borne in pairs. It is a popular wildflower and is occasionally grown for its young winged pods, which are edible.

Culinary Fare The asparagus pea is sometimes cultivated as a culinary herb, but is more often grown for its young pods, which can be cooked whole or pickled. The pods should be picked when very young, before they turn fibrous. Asparagus peas are best cooked simply: lightly steamed or boiled and dressed with butter, or with butter and a little cream. The pods offer small amounts of protein, iron, fiber, and carbohydrates.

Broad Bean, Fava Bean

Vicia faba

ABOVE: *Broad beans are shelled just like lima beans. The young and tender beans have a better flavor than older ones.*

Broad beans have been a staple food for millennia. Oddly, they were once regarded with mild superstitious dread by a number of ancient civilizations.

Historical Origins Broad beans are an ancient bean of Europe, West Asia, and North Africa, and are thought to have originated in the Mediterranean or southwestern Asia. Archaeologists have found the remains of broad beans in Neolithic sites in Israel dating to 6800–6500 BCE. Other remains in central Europe and the Mediterranean have been dated to 3000 BCE. Until haricot beans were introduced from the New World, broad beans were a staple food for the Ancient Egyptians, Greeks, and Romans. They spread to China some time after 3000 BCE and today are also cultivated in North and South America, the Sudan, and Uganda.

Botanical Facts Broad beans are hardy upright annuals bearing small, scented white flowers with black markings and large plump seedpods about 12 in (30 cm) long. The species belongs to the pea-flower subfamily of the legume (Fabaceae) family.

Culinary Fare In Italy, broad beans are eaten fresh in their pods with salt or with pecorino cheese. When larger and more mature they are best cooked, as they develop a tough outer skin that should be removed for best results. In Shanghai, Beijing, and the north of China, broad beans are used fresh and as sprouts. In Egypt the dried beans are the basis of *foul mudammes*, a traditional dish.

Cowpea, Black-eyed Pea, Yard-long Bean

Vigna unguiculata

LEFT: *The foliage, flowers, and pods of the yard-long bean. The pods are not actually a yard long; half a yard is more typical, and the plant is widely grown in Southeast Asia, Thailand, and southern China.*

Cowpeas are a popular legume cultivated for their dried beans and immature seedpods. They can be cooked in the same way as other beans and are especially popular in the Deep South of the United States.

Historical Origins The cowpea originated in Africa and was domesticated in West Africa around 3000 BCE. It reached India at some point after 1500 BCE and was also taken to Southeast Asia. In the sixteenth century the Spanish and Portuguese carried cowpeas to the New World. Cowpeas are a hot-climate plant and are currently grown in Africa, Central and South America, Southeast Asia, and the southern United States. They are a major food crop in Haiti.

Botanical Facts The cowpea has many common names and is the fresh bean from which the black-eyed pea is derived. It is an annual that comes in many forms—bushy, erect, prostrate, spreading, climbing, or twining. The flowers are often beaked and range in color from white through to light mauve, pink, and dark purple. The pods also vary in color from tan and pink through to red, purple, and nearly black. The pods also vary in size, with those of the type known as yard-long reaching up to 40 in (100 cm). Yard-long beans are grown exclusively for their long, immature green pods. The seeds are often kidney-shaped and can be colored buff, brown, red, black, or white. The white seeds contain a pigment confined to a small "eye"—these are described as black-eyed peas. The mature seeds, young pods, and leaves can be eaten.

Culinary Fare Black-eyed peas are synonymous with US "soul food" from the Deep South, where they are traditionally eaten on New Year's Day, as they are thought to bring prosperity and good luck. Sometimes a coin is buried in the dish and whoever receives it in their portion is considered to have their luck assured for the forthcoming year. Texas Caviar is another dish eaten in such a manner and comprises cold, cooked black-eyed peas with vinaigrette and chopped garlic. Another traditional dish is Hoppin' John, which is made from black-eyed peas cooked with rice, seasonings, onion, and sometimes fatty cuts of pork. The same cuts of meat are often cooked with black-eyed peas, to be then served as a side dish with a hot sauce. Dishes such as this are often accompanied by collard greens and cornbread.

ABOVE: *Black-eyed peas have a thinner skin than most beans, so require less time to cook and do not need to be soaked beforehand.*

Grains

Introduction to Grains

The classification "grains" or "cereals" refers to the edible seeds of grasses. In fact, the cereals are the most significant of the edible plants, as they are the most widely cultivated food crops in the world and provide the largest amount of caloric energy of any food. Despite the adage that "man cannot live on bread alone," in many developing countries carbohydrate-rich grains form almost the entire diet of the population. Even in more affluent societies, they represent a large percentage of the food consumed. The cereals can therefore be classified as a main staple of the human diet.

> "One type of rice nourishes one hundred types of people."
> Chinese Proverb

ABOVE: *White long-grain rice (in basket) and short-grain rice (in ladle) are two of the various cultivars of rice, classified by shape and texture, that are consumed around the world.*

WITHIN THE GRAIN

Grains are the edible part of cereal plants, and are actually complete fruits. Several plant products that are eaten in the same fashion as grains are often included in the cereal classification, though they are actually not the fruits of grasses but rather of broadleaf plants. These are the pseudocereals, and include buckwheat, amaranth, and quinoa, among others. Both cereals and pseudocereals are structurally similar. An outer seed coat consists of a layer of epidermis surrounding several thin inner layers, which include a thin, dry ovary wall. Next is the aleurone layer, which is very thin but rich in oils, minerals, proteins, vitamins, enzymes, and flavor. The center of the grain is the endosperm, which is a starchy mass of dead storage cells. At the base of the seed lies the embryo, also known as the germ, which is a rich source of vitamins, enzymes, oil, and flavor. While the dense mass of cells that is the endosperm is often the only part of the grain consumed by humans (the outer layers having been removed by milling and polishing, as part of the refining process), this part of the grain is actually the least nutritious.

SUPER STAPLES

Three types of grains have made up nearly 90 percent of the world's grain production in recent years, and have also represented more than 40 percent of the world's calories. These super staples are maize (an essential food of South America, North America, and Africa), wheat (the primary cereal of temperate regions), and rice (the preferred grain of tropical areas). Barley, the next most significant grain in terms of quantity produced worldwide, is largely used for malting to make humanity's beloved beers. Significant quantities of sorghum, millets, oats, rye, triticale, buckwheat, and quinoa are also grown worldwide, while other grains tend to have more localized significance. Among the important cereals, 3 bear fruits covered by tough husks or hulls; these are barley, oats, and rye. The others bear naked fruits that do not require husking prior to milling.

GRAINS IN HISTORY

Early humans in the temperate plains of the Near East (the region encompassing the Balkan states, Egypt, and Southwest Asia) fed on grains gathered from wild fields, easily harvesting enough in a few weeks to last an entire year. About 12,000 years ago, humans began to domesticate wild wheat and barley, and later rye and oats. They selected plants for seed size and ease of harvest, and over time cultivars developed. The most successful crops spread throughout Europe, northern Africa, and western and central Asia. Barley was adopted for its hardiness where growing conditions were poor; oats and rye were favored in cold wet climates; and wheat was favored for its gluten suited to bread making. Rice was domesticated at about the same time in southern Asia, and thrived in the hot wet growing conditions of the tropics and semitropics. The domestication of maize occurred slightly later in South America, where it was suited to the warm conditions.

CONTEMPORARY CULTIVATION

All contemporary grains are cultivated in a similar fashion. They are annual plants, thus each planting yields one harvest, but the timing of planting and harvesting varies by species. Wheat, oats, barley, rye, and triticale are cool-season cereals, ceasing to grow at warm temperatures, while most other grains are warm-season cereals. For all types, harvesting takes place after the parent plant has died, becoming brown and dry. Hand tools, such as scythes and cradles, were traditionally used for grain harvesting, but in the modern developed world the majority of harvesting is done more efficiently by machine.

Grains have been milled and refined since ancient times, though the practice has become more widespread in the wake of technological advances. Milling breaks up the grains, while refining removes the tough bran and germ from the endosperm—the bran and germ are made of a harder material that is easily sifted away. These processes make grains easier to cook, eat, and store. However, refining also removes almost all of the fiber, oil, B vitamins, and much of the protein of the grains, requiring many cereals to be artificially fortified with nutrients. Refined grains may be eaten whole, such as rice, or made into flour and used to make a plethora of baked goods and breakfast cereals. In recent times, many people have returned to eating unrefined grains as their health benefits become better understood.

ABOVE: *This fresco from an Ancient Egyptian tomb depicts men carrying a large basket of newly harvested ears of wheat. The grain harvest was a crucially important time in the Egyptian calendar.*

LEFT: *An assortment of cereals and pseudocereals. From left: oat ears and rolled oats, barley grains and barley flour, quinoa grains and quinoa flour, corn kernels and corn flour, rye grains and rye flour, wheat grains and wheat flour, buckwheat grains, and millet grains.*

Amaranth

Amaranthus caudatus

Amaranth is a nutritionally rich grain, cultivated since antiquity in the Americas and more recently throughout Asia and Africa. Once a staple crop of the Aztecs, it has also been an important supplementary crop. While the story of its historical diffusion is unclear, its potential to relieve contemporary hunger problems is apparent.

Historical Origins Wild amaranth seeds have long been gathered by Native American peoples, becoming a rich source of protein and vitamins to ancient hunting and gathering populations. Archaeologists date its domestication to 4000 BCE in Tehuacán, Puebla, Mexico, and by 2000 BCE it was well integrated into the Mexican diet. The Andean species was cultivated as far south as Pampa Grande, Argentina, while a third species has been found in present-day Arizona. Amaranth became a major crop of the Aztecs, who used it both for imperial tribute and as a significant dietary supplement to the less drought-resistant staple, maize. Its widespread cultivation throughout Asia is believed to predate 1492, but records are limited on the history of its diffusion.

ABOVE: *The bushy seed head and lush foliage of the purple amaranth make it a popular ornamental plant for the home garden.*

Botanical Facts Cultivars of amaranth are grown as green, grain, and ornamental. The plant grows from 1–10 ft (30 cm–3 m) tall, with broad leaves and large spiky seed heads that each bear thousands of tiny seeds. It is brightly colored in reds, yellows, purples, and oranges. Seeds are rich in protein and contain the essential amino acid lysine and more dietary fiber than any other grain, making it an excellent nutritional source. Amaranth can tolerate bright sunlight, heat, and adverse soil conditions. It has therefore long been an important supplementary crop in semiarid regions, though some cultivars also thrive in wet tropical conditions.

Culinary Fare Traditionally, the Aztecs consumed amaranth in many forms: as a toasted grain, a green vegetable, a popped grain (called *huauhtli*), and even as a tea. They also used it in sacred rituals, where the dough was baked into cakes shaped like deities and animals, a practice that has been assimilated to Christianity in parts of Mexico. Today Mexicans create *alegría,* a confection of popped seeds sweetened with honey or molasses, while a similar treat sweetened with sugar syrup is eaten in Nepal and China. In India and China the grain is made into cakes, while throughout Southeast Asia the leaves are commonly stir-fried as other greens. In the United States it is milled into flour and used to create health-conscious breads and breakfast cereals. Amaranth flour can be used to thicken soups, stews, and gravies.

BELOW: *Tiny yet nourishing, amaranth seeds are light brown or tan in color and are about the size of poppy seeds.*

Oats

Avena sativa

LEFT: *Oats are among the most nutritious of cereals, containing significant levels of Vitamin E, fiber, and fat.*

As the old song says, "mares eat oats and does eat oats," but this grain also features significantly in the human diet. In Wales and Scotland, oats have historically been the principal cereal crop. Highly valued for their nutritional content, they make the best porridge of all grains.

Historical Origins Wild oats appear as weeds among cultivated wheat and barley crops in Neolithic archaeological sites. Oat domestication did not occur until grain cultivation spread to Europe, where cooler, wetter, less sunny northern climates suited its growth. Definitive evidence of its cultivation in central Europe dates to the second millennium BCE, and the Chinese have been cultivating the grain since the first millennium CE. The Roman historian Pliny the Elder (CE 23–79) wrote that oats were grown in Europe for both human and animal consumption. The plant traveled to the Americas with both the Spanish and British, where it flourished in north–central North America.

Botanical Facts Oats were cultivated later than other grains because the wild plant did not easily lend itself to domestication. The wild grass contains single small grains atop small seed heads, which drop off on ripening, in the manner of a weed. Cultivated strains keep their seeds longer, enabling harvesting. It is a hardy plant, preferring moist, cool to temperate climates, but tolerating unfavorable growing conditions quite well.

Culinary Fare Although the majority of oats produced worldwide are still fed to animals, an increase in human consumption in North America from the late nineteenth century onward can be credited to Ferdinand Schumacher, who developed quick-cooking rolled oats, and Henry Crowell, who packaged the product under the retail brand "Quaker Oats." Today, oats are eaten primarily in the United States, United Kingdom, Australia, and New Zealand mostly in the form of oatmeal (porridge), commercial breakfast cereals, granolas, mueslis, and oatcakes. Rolled oats can also be used to make oatcakes and cookies. Oat flour inhibits rancidity, and is often added to products such as milk, bread, and vegetable oils to prolong their shelf life.

A Stomach Full of Oatmeal

While oats feature in many dishes aside from the standard oatmeal, perhaps their most infamous presence is in Scotland's national dish, haggis. The dish was made famous by Robert Burns in his poem "Address to a Haggis," which is commonly recited during celebrations with haggis on the menu. Haggis consists of a large sheep stomach stuffed with minced offal, oatmeal, fat, stock, and spices. In the medieval and early modern periods, it was eaten throughout Britain; today it is a dish that specifically celebrates Scottish culture.

Quinoa

Chenopodium quinoa

BELOW: *Quinoa is considered a complete protein because it contains all 8 of the essential amino acids required by the human body to aid digestion.*

Once the principal grain crop of the Andes region, quinoa (pronounced "keen-wa" or "kin-wah") has long played a significant ceremonial role and, today, remains a staple food of the people living in the area bounded by southern Colombia, northwestern Argentina, and northern Chile.

Historical Origins Quinoa was first domesticated around 5000 BCE, near Lake Titicaca in the Andes, where it is a native plant. It was the second most important staple crop of the Incas after the potato; they called it "the mother grain."

Botanical Facts Because it is not a member of the grass (Poaceae) family, quinoa is not a true cereal, though its grains are used in a similar fashion. It grows as a plant that resembles spinach, containing numerous clusters of small white, pink, black, yellow, or orange seeds on erect stems. The dehulled seeds measure ½–1¼ in (1–3 cm) long when dry, but when cooked they expand up to 4 times in volume and take on a pearly translucent appearance.

Culinary Fare The slightly nutty flavor and delicate texture of cooked quinoa affords it great versatility; it can be added to soups or stews, served like couscous, rice or millet, or ground into flour. Quinoa has a high content of unsaturated fats and is lower in carbohydrates than other grains. It has recently been recognized and adopted beyond its traditional range of cultivation.

Finger Millet

Eleusine coracana

Millet is the common term for cereals that do not belong to the wheat, barley, oats, maize, or rice genera. Finger millet gets its name from the fingerlike seed heads of the plant.

Historical Origins Native to Africa, finger millet was introduced to India during the first millennium BCE. Its wide geographical distribution and ancient history belie controversy over its first domestication, and archaeological evidence suggests that it may be the oldest domesticated tropical African cereal. It is currently cultivated in Africa, primarily in Uganda, Ethiopia, Malawi, Zambia, and Zimbabwe. In India, it is grown in the states of Andhra Pradesh and Tamil Nadu.

Botanical Facts Contemporary cultivated finger millet is divided into 5 races, with the major differences found in the number, size, and shape of its flowering branches. It is an annual plant that grows to 16–40 in (40–100 cm) in height in semiarid, tropical, and subtropical regions. Once harvested, the seeds can be stored for a very long time without insect damage, making it a popular crop in drought-prone areas.

Culinary Fare Known as *ragi* in India, finger millet is ground into flour and made into leavened *dosa* and flat *roti* breads. It is also cooked on the griddle as a flat bread known as *batloo*, or mixed with wheat and chickpea flour to prepare *dhebra*. In Africa, it is made into a gritty porridge, to which greens and other foods are added if available, and it is favored for brewing beer. Millet can also be added to soups and stews.

Buckwheat

Fagopyrum cymosum Perennial Buckwheat *Fagopyrum esculentum* Buckwheat
Fagopyrum tartaricum Tartary Buckwheat

Buckwheat is not a true cereal, as the plant does not belong to the grass (Poaceae) family, but rather to the knotweed (Polygonaceae) family along with sorrel and rhubarb. The grain, eaten widely across Asia, Europe, and North America, is actually a dry fruit. The plant's hardiness has made it a popular crop, while the fruit's high nutritional value and nutty flavor have earned it a place in many traditional dishes, modern health foods, and medicinal treatments.

Historical Origins Native to Manchuria, China, and Siberia, buckwheat was likely cultivated in Japan as early as 5000 BCE, and certainly cultivated in China by 1000 BCE. During the fourteenth and fifteenth centuries it traveled from the Far East to Europe via Russia and Turkey, and was later introduced to the New World by early Dutch settlers. It proved a popular crop in Europe, growing hardily where other grains would not. Its nutty flavor enhanced porridge and pancakes. Production peaked in the early nineteenth century, and today, most of the world's crop is grown in former Soviet Union countries.

Botanical Facts
The various cultivars of common buckwheat (*Fagopyrum esculentum*), which accounts for over 90 percent of the world's production, can be divided into 2 groups. One is characterized by tall plants, late maturation, and photoperiod sensitivity (only flowering when the day length is short), and tends to grow in eastern and southern Asia. The other features small plants, early maturation, and is insensitive to photoperiod (flowers all year round), and is found in northern China and Europe. In general, the buckwheat plant grows up to 5 ft (1.5 m) tall, has broad arrow-shaped leaves, a small root system, and densely clustered flowers that depend upon cross-pollination. The triangular kernel measures ¼ in (6 mm) across and is comparatively rich in minerals, starch, protein, fatty acids, flavonoids, and the amino acids lysine, arginine, and tryptophan.

ABOVE: *Buckwheat kernels do not contain gluten and are used in breads and cereal products for people who cannot tolerate wheat.*

Culinary Fare In Europe, buckwheat is most commonly milled into groats and used to make oatmeal, added to hamburger, eaten fresh with sour milk, or used as a component in dumpling fillings. It is also milled into flour and used to make breads, biscuits, confectioneries, and breakfast cereals. In North America it is mainly found in premixed products: as flour in buckwheat pancake mixes; or as groats in the Jewish dish *kasha*, and in soup mixes. Buckwheat pancakes are known as *blinis* in Russia and *galettes* in France. In Asia the flour is used to make noodles.

LEFT: *Buckwheat grows in Japan. The grain is milled into a flour and mixed with wheat flour and water to make soba noodles, a staple of Japanese cuisine.*

Barley

Hordeum vulgare

Barley is the oldest cultivated cereal in the Near East and Europe, and for many centuries was the most important food grain in the ancient world. Today, it is primarily grown for animal feed and to make beer and whiskey, but it is still used in many dishes throughout the world.

Historical Origins Wild barley has long thrived throughout the Near East, and its remains are found at nearly all archaeological sites across the region, even those predating cultivation. Gradually, throughout the Neolithic period, the ratio of domesticated barley to its wild counterpart began to increase, and about 9,000 years ago its cultivation spread throughout the Mediterranean coast, the Balkans, temperate Europe, and central and Southeast Asia. Though its domestication followed that of wheat, its geographical range is more extensive and it can tolerate more diverse ecological conditions. Early domesticated barley was used as human food, for brewing beer, and as animal feed, but it is unclear which of these was its primary function. Barley was also among the "7 species" of crops sacred to the Ancient Israelites, and also featured prominently in Ancient Greek and Egyptian rituals.

Botanical Facts Native to the Mediterranean and central Asia, barley is an annual grass with long, narrow, downy leaves. The flowers are grayish green tinged with purple, and grow in dense narrow spikes. The chief cultivar of the ancient world was "6-rowed," which seemed to give a better yield, though today the "2-rowed" type is preferred. The domesticated plant has also been engineered so that its indigestible husk falls away easily, while the wild plant has seed leaves that enclose and protect the grain.

Culinary Fare Because barley could grow easily in a variety of climatic and geological conditions, it was once a staple food in many communities. However, it contains much less gluten than wheat, and therefore barley bread is much denser, coarser, and darker. Thus, as wheat cultivation improved, barley fell out of favor. Today we may taste pearl barley in soups or find barley flour in health food stores, but our most common contact with the grain is in fermented form. After water, barley is the key ingredient in beer. Barley grains are germinated to create fermentable sugar, then either dried or roasted to create malts of varying darkness, and fermented in water.

RIGHT: *This crop of barley is most likely to be used for brewing beer. Barley has the shortest growing season of any grain, maturing in about 90–100 days.*

Rice

Oryza sativa

Rice is the staple food for about half of the world's population. In many countries, particularly those of Southeast Asia, people often obtain up to three-quarters of their total caloric intake from this grain. Even in areas of the world where rice is not the principal food, rice dishes are firmly embedded in the culinary canon.

Historical Origins Native to tropical India, northern Indochina, and southern China, rice was most likely domesticated independently in several places. The short-grain cultivar was grown in the Yangtze River valley in China around 7000 BCE, with the long-grain cultivar following later in Southeast Asia. It traveled west to Persia, where it was adopted by the Arabs, then it made its way to Europe. It was widely cultivated in Moorish Spain in the eighth century, and reached northern Italy and France via Sicily and Spain by the fifteenth century. The Spanish and Portuguese brought rice to the New World in the seventeenth century.

Botanical Facts The more than 100,000 distinct cultivars of rice worldwide are divided into 2 subspecies. *Indica* rices, characterized by long firm grains and a large amount of amylose starch, are usually grown in low-altitude tropical and subtropical areas, and include the long-grain rices popular in China, India, and the United States. The shorter stickier *Japonica* rices are much lower in amylose starch and grow in upland tropical and temperate climates. Medium-grain Mediterranean rices are of this subspecies, as are the clumpy short-grain rices popular in northern China, Japan, and Korea. Most rice is milled to remove the outer bran and germ, resulting in polished white grains that can be stored for months, but what is commonly called "brown rice" is simply any variety of rice that has not been milled.

Culinary Fare Rice is generally cooked in warm liquid; water softens the grains, and heat thickens the starch granules. Where rice is the staple of the diet, it is usually cooked in plain water to produce intact, tender, glossy, white grains. In areas where rice is more of a luxury, it often features in elaborate preparations with many rich ingredients, such as Indian *pilau*, Iranian *polo*, Italian risotto, or Spanish paella. Rice can also be made into a gluten-free flour, ground into a powder (which is eaten as a condiment in Vietnam and Thailand), cooked into noodles, and made into pastries or pudding.

ABOVE: *Some of the most popular rices eaten today are, clockwise (from left), basmati (raw and cooked), long grain, brown, wild, red camargue, and arborio.*

Sushi Rice: Flavor Follows Function

Japanese sushi rice is seasoned with rice vinegar and sugar, producing a mildly sweet and tangy taste. However, this treatment turns out to have a purpose beyond that of flavor. Raw rice usually carries dormant spores of the bacterium *Bacillus cereus*, which sometimes survives cooking. Left for a few hours at room temperature, the spores germinate and the rice becomes riddled with toxins. Since sushi is served at room temperature, its acidic dressing has antimicrobial properties, making it safe to eat.

Proso Millet

Panicum miliaceum

Proso millet is widely grown throughout the world, and is also known as common millet, hog millet, Indian millet, or broom corn (since the seed head spreads in an untidy fashion).

Historical Origins Native to central China, proso millet has been grown for over 5,000 years. Various cultivars can be found throughout Asia, the Middle East, and far eastern Europe, including Bangladesh, Pakistan, Afghanistan, Turkey, Hungary, Russia, Japan, Iran, and Iraq. It was widely cultivated as a cereal throughout southern Europe for about 3,000 years but has now been almost entirely replaced by wheat.

ABOVE: *Proso millet is the most widely traded millet on the world market.*

Botanical Facts Requiring the least water of any of the major cereals, proso millet is well suited to dry areas. It is an annual grass that grows to about 36–40 in (90–100 cm) in height, with upright clumps of bright green arching leaves. It produces clustered heads of small seeds colored white, cream, yellow, orange, or brown.

Culinary Fare In China, proso millet was once among the sacred 5 grains, and though held in lower esteem than rice, it was still widely consumed in the north, where rice was not cultivated. It is most commonly prepared as a porridge in this region. Because it contains no gluten, it is often sold in health food stores for people with gluten allergies.

Pearl Millet

Pennisetum glaucum

ABOVE: *The stiff compact panicles, or earheads, of pearl millet are a distinct feature of the plant.*

Also known as bulrush millet, pearl millet is the second most widely grown native African cereal. It is the main grain crop in the Sahelo-Sudanian zone of West Africa and is a staple of the diet in this area. It is also grown in Zambia, Zimbabwe, Namibia, Angola, Pakistan, and northwestern India.

Historical Origins Native to the arid and semiarid tropical regions of Africa and Asia, pearl millet was most likely domesticated along the southern edges of the Sahara and dates back 3,000 years. The cultivated cereal reached India about 2,500 years ago.

Botanical Facts Pearl millet is an important crop because it consistently produces a harvest, even with low rainfall and high soil temperatures. A member of the grass (Poaceae) family, the adaptable plant grows quickly (up to 3–6 ft/0.9–1.8 m) with adequate moisture and heat, and can survive short periods of drought by becoming dormant in order to resume growth when water is again available. There are numerous wild and cultivated variations of pearl millet, which differ in panicle length, seed size, seed color, and plant height.

Culinary Fare Pearl millet can be baked into both leavened and flat breads, boiled or steamed as a porridge, used as an ingredient in couscous, and brewed into alcoholic beverages. In Namibia, where pearl millet is commonly known as *mahangu*, it is ground and made into a stiff oatmeal called *oshifima*, or fermented to make beverages called *ontaku* or *oshikundu*.

Rye

Secale cereale

Once the predominant bread grain for much of the population of northern Europe, rye has in many places been replaced by wheat today. However, the taste for rye persists in eastern Europe, Scandinavia, and North America, and distinctively flavored and colored rye breads and crackers are still consumed among these populations.

Historical Origins Wild rye is indigenous to the cool dry mountainous regions of Turkey, Iran, and Caucasia (a region including southern Russia, Georgia, and Azerbaijan). The species seems to have developed as a weed in wheat and barley fields in the Near East, often turning out a larger yield than the intentionally cultivated grains during unfavorable growing seasons and later earning it the nickname "wheat of Allah." Rye cultivation did not gain importance in Europe until the Roman age, when population increases necessitated the agricultural use of marginal fields and winter crops for which rye was well suited. Rye became a primary crop in northern and central Europe during the Middle Ages, and later traveled to the New World where it thrived in North America.

Botanical Facts Rye grows well in cool and mountainous climates, and is able to grow in less fertile soils than wheat. The plant is an annual that grows up to 6 ft (1.8 m). Its seed head contains 2 rows of large husks with stiff bristles surrounding small, green-gray seeds.

Culinary Fare Rye flour is most commonly made into bread, particularly in northeastern Europe, Scandinavia, and North America. Jewish immigrants popularized its use for deli sandwiches in the United States and Canada. Rye can also be made into flakes that look similar to rolled oats. It is fermented to make alcoholic drinks, such as American rye whiskey.

ABOVE: *Ripe rye ears are often milled into a flour to make rye bread, which is dark brown in color with a rich and hearty flavor.*

Foxtail Millet

Setaria italica

This most important of millets in east Asia, foxtail millet is a primary food crop in the dry areas of northern China.

Historical Origins This grain was first cultivated in China in the sixth millennium BCE, along with common millet, and became the predominant grain of the ancient Yangshao culture. Archaeological evidence for its presence in central Europe dates back to the second millennium BCE, and about 600 BCE in the Near East.

Botanical Facts The plant grows as a grass, reaching 4–7 ft (1.2–2 m) in height. The seed head of the plant is its namesake, as it resembles a fox's tail in shape and color. The small seeds are encased in a papery hull, and vary greatly in color.

Culinary Fare Foxtail millet is usually cooked like rice. In southern India, boiled foxtail millet is called *korrannam*, and is eaten with curries or pickles, much like rice, or it can be cooked with spices and vegetables in a pressure cooker. Foxtail millet is available in whole grain or flour form in many Western health food stores, and can be added to stews, salads, and a variety of baked goods.

Sorghum, Great Millet

Sorghum bicolor

ABOVE: *This vast sorghum field is in Kansas. It is the third most important grain grown in the United States, mainly for export.*

Sorghum is the most significant of the native African cereals. It is a staple in the diet of people living in the drier parts of northern Africa and India, and is grown elsewhere in the world as animal feed.

Historical Origins Species of sorghum are native to tropical areas of all continents, including Australia. The early relatives of contemporary commercial sorghum are native to areas south of the Sahara extending into the Guinean zone, and its cultivation dates to between 4000 and 3000 BCE in Ethiopia. From there it was disseminated to western Africa and throughout the Middle East, India, and Asia. Sorghum was brought to the New World in the seventeenth century by African slaves. Today, it is an important food crop in Africa, Southeast Asia, and Central America. Its ability to resist extreme drought and heat makes it particularly suited to arid regions.

Botanical Facts Belonging to the grass (Poaceae) family, sorghum has a robust, usually upright stem and features narrow straplike leaves and feathery seed heads. The color and size of the plant vary. It may reach as tall as 20 ft (6 m), but it is also grown in dwarf cultivars for ease of machine harvesting. The white-seeded cultivars are most commonly consumed as food and those with red seeds are often brewed into beer.

Culinary Fare When eaten as a grain, the flavor of sorghum is similar to buckwheat. Its flour lacks gluten, so it is most commonly prepared as a porridge or used to make couscous. Indians use sorghum meal to make unleavened breads such as *chapati*, *bhakri*, and *roti*. In South Africa, a porridge prepared with sorghum meal is known as *mabele* and is eaten with soured milk or boiled greens. A particular group of cultivars in the *S. saccharatum* species are grown in order to use the sap in their thick stems to make syrup. This sweet syrup is made up of about one-third sucrose, one-third dextrose, and one-third fructose, so it is not suited to making solid sugar. In China, the syrup is partly evaporated to produce dried brown strips, while in the southern United States sorghum syrup is produced as a popular and cheap alternative to maple syrup. Sorghum is also used to produce alcoholic beverages; in China it is distilled into *maotai* and *kaoliang*, while in Africa it is brewed into beer.

ABOVE: *Brewing beer from sorghum grains. The beverage is popular in South Africa, particularly in Zulu communities.*

Triticale

x *Triticosecale* hybrids

Triticale is a cross between wheat and rye, and its name is a combination of the botanical names for the 2 plants (*triticum* and *secale*). As it is easier to grow and more nutritious than wheat, it holds great potential to address particular problems within the contemporary cereal industry, and research is still being conducted to further enhance its position as a commercial crop.

ABOVE: *The triticale cultivar 'Salvo' has the large seed heads typical of most cultivars of this wheat-rye hybrid.*

Historical Origins Triticale is a modern cereal grain, first developed by the Scottish botanist Alexander Stephen Wilson in 1876. Like natural hybrids between wheat and rye, Wilson's seedlings were sterile, and it was not until the 1930s that advances in the field of genetics allowed for chemical doubling of chromosome numbers to create a fertile version. By the mid-1950s, the plant was being intensively bred by scientists at the University of Manitoba in Canada, and in 1970 the first commercial cultivar was sold. Today, the primary producers of triticale are Germany, France, Poland, Australia, China, and Belarus. While it is well established as a feed crop, research is currently being conducted to advance its potential for human consumption and to investigate its use as an energy crop for producing bioethanol.

Botanical Facts Triticale is a commercially attractive cereal crop, as it combines wheat's excellent bread-making properties with the hardy growing ability of rye. It can resemble either of its parents (depending upon the particular cultivar), but as a rule it appears as a tall grass with large seed heads. Today, most commercially available triticale is a second-generation hybrid—a cross between 2 types of triticale. It is nutritionally superior to wheat, with a higher content of the amino acid lysine, but it is slightly weaker in gluten content.

Culinary Fare Though triticale is well established and economically significant as an animal feed crop, its place in the human diet is still being negotiated. Grains and products made with it, such as flours and breakfast cereals, can be found at health food stores. It has also been recognized as having potential to improve production and nutrition in developing countries, because it features the high quality and yield of wheat along with the environmental and disease tolerance of rye. Having a nutty and naturally sweet flavor, triticale can be used in various products such as pastas and breakfast cereals, and contains more protein than wheat. However, it requires different milling techniques than wheat, necessitating industrial adaptation in order for it to become more widespread.

Wheat

Triticum aestivum Bread Wheat, Common Wheat · *Triticum dicoccon* Emmer Wheat
Triticum durum Durum Wheat · *Triticum monococcum* Einkorn Wheat
Triticum sativum Winter Wheat · *Triticum spelta* Spelt

ABOVE: *Ripe bread wheat has bearded seed heads of oval kernels that are usually yellowish brown in color.*

RIGHT: *Einkorn wheat, domesticated 9,000 years ago, is one of the earliest forms of cultivated wheat.*

With a production second only to maize among the world's cereal crops, wheat is one of the most important human foods. There are over 30,000 known cultivars, the majority of which are used to make flours for breads, pastries, and pastas.

Historical Origins Wheat originated in the Fertile Crescent (a region in the Middle East including the Levant, Ancient Mesopotamia, and Ancient Egypt), and was among the first food plants to be cultivated by humans in the early days of agriculture. It was first domesticated near Diyarbakir in Turkey, through a process by which the planting and harvesting of wild grains led to favorable mutations, which were then selected for their durable ears, larger grains, and spikelets that remained on the stalk until harvesting. During the Neolithic period, cultivated wheat spread throughout the Mediterranean, becoming the most important cereal among these ancient civilizations. By 5,000 years ago it had spread to Ethiopia, India, Britain, Ireland, and Spain, and shortly after it reached China. As technology for grain cultivation improved, such as the use of horse-drawn plows, more and more wheat was grown. Between the Middle Ages and the nineteenth century, wheat was largely replaced as a staple in Europe by less versatile, but hardier, grains and potatoes. However, with the advent of industrialization and simultaneous advances in agricultural production, wheat yields increased and have continued to do so up to the present day. Wheat traveled to North America in the early seventeenth century and reached the Great Plains by the mid-nineteenth century, and today the United States is the world's third largest producer of this grain (behind China and India).

Botanical Facts Wheat is a member of the grass (Poaceae) family. It is a tall plant with spiky seed heads containing small, light-colored grains surrounded by a reddish brown bran layer. Historically, several different wheat cultivars have been grown in various parts of the world; today, they can be classified into

3 categories based upon the cultivars' genetic makeup. Einkorn wheat is the simplest and was the first to be cultivated; it is a diploid species, meaning it contains 2 sets of chromosomes. The second group contains 4 sets of chromosomes and evolved from a mating of wild wheat with wild goatgrass. Included in this tetraploid species are emmer and durum wheats, the 2 most significant varieties used by ancient Mediterranean civilizations; durum wheat is still widely cultivated today. When a tetraploid species mated with a goatgrass about 8,000 years ago, the 6-chromosomed hexaploid type arose. Contemporary bread wheat and spelt belong to this last category, and the extra chromosomes are thought to contribute to the elasticity of gluten proteins, as well as the agricultural diversity and hardiness of modern wheat.

Culinary Fare The primary reason that wheat has become the most prized grain in the Western world is its high gluten content. When wheat flour is mixed with water, gluten proteins bond, forming a stretchy mass that expands with heat and accommodates gas bubbles produced by yeast, giving us raised breads, cakes, and other baked goods. This preparation is so significant that one species of wheat is now known by its purpose: bread wheat. Perhaps the second most common use for wheat is as the key ingredient in pasta, for which durum wheat ground into semolina flour is used. Another way to eat wheat is to cook its whole grains, which the Italians do with faro wheat in a dish similar to risotto.

Wheat can be milled into flour with or without the outer bran; removing the bran creates white flour while including the bran produces whole-wheat flour. The germ itself is often sold separately and is added to many health foods because it is a good source of protein, oil, and fiber. Another preparation of wheat is known as bulgur or burghul, an ancient product still made in North Africa and the Middle East. Bulgur consists of whole-wheat grains cooked in water, then dried and milled to remove the bran and germ but leaving the chunky endosperm. This highly nutritious remainder is then boiled or steamed and eaten like rice or couscous (the latter is a wheat product as well, actually a very tiny pasta made of semolina flour). Throughout the Arab world green wheat grains are prized for their unusual sweet flavor. Called *frikka*, it consists of young grains that are charred and thrashed and eaten fresh or dried.

ABOVE: *Whole-wheat flour (top) contains more protein and fiber than white flour (bottom), but it also has a shorter shelf life.*

Wheat Meat

Sometime around the sixth century, noodle makers in China discovered that gluten proteins could be separated from the rest of the flour by kneading a dough in water until the starch washed away and only the chewy gluten remained. By the eleventh century this product was known in Chinese as *mien chin*, the "muscle of the flour" (*seitan* in Japanese). This plant-based protein became an important ingredient in Buddhist monasteries, as monks were required to follow a strict vegetarian diet. Recipes have been found dating to this period for imitation venison, jerky, and other meat products. This tradition continues today in the East, and has also been adopted among vegetarians in Western countries to make a wide variety of high-protein meat substitutes.

A chef strains pieces of seitan *before preparing them to eat.*

Maize, Sweet Corn

Zea mays

Maize is a cereal commonly known in the United States, Australia, and Canada as "corn," while in Britain this latter name is a general term for any staple grain. Native to the Americas, it is the largest crop grown on these 2 continents, and is widely cultivated throughout the rest of the world.

Historical Origins The maize species originated in Central and South America, evolving from a wild ancestor that disappeared in prehistoric times. Wild maize with very small cobs seems to have been sown deliberately by people living in Tehuacán, Puebla, Mexico, in about 5500 BCE, and similar evidence has been found at Bat Cave in New Mexico, which dates to 4500 BCE. Other remains have been found in Peru dating to 1000 BCE, by which point the South American continent was inhabited by a great number of mobile people. Maize therefore seems to have been selected and disseminated by human intention. It has long been accepted that maize was distributed to the rest of the world by Christopher Columbus, but controversy has arisen recently as etymological and archeaological evidence suggests that the plant may have been under cultivation in Asia, Africa, and Europe at a much earlier date. In any case, by the sixteenth century maize had gained staple status beyond its native soil.

ABOVE: *Aztec farmers harvest maize in this illustration from the Florentine Codex, c. 1570.*

Botanical Facts Maize is a type of grass similar to millet and sorghum, but its seed heads are distinguishably larger than those of its relatives. These seed heads are what we know as cobs or ears, and can grow up to 24 in (60 cm) in length. Surrounding each cob are even-numbered rows of grains or kernels, as many as 1,200 per ear. All modern cultivars are covered by a husk of long leaves, which protect the grains and make the domesticated plant convenient to grow and harvest. This husk, however, prevents the plant from distributing its seed, thus all contemporary maize is dependent upon human cultivation. The grains themselves are coated with a hard or soft material colored white, yellow, orange, red, brown, purple, blue, or black.

Most maize consumed by humans today as "corn on the cob" is of the yellow or white sweet corn cultivar. Three other cultivars are widely grown as well: flint corn, which has small grains with hard skins and is favored by Native Americans; flour corn, which features

Unlocking Nutrition

When maize was disseminated to the Old World, it was embraced enthusiastically for its growing properties and taste. However, many of the farmers who adopted the grain as a mainstay of their diet began to suffer from a disease of niacin and protein deficiency known as pellagra. At first this malnutrition seemed mysterious, as it did not affect Native Americans who lived on maize. Eventually it was discovered that native maize eaters employed a process known as nixtamalization, in which the corn is treated with an alkali substance (such as lime or ash) in order to chemically release the niacin and amino acids contained within corn, thereby making its proteins more biologically available. Furthermore, Native Americans had learned to eat relatively low-protein maize in conjunction with protein-rich beans to ensure more complete nutrition. Pellagra disappeared once alkali was applied and a more balanced diet encouraged.

big starchy grains that are often blue in color and commonly used to make cornmeal in South America; and dent corn, which grows across the "corn belt" in the United States and is used primarily for animal feed.

Culinary Fare Aside from the sweet-corn cultivar mainly eaten as a vegetable, there are 2 major categories of maize preparations: those based on whole grains, and those based on ground grains. Among the whole grain group, popcorn is the most popular preparation, and is deeply embedded in American snacking culture. Whole grains are also soaked or cooked in a lime or lye solution and made into dense chewy hominy, or further fried into corn nuts. Ground corn can either be dry or wet milled. Dry-milled corn is categorized by texture: grits are the coarsest and are cooked into a popular breakfast oatmeal in the American south; cornmeal is of medium graininess and cooked into mush, polenta, johnnycakes, corn breads, and other baked goods; and corn flour is finely ground and used exclusively for baking. Wet-milled corn is first cooked in an alkali solution, then steeped, cooled, and stone ground into masa dough (which can be flash-dried into *masa harina* for storage purposes). Masa is baked into thin tortillas, fried into corn chips, or steamed inside corn husks into tamales. Outside of the Americas, maize has also been well incorporated into many of the cuisines to which it was more recently introduced; for example, the cornmeal dish polenta, common in Italian cooking, and the Ghanaian dish *kenkey*, which is made of fermented balls of maize steamed in its own husk (or banana leaves) and served with fish or meat.

ABOVE: *Cobs of yellow sweet corn are cooked and eaten as "corn on the cob," or the kernels are canned or frozen.*

BELOW: *Male flowers of maize plants are at the top of the stems, while females are in the leaf axils. The cobs emerge from the fertilized female flowers.*

Nuts

Introduction to Nuts

Invaluable foods since the dawn of time, nuts and nut oils are superb ingredients for the cook. They are valued not only for their nutritional excellence, but for the glamor and richness they add to both sweet and savory dishes. Nuts are a globally important food crop, with many now grown far from where they originated. A tenth of world production is from the United States and includes almonds, pecans, and pistachios. Exports from China and Turkey are growing, almonds still come from Spain and Italy, and Australia is predicted to be the second biggest nut producer later in the twenty-first century.

> *"No man in the world has more courage than the man who can stop after eating one peanut."*
> Channing Pollack (1880–1946)

BRAIN FOOD

Nuts first appeared at the end of the Cretaceous period when the dinosaurs were in decline. At that time, almonds, hazelnuts, and walnuts, all rich in oil and omega-6 essential fatty acids (EFAs), were among the emerging flowering and seed-bearing plants. A diet that combined the omega-6 EFAs of nuts with the omega-3 group of EFAs in seafood, in the vital ratio of 1:1, enabled new placental animals to develop. The ideal environment for the new species of mammals to maintain brain size while growing larger was where land and water were in close proximity. The nut-bearing plants in this area made it replete with foods that were abundant in the fatty acids which were so important to neural development.

BELOW: *In the Middle Ages, the nutritious qualities and health benefits of nuts were widely known. Nuts were sold in the streets and by apothecaries. This illustration from a fifteenth-century manuscript shows a seller of medicinal nuts carefully weighing and preparing pieces of coconut for sale.*

DIGGING UP THE PAST

Archaeology and ancient writings reveal the history of the best-known European nuts. Food remains excavated in the Iranian sites of Deh Luran at Ali Kosh (7500–5600 BCE) and Tepe Sabz (5500–3700 BCE), when analyzed using modern nutritional tables, showed that almonds and pistachios provided more calories than any other food. The calories were double that of the lentils and wild legumes that were second on the list. Almonds and pistachios also provided more protein, weight for weight, than goat, sheep, or pig meat. In addition, only mussels, turtle, and freshwater crabs contained comparable amounts of calcium.

Almonds were found in the Neolithic level at Knossos in Crete, and 30 almonds were found in the tomb of Tutankhamen in Egypt. As they are not native to Egypt, these must have been imported from Turkistan, a region in central Asia, as they were to China, where they have been cultivated since late in the Tang dynasty (CE 618–907). Both almonds and pistachios are also mentioned in the Bible.

Other nuts that have an ancient lineage include chestnuts and walnuts, which appear frequently in Classical Greek and Roman gastronomic texts. Centuries later in France during the ninth century, Charlemagne listed chestnuts as trees to be planted in imperial gardens. In 1765, when excavations began in Pompeii, Italy, the carbonized remains of a meal of eggs, fruit, and walnuts that had lain undisturbed since August 24, CE 79 were discovered at the site.

NEW OLD NUTS

Spanish explorers called the Brazil nuts they first saw in South America "almonds of the Andes." When Brazil nuts first arrived in England in 1836, they were marketed as the "king of nuts," and imported in great quantity. By the 1860s, they were equally popular in North America.

Cashews and peanuts originated in the Americas, and macadamias in Australia. By the end of the sixteenth century, the Portuguese had planted cashews in East Africa and Goa where they thrived and eventually spread to the Malabar Coast and southwestern India. Macadamias were first successfully cultivated commercially in Hawaii, where they were introduced in the 1880s, only later to be cultivated in their homeland.

Peanuts were grown in pre-Inca Chimu in Peru, long before Cortez and Columbus saw them in Mexico and Haiti. As they had with the cashew, the Portuguese took them to their colonies in Africa, India, and the Far East. Peanuts were sometimes the only food on slave ships because they kept well on the long voyage from Africa to ports in the southern United States and the Caribbean. These less than auspicious circumstances did nothing to diminish the peanut's almost mythic popularity in the United States, overtaking native hickories, pecans, and pine nuts.

MORE THAN ONE NAME

Nuts are the seed of a fruit with a thick, hard shell (pericarp). "True nuts," with a cap at the stem, are oak acorns, chestnuts, hickory, and hazel. Drupes are almonds, coconuts, pecans, and walnuts, as are peaches, plums, and cherries. Others are seeds, including Brazil nuts, cashews, and pine nuts. Peanuts, peas, and beans are legumes.

NUTS IN COOKERY

Nuts feature in the earliest known recipes from Ancient Greece and Rome, the Orient, and the Middle East. Almonds were hugely important in Europe during the Middle Ages, and were particularly useful when animal products were forbidden during church-regulated periods of fasting.

LEFT: *The peanut is also known as a ground nut because it grows from a flower under the ground. Peanuts are eaten raw straight from the pods or in a variety of foods, such as peanut butter.*

Nuts native to Southeast Asia and the Pacific, such as candlenut, sawari, and water chestnut, are authentic ingredients in the cuisines of these regions. These lesser-known nuts are now becoming more widely available as global transport, travel, and a dramatic shift in demographics has increased the demand for more exotic flavors.

Nuts are full of health-promoting nutrients. Their freshness is critical to their flavor. Once they are shelled, they are best kept cold to prevent them from becoming rancid.

BELOW: *Chestnuts are known as a "true nut." They can be peeled and eaten fresh, boiled or roasted, or ground into flour for use in baked goods.*

Candlenut

Aleurites moluccana

The candlenut is an evergreen forest tree that is of particular importance in its native regions from tropical Asia to islands in the western Pacific. Most parts of the tree are used, in particular the nut, which is so oily it can be burned like a candle.

Historical Origins The candlenut tree has a long and wide-ranging history. Known in Hawaii as *kukui*, it is the state tree—a symbol of enlightenment, protection, and peace. In ancient Hawaii, the nuts from this tree were burned to provide light. Today, Hawaiian *kukui* oil is sold as a treatment for skin ailments such as burns, chapped skin, and other minor problems. In traditional medicine, candlenut oil has been used for many years in the tropics as a purgative and hair restorative, and in Southeast Asia it is still used externally for headache, fever, and swollen joints. In addition, black dye is made from charred shells and soot from burned candlenuts or their roots. Today, ripe candlenuts are pounded into a paste for shampoo and soap in Tonga, while the bark, leaves, and oil are used medicinally in Japan, Malaysia, Java, and Sumatra. Raw kernels can be used to polish and waterproof wooden bowls. And in Hawaii, fishermen spit the oil from a chewed nut into the water, in order to clear it enough for fish to be seen.

Botanical Facts This tropical nut has a number of names: candleberry, Indian walnut, *kemiri*, *kukui*, Tahitian walnut, and varnish nut. The inedible fruit of the tall evergreen is round, and contains one or two waxy kernels. The thick, fleshy husk is off-white when young, and black when the nut is mature. The nut is edible, but most are toxic when raw, and some contain cyanide.

Culinary Fare The nuts are primarily grown commercially for their oil, but they do have culinary uses. The raw nut has a powerful laxative effect and is always roasted or fried before eating. In Indonesia and Malaysia, it is ground with other flavorings, such as galangal, garlic, and ginger, to flavor and thicken sauces. Brazil nuts or macadamias have been suggested as a substitute, but candlenuts are more astringent, with an agreeable bitterness lacking in the others. A cooked candlenut can be rubbed around a frying pan to prevent food from sticking.

In Hawaii, *kukui* nuts roasted in hot ashes are shelled and pounded with salt to make the condiment *inamona*. This is an essential ingredient of the sauce that is mixed with chunks of raw tuna for the appetizer called *poke*, the Hawaiian version of sashimi. The appetizer is a traditional feature of a *luau* feast, but it is also extremely popular to serve at home.

BELOW: *The fruit from the candlenut contains nuts that should not be eaten raw, but are delicious and nutritious when roasted. They are often ground into a paste and used in curries and Indonesian dishes.*

Cashew

Anacardium occidentale

The cashew apple is the fruit of a large, fast-growing evergreen shrub or tree, native to tropical America. India now produces half of the world's supply, with Brazil the second largest producer, followed by Indonesia, Mozambique, and Vietnam. Most cashews are grown for the nuts—more than half of which are sold roasted and salted. The cashew nuts we eat are not truly raw. Roasting removes their caustic oil and dries them out.

Historical Origins The name "cashew" comes from what Portuguese explorers heard as "caju" when Native Americans were really saying "acaju"—meaning to pucker the mouth. The acrid substance that lies between the 2 shells of the nuts is so potent that it blisters the skin and can even be used as a wart and corn remover.

The French naturalist André Thévet saw cashews in northern Brazil in 1558, growing on land that divided the territory claimed by Spain from that claimed by Portugal. He wrote that this land was "far too good to belong to the cannibals" as it had great numbers of the tree called "acajous." The fruit he described as hardly edible. But, he wrote, the nut that hung at the base, which was "as big as a chestnut and the shape of a kidney," was a different matter: "as to the kernel therein, it is excellent to eat when lightly cooked."

Botanical Facts This perennial tree is related to the mango and pistachio, as well as poison ivy, oak, and sumac. It is very hardy in extreme tropical heat and can live for 30 or 40 years. The rough bark contains resin and acrid sap. The branches are often twisted and crooked, and the lower ones resting on the ground take root, giving the cashew an impressive spread. Its fragrant flowers are a coppery pink color. The comma-shaped nuts grow outside the bright red or yellow fruit, dangling beneath the lower end of the cashew "apple," which actually more closely resembles a pear in shape.

Culinary Fare Cashews can be substituted in any recipe that calls for nuts. Cashew butter is delicious and is easily made at home. In Chinese cooking, cashews are often added to a stir-fry, and after quickly frying them in a little oil, they make an excellent addition to salad or vegetable dishes. Mixing cashews with wasabi paste and roasting them in a medium oven for 10–15 minutes, stirring occasionally, transforms them into an exotic snack. In Brazil, the juice of the cashew apple is drunk fresh, or fermented into *cajuada*, a Madeira-like wine. Also in Brazil, the pulp of the cashew apple is used for preserves, jellies, syrups, and a vinegar called anacard.

Cashews are a good source of iron, potassium, and zinc. They contain protein, and are high in fat but low in saturated fat. They have no cholesterol, and are slightly higher in carbohydrate than most other nuts, with the exception of pistachios.

ABOVE: *Cashew nuts have a delicate flavor that is enhanced by dry roasting. They are also popular ground into flour and used as a paste in curries and stir-fried dishes.*

LEFT: *The kidney-shaped cashew nut grows outside the cashew fruit, which is commonly called a "cashew apple." The fruit falls to the ground when it is ripe and the nut is gathered for processing. Although edible, the fruit is usually discarded.*

Peanut

Arachis hypogaea

The peanut (also known as groundnut) is the only nut that is one of the world's leading food crops. Only the soybean contains higher levels of protein, but peanuts have the advantage of being cheaper to produce. In India, China, Southeast Asia, and Africa, peanuts yield the highest protein per acre of any food. Unlike the chickpea, which has the third highest protein level, peanuts grow underground, which makes them safe from pests such as locusts.

ABOVE: *When peanuts are harvested, the whole plant, usually including the roots, is dug up from the soil. This is then left on the ground to dry for 2 or 3 weeks before the peanut pods are separated from the rest of the bush.*

Historical Origins Numerous species are native to South America—from the Atlantic coast to the Andes. Representations of peanuts on pre-Inca Chimu pottery in Peru and string bags found in Inca graves containing peanuts, maize, beans, chili peppers, and coca suggest that they may have been the first to cultivate the peanut. It was via the Portuguese that the peanut traveled from Brazil to Africa, the Far East, and India, which is now a major producer.

Slaves arrived in southern American ports from Africa on ships provisioned with peanuts. One of the many names for the peanut is "goober pea," which comes from the central African Bantu name *nguba*. Another tribal name is *pindar*, but Southerners also call them ground nuts, ground peas, and monkey nuts. Little known in the northern American states before the Civil War, peanut production more than doubled within five years of the Confederate Army's surrender at Appomattax in 1865.

In the United States, peanuts were sold as a snack at Barnum's traveling circus, a Victorian confection of sugar-coated peanuts was called Boston Baked Beans, and during the depression of the 1930s, the Planter's peanut company created a peanut bar called Nickel Lunch. These days, more US peanuts go into peanut butter than are roasted in the shell, salted and roasted, or made into candy.

Botanical Facts The peanut conforms to the dictionary definition of a nut as being a fruit that consists of a hard or leathery shell enclosing an edible kernel, even though it is not botanically a nut. It is botanically classed as a legume—a pulse like peas, beans, and lentils. The peanut grows on an annual bush, the pollinated flowers of which are pulled underground as the peanuts develop. The species name *hypogaea* means growing beneath the ground.

Culinary Fare Like all nuts when they are young and spanking fresh, peanuts are a seasonal delight. Before the peanuts are fully mature, they are washed and boiled in their hulls in salted water. The kernels are gelatinous, tender, and sweet. They are particularly relished in the peanut-growing south of the United States. In Florida, the old name for peanuts was "pinders" while in Georgia they were called "goobers." The Civil War marching song about eating goober peas is said to have been sung by both sides and is still a classroom favorite. In the United States,

peanuts ripen at harvest time and hogs, let loose in the fields to root up the vines, grow fat on the rich, oily kernels in a few weeks. Virginia's celebrated Smithfield ham is traditionally made with peanut-fed pork. In some parts of the world, immature peanuts are prized and some of the crop is harvested before the peanuts are ripe.

Peanuts are used all over the world in everything from soup to desserts. Where peanuts were introduced in Southeast Asia they came to be used in a variety of ways.

One of the most familiar Southeast Asian dishes, the satay, originates in Indonesia and Malaysia, and consists of pieces of meat, poultry, or fish on skewers, covered in a chili-spiced peanut sauce and chargrilled.

Chinese peanut brittle, sharp with black pepper, is an unusual variation of the popular American candy. Salty peanuts, boiled in their shells are sold on the street in southern China. In a restaurant, the soft kernels often arrive with a few pickles as a complimentary snack while waiting for the menu.

Peanut oil has a higher smoking point than almost any other cooking oil and is ideal for stir-frying, in which a very high heat is needed, and is used widely in Chinese cooking. It contains a natural antioxidant that makes it resist rancidity. Most of the fat in peanut oil is polyunsaturated.

ABOVE: *Peanuts are grown on a bush that turns itself downward into the soil after flowering in order for the fruit to mature. There the peanut pods remain for 4–5 months until they ripen.*

ABOVE: *Peanut oil is popular for use in cooking because it does not impart a strong flavor to the dish. It is perfect for stir-fries as it can reach high temperatures without burning.*

Saltwater Therapy

When immature peanuts are not available, try the following to achieve something resembling their flavor. Rinse unshelled raw peanuts, as fresh as possible with firm shells, cover with water in a large saucepan, add sea salt until it tastes like seawater, bring to the boil, partially cover, lower the heat, and simmer until the peanuts are soft and tender, about an hour. The shelled nuts should taste quite salty and will absorb a considerable quantity of water, which gives them an exquisite texture.

Brazil Nut

Bertholletia excelsa

ABOVE: *Brazil nuts are a good source of protein, and in herbal folk medicine, the husks of their seed pods are sometimes made into a tea to cure stomachache.*

Most Brazil nuts come from wild trees that can live only in a healthy rain forest. These trees are evergreen Titans, originating in the Amazon region of South America. Taller and wider than other forest trees, they can grow to a height of 150 ft (45 m). Exports of unshelled nuts exceed that of the more expensive shelled nuts.

Historical Origins The Spanish first saw Brazil nuts in the sixteenth century, and in 1633, Dutch traders sent "wild oil fruits" from the Amazon to the Netherlands. By 1818, Brazil nuts were being shipped to Europe in increasing quantities. The first official United States customs entry for Brazil nuts was in 1873 when 4 million lb (1.8 million kg) of unshelled nuts were imported. The price of just below five cents a pound stayed roughly the same until 1940. In modern times, deforestation has resulted in a severe decline in production.

Botanical Facts The trunk of the tree is straight, with branches only at the top where they can spread to more than 100 ft (30 m). The leaves are just as impressive. They are dark green, shiny, and enormous—6 in (15 cm) in width, and twice as long. After the clusters of lemon-yellow flowers at the end of the branches fruit, it will take about 14 months for the fruits to mature, when they weigh 2–4 lb (1–2 kg). Inside the woody outer husk is a second tough shell containing 12–24 closely packed nuts. When ripe, fruits fall to the ground. A huge, heavy fruit falling from 100 ft (30 m) can cause serious injuries to anyone unlucky enough to be in its path. It takes 12–15 years for the tree to bear fruit, and 30 years for a new plantation to show a profit. A mature tree can produce between 250 and 500 lb (115–225 kg) of nuts one year, but the yield the following year is likely to be poor. Harvesting the nuts is laborious. The fruits are gathered immediately after they fall and the nuts extracted by hand.

Culinary Fare Brazil nuts are very rich and creamy in flavor. With a fat content of roughly 67 percent, they can be ground to use as an excellent vegetarian substitute for suet when making suet puddings or pastry. They are extremely popular dessert nuts, although their hard, tough shells make cracking them a challenge. Unshelled nuts should always be stored in a cold place to keep them from becoming rancid. They are most often used in cakes and biscuits, but because they are so nutritious they are a good addition to meatless dishes and also add extra flavor and texture. Brazil nuts can be grated or chopped and mixed in with other ingredients, or used as a crunchy topping.

RIGHT: *Although Brazil nut trees are enormous, they grow very slowly. They have a long life, with a number of trees living for between 500 and 800 years.*

Breadnut

Brosimum alicastrum

The breadnut is also called the Mayan nut, snakewood, and in Spanish-speaking Latin America, *ramón*, from the word "to browse or forage." The fruit, seed, and leaves of the breadnut are edible, and the tree is listed as an underutilized rain-forest food resource for humans and animals.

Historical Origins The Maya civilization of southern Mexico, Guatemala, and northern Belize arose from about 1800 BCE. The Maya ate breadnuts raw, dried and ground into flour, or mixed with honey. Today, breadnuts grow wild among the ancient ruins in Yucatán in Mexico and Tikal in Guatemala.

Botanical Facts This tropical evergreen giant, which is native to the tropical and subtropical Americas, can reach a height of 120 ft (36 m) with a straight trunk that grows up to 3 ft (0.9 m) in diameter. A single mature tree can produce over 150 lb (70 kg) of nuts in a year. It is a member of the Moraceae family, which also includes figs and mulberries. This fast-growing tree grows best in areas with a dry season but can also tolerate some humidity. The fallen nuts and leaves, which contain about 10 percent protein, provide excellent forage for livestock, such as cattle and pigs. The small fruit, which is peeled by hand, is yellow when ripe and contains a single, edible seed. Seeds are washed and dried before storage.

Culinary Fare Roasted breadnut seeds have a nutty cocoa-ish flavor and are often made into a drink. The seeds are boiled and ground into a bright green dough and can be formed into patties and fried, or diluted in milk and sugar. Breadnut flour is gluten-free.

Canarium Nut

Canarium indicum

The canarium, or galip, has the potential to become an important crop. Around 100 species of evergreen and deciduous *Canarium* trees are native to tropical Africa, southern Asia, Australia, Mauritius, India, Indonesia, and the Philippines. Every part of the tree is used—bark, trunk, resin, leaves, young shoots, fruit flesh, kernels, nut oil, and shells for timber, firewood, and traditional ceremonies and medicine.

Historical Origins This species is native to Papua New Guinea, and has been introduced into Melanesia and Polynesia. Evidence is emerging of the antiquity of canarium nut production and its role in the trade and exchange of clay pots and obsidian on the north coast of Papua New Guinea. Recent finds of obsidian in the Sepik region indicate canarium nut links with Manus province, first colonized more than 20,000 years ago.

Botanical Facts Many tropical and subtropical trees of the genus *Canarium* bear edible nuts and fruits. The canarium is the best known. The fruits are borne on a forest giant that can reach 130–165 ft (40–50 m). They are oval drupes with thin flesh and a hard shell enclosing a large, oily seed. In relation to the size of the tree, some fruits are relatively small, not unlike green almonds. The flowers appear frequently on young shoots and the fruits ripen over a long period.

Culinary Fare Canariums are a popular snack eaten raw. Roasted and dried, they are used in cooking for soups and staple root dishes. In the Solomon Islands they are eaten with eggs of the megapode, a native bird. Canarium nuts make a nutritious infant food when crushed, strained, and added to milk.

ABOVE: *The kernel of the canarium nut is an important source of protein and fat in some South Pacific countries, particularly Papua New Guinea, Vanuatu, and the Solomon Islands.*

Pecan

✳ *Carya illinoinensis*

RIGHT: *Pecans flower in the late spring, and the nuts develop over summer until their shell dries, hardens, and breaks open.*

BELOW: *Pecan nuts have a thin shell that is removed easily. They are great as a snack, but are also tasty in both sweet dishes such as pies, and savory dishes such as stir-fries.*

The pecan is the most important native North American nut. Pecans are related to hickories and walnuts, although the sweet, rich kernels of pecans are fatter and less deeply wrinkled, and the shells are smooth. In commercial production, selected cultivars are generally used. The trees begin to produce nuts after 6–10 years, and are expected to be profitable for at least 200 years. Mature trees have been known to produce 900 lb (410 kg) a year. The pecan tree is a long-lived tree, and some specimens in the southeast of the United States are believed to be more than 1,000 years old.

Historical Origins Native Americans used pecans extensively. The Natchez called it *pacane*, a name that was later adopted by the French in Louisiana. Thomas Jefferson, a keen gardener, transplanted several hundred pecans from the Mississippi Valley to Monticello. He gave some to George Washington who, on

March 25, 1775, planted them at Mount Vernon, where three of them were still alive in the twentieth century. Wild pecans grow prolifically in Texas and are the state tree. The first commercial pecan plantation of 400 acres (160 hectares) was planted in 1880 in Brownwood, Texas.

Botanical Facts When conditions are ideal, the large, elegant, deciduous pecan tree can grow to 100 ft (30 m), and the trunk size and the spread of its branches is equally impressive. It requires deep, well-drained soil, a good supply of water, and a frost-free growing season of several months followed by a cool season for the nuts to develop. Each cultivar responds to a slightly different climate. Because pecan wood is brittle, it can be adversely affected by strong winds, and even rough harvesting can be damaging to the fragile branches. Although both male and female catkins occur on the same tree, self-pollination is rare, and pollination most often comes from another tree.

Culinary Fare Pecans are most often used in cakes, pies, and candies such as New Orleans pralines, butter pecan ice cream, and the celebrated pecan pie. They can be used instead of walnuts in most dishes because they are high in fat, but very little of it is saturated fat. They are a fine source of iron, magnesium, potassium, and zinc, and are also rich in protein, and have significant amounts of the B vitamins. Pecans contain powerful antioxidants, and are said to be effective in lowering cholesterol and aiding weight loss by increasing metabolic rates.

North American Nut

For many native North Americans, the impressive pecan tree was a symbol of the Great Spirit. Its fruits were so valued that furs and other goods were traded for pecans in Spanish Florida. A true American nut, the pecan is the second most popular nut after the peanut in the United States, which produces 80 percent of what the world consumes. Related to the walnut, its flavor is often said to be similar, but pecans are sweeter and creamier. The original Native American names (which also referred to hickories) meant a nut with a shell so hard that it had to be cracked with a stone. In their natural state, pecan shells are an attractive mat light brown with deeper colored random markings. For commercial and cosmetic reasons, most pecan nuts are tumbled and polished to make them more uniform and shiny.

Shagbark Hickory

Carya ovata

Of all the many hickories that are native to North America, the most abundant and esteemed is the shagbark. The single creamy kernel is similar in flavor to pecans and resembles the Persian walnut in shape. The only hickory with commercial importance is the pecan; shagbarks are a local delicacy. The nuts begin to ripen in October, and in December most have fallen from the trees without their shells, when they are eagerly gathered by squirrels and humans.

Historical Origins Hickories once grew all over Europe but disappeared during the last Ice Age and survived only in North America. Early settlers in Virginia shortened the Algonquin name *pohickery*. Thomas Harriot writing in 1588 mentioned the creamy nut "milk." Hickories were pounded, added to water, then strained to be made into a type of broth. This added considerable nutrition to corn cakes and sweet potatoes.

Botanical Facts Hickories are members of the walnut family (Juglandaceae). The shagbark grows to a height of 60–80 ft (18–24 m) with large branches and an open crown. It is a deciduous tree with dark green, smooth, glossy leaves. The yield of nuts is not large, and a seedling may not produce a crop for 15 years or more. The bark of mature trees is ragged and shaggy.

Culinary Fare Shagbark hickories can be eaten raw. A syrup made from the bark is now a niche gourmet item in the United States. Sharon Yarling and Gordon Jones have been making the "original and only" syrup from a 200-year-old recipe since the 1990s. The flavor is described as complex, smoky, nutty, and less sweet than maple syrup.

BELOW: *The nuts of the shagbark hickory contain slightly more carbohydrate than protein, moderate fat, and are a good source of magnesium and thiamin.*

Sawari Nut

Caryocar nuciferum

The sawari nut, or butternut, is the edible, nutlike seed of a large evergreen tree that grows in rain forests in tropical northern South America. It also goes by the names of souari, swarri, and souri—all stem from the Cariban Indian name *sawarie*.

Historical Origins Once an important food in pre-conquest Amazonia, sawari nuts are now a culinary curiosity. The fine-flavored nuts, said to be even sweeter than almonds, were sold as a luxury item in London and Paris in the late 1940s.

Botanical Facts The sawari is one of the largest trees in the primeval forest. The grayish brown fruit of the sawari tree is as large as a coconut with as many as four large kidney-shaped nuts. The pure white kernel is soft and aromatic with a rich, oily flavor, surrounded by edible flesh. Sawari nuts are seldom successful as a commercial crop, and are an underutilized food. The seeds have a short viable life, do not dry well, and cannot withstand low temperatures. Because they are protected by their hard, thick, woody shells, the kernel containing the embryo can remain on the ground for months at a time, and germination can take up to a year.

Culinary Fare The nuts can be eaten raw or roasted, and the oil extracted from them is used for cooking. In Guyana they have a medicinal use, as an anti-fungal. The nuts contain more than 60 percent fat, like the butternut walnut (*Juglans cinerea*). This makes them nutritionally valuable in underdeveloped regions.

Chestnut

Castanea sativa

RIGHT: *Chestnuts are housed inside a prickly husk, which splits open when the nuts are ripe. The nuts are large, shiny, and a dark brown color. Fresh chestnuts are delicious roasted over hot coals or in the oven.*

Native to southern Europe, the chestnut is a noble tree and can live to a great age. The most familiar chestnut is probably the plump cultivated variety in its sublime candied form, *marron glacé*. Wild chestnuts are smaller and leaner but nevertheless the so-called "bread tree" sustained poor mountain communities for centuries. The chestnuts obligingly fall to the ground when ripe, keep well, and can also be ground into flour. English speakers would describe fasting as living on bread and water, but for French speakers, it is water and chestnuts.

Historical Origins Chestnuts have been a valuable food for humans since prehistory. Fossil remains in Europe, Greenland, and Alaska suggest that chestnuts were consumed by early hominids with other nuts, fruits, seeds, roots, and leaves.

The chestnut was known and valued in ancient Greece and Rome. The Roman writers Virgil and Pliny mention it, and there is a recipe for a highly seasoned chestnut sauce to be mixed with boiled lentils in *Apicius*, a Roman collection of recipes. In the seventeenth century, Giacomo Castelvetro suggests a number of appealing ways to cook them.

In Gloucestershire, England, a tree known as the Tortworth Chestnut was already a century old when it was recorded in boundary records during the reign of King John in the late twelfth century. It was 50 ft (15 m) in circumference in 1720, and is still alive today in St. Leonard's churchyard, Tortworth.

Botanical Facts The family Fagaceae includes the cup-bearing chestnut, oak, and beech. The chestnut is a magnificent spreading tree that provides rot-resistant timber as well as good crops of its versatile nut. Chestnut blight, *Endothia parasitica*, a canker-forming bark disease, destroyed most of the native American chestnuts within 50 years after it first appeared at the turn of the twentieth century. Extremely contagious, the microscopic fungal spores are both wind-borne and carried by birds. European chestnuts are also vulnerable to the disease, although the Japanese and Chinese varieties appear to be more resistant.

Culinary Fare Fresh chestnuts are in season only from autumn to early winter. Dried, vacuum-packed, or tinned, the puree and flour are all handy pantry ingredients. Chopped chestnuts are a good complement to brassicas such as Brussels sprouts and cabbage. The delectable Italian confection known as *Monte Bianco* is a "mountain" of chestnuts cooked in milk, flavored with cocoa, vanilla, and rum, pushed through a coarse-bladed food mill, then covered with snowy whirls of lightly whipped cream.

Hazelnuts

Corylus avellana Hazelnut *Corylus maxima* Filbert

These nuts have many different names—cob, filbeard, hazlenut, witchales, halse—and myriad associations with myth and magic. The distinctions between them are blurry, but the filbert is most likely the parent of most cultivated varieties. By any name they are delicately sweet.

Historical Origins The tough little hazels were among the first shrubby trees to spread north over Europe after the last Ice Age. It was so extensive that more hazel pollen grains than any other trees combined were found in the peat layer dated to 75,000–5500 BCE. Charred hazelnut shells have been found in Mesolithic and Neolithic sites in northern Europe. The tree was the dominant form of vegetation in the British Isles and much of Scandinavia.

There were two classical names for these nuts, *nux pontica* and *nux abellana*, both honored, according to Virgil, more than vine, myrtle, or bay. Mercury's winged wand with two entwined snakes was hazel, and it is still a symbol of communication and the wood of choice for divining rods.

For the pagan Celt, it was the tree of knowledge and a potent ally in the struggle against the forces of darkness. The hazel was one of the protective plants taken inside the house on Midsummer's Eve when the barrier between the natural and supernatural worlds was most fragile.

The feast day of St. Philibert, who died in 684, is August 20, when filberts are starting to ripen. However, having a saint's name did not erase the pagan associations and fairy lore that surrounded the nut.

Botanical Facts In 1942 the American committee on horticultural nomenclature decided that the name "filbert" would be used for the genus *Corylus*. Hazelnut is still the name most used internationally. The dainty hazel with its catkins borne on bare branches is a welcome harbinger of spring in English hedgerows. It requires mild winters, warm springs, few late freezes, and cool summers with a body of water nearby to help moderate the climate. The slow-growing cultivar *C. avellana* 'Contorta' is a popular garden plant. Its twisted branches are particularly beautiful in the winter garden and in flower arrangements. There is also a purple-leafed cultivar, *C. maxima* 'Purpurea'.

Culinary Fare Most commercially grown hazelnuts are used in confectionery. One of the most refined types is praline, a powder of caramelized crushed almonds, hazelnuts or both. In Europe, Nutella and similar chocolate-hazelnut spreads are very popular, as is the hazelnut liqueur Frangelico. Hazelnuts are the richest nut source of folic acid and also contain protein, fiber, and niacin and many other vitamins.

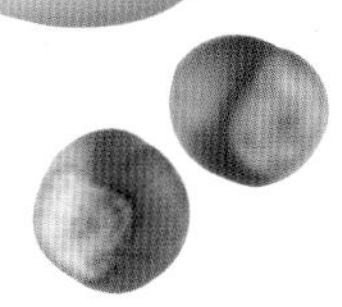

ABOVE: *This bowl contains a mixture of both species of hazelnut. The husk of* Corylus maxima *is longer than the seed and fully encloses it. The husk of* C. avellana *is shorter than the seed and only partially encloses it.*

LEFT: *The hazel has light green, oval-shaped leaves that have a ragged appearance. The nuts mature in summer and are harvested in late summer or early autumn.*

Walnuts

Juglans cinerea Butternut Walnut *Juglans nigra* Black Walnut *Juglans regia* Persian Walnut, Walnut

RIGHT: *The oval fruit of the walnut has a softish outer green husk that changes to a dark brown color when it is mature. The fruits are ready for harvesting when the hull gives under gentle pressure. The nuts should be hulled immediately after they have been harvested.*

All walnuts bear the name of the Roman sky god Jupiter, from the Latin *iovis glans*, or "Jupiter's nut." Walnuts may originally have come from Persia, where in the classical world, the "Persian" or "royal" species was considered the finest. The nuts and their perfumed oil are of enormous gastronomic importance.

Historical Origins Indigenous to eastern Asia, southern Europe, and North and South America, walnuts have been found in prehistoric Iron Age sites in Europe, mentioned in the earliest Chinese and classical texts, and the Bible. From Persia, the walnut may have traveled from Greece to Asia and Italy. Roman armies took it to France and Britain. "Gaul nut" or *wealh* (the Anglo-Saxon word for foreign or alien) may be why they are called walnut in English.

Botanical Facts An elegant, aromatic tree, the walnut can reach 100 ft (30 m) in height but has a reputation for not bearing fruit. It is self-fertilizing, but a single tree with male and female flowers at different times sulks and is unproductive. The fruit is a green drupe with a wrinkled seed. The primary difference between the species is the hardness or brittleness of the shell and the fullness of the kernel; thin for eating, thick for its wood. The black walnut is a North American native. The butternut has a fine flavor, but a very thick shell and is not particularly commercially viable. All parts of the tree are used: the wood for fine furniture, husks and roots for dye, leaves for medicinal tea, with tonka bean and spices for pot pourri, and the oil as a painting medium.

Culinary Fare Walnuts are a most excellent dessert nut, much used in baking. Finely crushed walnuts are used for sauces and soups. Georgian and Armenian dishes that combine them with pomegranate are superb. Fresh walnuts are peerless. Pickled green walnuts are very popular in England; in France they flavor wines and liqueurs. The best walnut oil is incomparable. Persian walnuts contain calcium but not so the American variety, which retains its flavor better when cooked.

Good for the digestion and said to be an antidote to dyspepsia or poison, walnuts contain significant amounts of fiber, protein, calcium, iron, magnesium, potassium, and zinc. They are a good source of omega-6 fatty acids.

Brain Food

In the ancient Doctrine of Signatures, the system used by herbalists for centuries, a plant's resemblance to an afflicted body part was an indication of its therapeutic value. "The kernel hath the very figure of the Brain, and therefore it is very profitable..." wrote William Cole of walnuts in 1657. Now, recent scientific studies have suggested that eating walnuts promotes the development of more than three dozen neural transmitters that are essential for brain function.

Macadamia

Macadamia integrifolia

Macadamias are the only Australian native food plant known all over the world. The sweet, rich little nut is a valuable crop. More than half come from Hawaii, with Australia the second largest producer. Like pine nuts and pistachios, macadamias are costly compared to other nuts but as demand and production steadily increases, their price may come down.

Historical Origins One of the macadamia's many Aboriginal names is *gyndl* or *jindilli*, called *kindal kindal* by early European settlers. It was also called the Queensland nut, but in 1858 those found in the Moreton Bay area of Queensland were named after Dr. John MacAdam of the Philosophical Institute of Victoria, a name that lives on. The macadamia was introduced in Hawaii in the 1880s, and after much experimentation and plant selection, thousands of acres were devoted to macadamia plantations.

Botanical Facts The spreading, long-lived, rainforest evergreen is indigenous to the slopes of the Great Dividing Range. Most macadamias bear self-pollinating fruit and quickly hybridize when other cultivars are grown nearby. They will produce nuts in 6–7 years in good growing conditions. The leaves are similar to those of holly, and make a useful seasonal decoration in the tropics.

Culinary Fare Rich, smooth, and unctuous, with a unique flavor, macadamias are not only an ideal nibble but also a luxurious ingredient for the cook. They have a particular affinity with other tropical ingredients such as mango, pineapple, coconut, and palm sugar. When crushed, they can be used as a coating for poultry or fish.

ABOVE: *Macadamias are high in fat, but low in carbohydrate. When finely ground, they make a delicious addition to a biscuit base for cheesecake.*

Pachira Chestnuts

Pachira aquatica Malabar Chestnut *Pachira insignis* Wild Chestnut

The trees of the Malabar chestnut and the wild chestnut look very different, but their botanical names are often used interchangeably. In order to thrive, they require a warm climate and a well-drained, moist position in full sun. All parts of the tree are useful.

Historical Origins On the west coast of America and Hawaii, the smaller-fruited species is called Malabar, and in the eastern United States that name refers to the larger Guiana chestnut. Both are native to tropical rain forests of Central and South America and Mexico, growing in swampy regions near estuaries and lake shores. In these regions, the bark is used in traditional medicine for anemia, low blood pressure, fatigue, and as a strengthening tonic.

Botanical Facts Pachira chestnuts are medium-sized, fast-growing evergreens that are grown ornamentally, outdoors, and as house plants, for their enormous fragrant flowers, spectacular white exploding clusters, and their edible nuts. Inside the large brown fruit are several tightly packed nuts, which closely resemble the European chestnut. When enlarged, they burst open the pod.

Culinary Fare Eaten raw, pachira chestnuts are said to be rather similar to peanuts, and stored in a cool, dry place will keep for months. After being roasted, boiled, or fried, their flavor is more like cashews or European chestnuts. They are also ground into flour for baking. The leaves and flowers are also edible when cooked like vegetables.

Pine Nuts

✳ *Pinus koraiensis* Korean Pine ✳ *Pinus pinea* Italian Stone Pine

BELOW: *Unshelled pine nuts last longer than shelled nuts. They can be added raw to meat and fish dishes. A sprinkling of quickly fried pine nuts adds a note of grace to almost any dish.*

The luscious, fat little seeds of a wide variety of pines enjoy a culinary reputation disproportionate to their size. The most common species is harvested from the Korean pine. Exported in quantity from China, pine nuts were praised for their delicacy of flavor during the Tang dynasty (CE 618–907).

Historical Origins The stone pine, native to the Mediterranean, has been a prized food since ancient times. Sacred to the god Neptune, pine nuts preserved in honey are mentioned by the Roman writer Pliny. They were found at Pompeii, and they provisioned Roman legions in Britain where shells are found in rubbish dumps surrounding Roman encampments of the first century CE. In China, pine nuts, but no tree pollen, were found in Neolithic sites in Shensi. The two pinecones that were found in a twelfth-dynasty Egyptian tomb may have been imported from Lebanon.

Botanical Facts Of the 100 or so "true" pines worldwide, only 12 in the Northern Hemisphere bear nuts that are thought worth gathering. Some that have a strong turpentine taste are admired locally. Each pinecone, which can take as long as 3 years to mature, contains more than 100 narrow seeds. Each has a tough brown covering that is removed to reveal the sweet nut within.

Culinary Fare Recipes using pine nuts occur in the earliest collections, and many could easily be made today. Recipes still made include all of the sauces *alla Genovese*, which include pesto, and the sweet Spanish pine nut paste *piñonate*, or the *pinocchiato* of the fifteenth century.

Pistachio

✳ *Pistachia vera*

RIGHT: *Pistachio nuts are an excellent source of protein, potassium, phosphorus, and magnesium, and they contain no cholesterol.*

A pistachio is an elegant, dainty, green nut, which is not only decorative, but also finely flavored. The Queen of Sheba is said to have kept all the pistachios in Assyria for herself and her court. A tree in Kerman, Iran, is believed to be more than 700 years old. This region is still an important pistachio producer, and it was a Kerman cultivar introduced in 1929 to California that is the main variety grown there.

Historical Origins Pistachios have been enjoyed for thousands of years. In 1965 carbonized pistachios were found in an early Neolithic settlement in Jordan, carbon-dated to 6760 BCE. Paleolithic underground storage pits in the Zagros mountains of Iran contained pistachios, acorns, and snails. Eleven thousand years ago, oaks and pistachio trees reforested the foothills after the last Ice Age.

Botanical Facts The fruit, of which the pistachio nut is a seed, grows in clusters that resemble grapes. When the conditions are right, the shells split open. The nuts at this stage are described as "laughing," and are easier to eat and to sell.

Culinary Fare Everything made with pistachios is grand: ice cream, baklava, Turkish delight, halva, stuffings, or coatings for meats. A simple bowl of salted and roasted pistachios is an elemental way of pleasing everyone.

Almond

Prunus dulcis

The beautiful almond has no rival for its place as the premier culinary nut. "The possibilities of almonds are varied; they can be used to change a commonplace dish into a delicacy" wrote Mrs. C. F. Leyel in *The Gentle Art of Cookery* in 1925. They are the most widely grown nut, are inexpensive, and essential for good cooks.

Historical Origins The almond probably came originally from the Near East and spread to the Mediterranean basin. Hybrids from wild species were domesticated during the Bronze Age in Greece and Cyprus. The almond is mentioned a number of times in the Bible: Aaron's rod bloomed and "yielded almonds" (Numbers 17:8). They were brought to England by Roman armies in the first century CE. In the Middle Ages almonds were an important food during church-regulated periods of fasting, an essential source of calcium when animal foods were forbidden. Early cookbooks, however, contain a huge proportion of recipes with almonds, not all of them meatless. A household account of August 29, 1265 records that 12 shillings and 6 pence was paid for 60 lb (27 kg) of almonds and 12 pence for a pair of shoes for Petronilla the laundress. Roughly the same price comparison could be made today.

Botanical Facts The almond is a member of the rose family, closely related to peach, plum, and apricot. Delicate white or pinkish flowers bloom in early spring on its bare branches. It grows in the Near East, southern Europe, California, South Australia, and South Africa. The downy outer hull of the almond fruit, which resembles a miniature peach, splits when it is mature. The almond is the pit inside the flesh, which, unlike its near relatives, is the edible part of the fruit.

Culinary Fare Almonds are beyond compare when fresh, delicious still after drying, and useful in all manner of dishes. Eaten raw, they are satisfyingly healthy. They are indulgent when carefully cooked in rather a lot of olive oil, lightly salted while still warm, and far better than store-bought. Almond oil, essence, and milk are also splendid ingredients. Almonds are particularly nutritious. They contain significant amounts of protein, calcium, iron, magnesium, potassium, zinc, niacin, and riboflavin.

ABOVE: *The hull of the almond splits open while the nut is still on the tree. When this occurs, the nuts are harvested by knocking them to the ground.*

Romance and Luck

In Greek mythology, when the Thracian princess Phyllis was jilted at the altar and later died by her own hand, the goddess Athena transformed her into an almond tree—*phylla* in Greek. Her lover returned, but too late, and as he embraced the branches of the forlorn tree on her grave it burst into bloom. In the Muslim world, almond flowers represent hope. For centuries in Europe, the almond has been a good luck symbol. Sugared almonds are still distributed at celebrations such as weddings and baptisms. A "butter" of almonds, sugar, and rosewater to be eaten with violets was recommended by Nicholas Culpeper in 1652 as being "very wholesome and commodious for students for it rejoiceth the heart and comforteth the brain."

Herbs

Introduction to Herbs

RIGHT: It is not necessary to have a garden or even much space to grow herbs. The smallest sunny spot is ideal for growing a few kitchen herbs, ready to pluck fresh from the pot.

Herbs are the stuff of myth, lore, and legend, used by every culture since before recorded history. The botanist defines an herb as any plant without a persistent woody stem that dies back to the ground each year—an herbaceous plant. The herbalist—a person skilled in collecting, growing, or using herbs—defines an herb as any plant used for flavoring or for its fragrant or medicinal qualities.

We encounter herbs daily. They are used for flavor and aroma in every category of household product. Almost every kitchen has a collection of herbs and spices at hand for the potential to transform each meal. Many canned goods, packaged baked goods, and alcoholic and non-alcoholic beverages have extracts, essential oils, or oleoresins of herbs added to tweak flavor or enhance fragrance.

Often used in parts-per-million quantities, herbs and their derivatives in packaged products account for nearly half of the volume of herb use in the modern world. In addition, every soap, detergent, and cleaner in the household contains fragrances derived from herbs, either directly or mimicked by laboratory chemists. When brushing your teeth, the toothpaste is probably flavored with peppermint essential oil or menthol that was derived from the essential oil. If the menthol is synthetic, the chemist's starting material is likely to be the essential oil of thyme. Herbs touch our lives in many ways. We can appreciate them not only for their sensory qualities but also for their history, origin, botany, and biology.

> *"Those herbs which perfume the air most delightfully, not passed by as the rest, but, being trodden upon and crushed, are three; that is, burnet, wild thyme and watermints."*
>
> Francis Bacon (1561–1626)

HERBS IN HISTORY

Herbs are inextricably woven into human history, whether used for flavor, fragrance, or medicine. Following publication of the first printed book in 1455—the Bible—came printed herbals. The Bible, Koran, and other texts from virtually all religions describe the ritual or practical uses of herbs. The earliest works on Chinese herbal traditions can be traced back at least 5,000 years. Ayurveda, incorporating the herbal traditions of India, has origins dating to 7,000 years ago. Ancient Egypt was famous for its unguents, balms, fragrant oils, incense, and perfumes.

Garlic served as health food to give strength and prevent disease among the workers who created the pyramids. Today's garlic is not known from the wild; rather it is a cultigen—a plant that has evolved over a 7,000-year period. It is a creation of humans. Garlic appears in the Vegetables chapter, with its relatives, the onion and leek.

Kava (*Piper methysticum*), used by Pacific Islanders for thousands of years to make a ceremonial beverage, is also not known from the wild. It never produces seeds and does not spread on its own. Kava is entirely dependent on humans for its reproduction.

When Christopher Columbus sought passage to India in search of a new trade route for herbs and spices, he stumbled across the American continents, and instead of returning with the black pepper he sought, he introduced red, or chili, peppers to the Western world. Cayenne (*Capsicum* species) is now the most widely consumed flavoring ingredient in the world.

In the modern kitchen, herbs are associated with fragrant plants that add subtle or dynamic flavor dimensions to food. Most herbs currently used in Western culture are plants that originated in the Mediterranean region, where the sun-baked, dry, rocky soils forced plants through thousands of years of evolution to produce chemical compounds called secondary metabolites to help them survive. Secondary metabolites are plant compounds that are not necessary for essential nutrition. They are aromatic compounds, bitter compounds, or perhaps poisonous substances that dissuaded animals from eating the leaves or prevented insects from attacking the plant. In the case of most of the plants that we call herbs, these are essential oils (also known as volatile oils), which are responsible for the plant's flavor and fragrance.

ABOVE: *Many herbs can be dried for later use, taking advantage of good seasonal harvests. Air dry by hanging them in bunches upside-down. Once dried, store in airtight containers in a cool dry place away from light.*

TWO LARGE HERB FAMILIES: MINT AND PARSLEY

Many herbs used for culinary purposes originate from just a few botanical families. Most notable are the Lamiaceae—the mint family (also known as Labiatae) and the Umbelliferae or parsley family (also called Apiaceae). The mint family is perhaps the most important herb family, with over 250 genera representing nearly 7,000 species. This botanical family represents less than 2 percent of the world's flowering plants, yet over 25 percent of plants commonly called herbs belong to it. Lemon balm, basil, peppermint, spearmint, oregano, sage, thyme, marjoram, savory, and rosemary are just a few of its representatives.

Mint family members produce oil glands, often on the underside of leaves, scattered along the stems, and largely concentrated on the flowers. Stroking or bruising the leaves releases the essential oil, delighting the palate with the flavors within.

Similarly, the parsley family members—particularly seed crops such as dill, angelica, fennel, lovage, sweet cicely, and parsley—have oil glands in the leaves, or seeds (which, technically, are fruits) with oil tubes holding the essential oils that impart the flavor and fragrance we enjoy.

LEFT: *Fresh mint is an excellent palate cleanser, as is parsley. All mints have bold pungent aromas that awaken the tastebuds. Fresh herbs have a different flavor to dried and add a special piquancy to salads and other cold dishes.*

Chives

Allium schoenoprasum Chives   Allium tuberosum Garlic Chives

RIGHT: *Chives are very easy to grow, either in the garden or in containers. Harvest the fresh leaves as needed and within days a new crop will appear.*

With its subtle onionlike piquancy, chives are among the easiest to grow of the fresh herbs for the kitchen. Their cousin, garlic chives, impart a more punchy but light garliclike flavor.

Historical Origins "Chives" comes from the French *cives*, derived from *cepa*, the Latin word for onion. Chives originate from the colder regions of the Northern Hemisphere, including Europe and North America. Historically, they are associated with English, French, and German gardens.

Garlic chives, also known as Chinese chives or oriental chives, were unknown to Europeans until the seventeenth century. Originating in Southeast Asia, they are grown from Nepal to Japan, where they have been a common herb since ancient times.

Botanical Facts Related to the lily, chives belong to the genus *Allium*, which is best known for onions, garlic, and leeks—and these are just a few of the nearly 700 species it contains. Common chives are a hardy, bulbous, perennial plant. The round, hollow, spearlike leaves, 6–8 in (15–20 cm) long, arise from the crowded bulbs in tufts. The purplish flowers are borne in crowded globular flowerheads. Garlic chives have flat fleshy leaves, from 8–15 in (20–38 cm) high. The thick round flower scapes produce white flowers in a globular head. Left to go to seed, chives may become weedlike in the garden.

Culinary Fare Clipped short before the flowerheads appear, the leaves can be enjoyed throughout the growing season. The marriage of chives and potatoes via sour cream is their best-known use. Otherwise add their colorful flowerheads to help brighten up a salad. Chives are almost a necessity in omelets. They can also eaten with cottage cheese, cream cheese, or mixed into softened butter. Chopped fresh garlic chives are excellent with egg dishes and stir-fries alike.

A Delightful Delicacy for Chinese New Year

The Chinese New Year, also called the Spring Festival or Lunar New Year, begins on the first day of the first lunar month in the Chinese calendar, typically in early February. A delicacy at feasts for this, the most important of traditional Chinese festivals, is *jiu cai huang*—literally "yellow everlasting vegetable"—the blanched leaves of garlic chives; ("*huang*" means yellow). In China, fresh garlic chives are called *jiu cai*, "everlasting vegetable" as they stay green and fresh through all but the coldest winter. Healthy rootstocks are kept in darkened greenhouses in Beijing. In Chengdu, as leaves appear, sand is gradually hilled up around the plants to stop the sun reaching the leaves, blanching them white-yellow. In south China, earthenware pots are placed over the plants to blanch the leaves. Bundles of the delicate blanched leaves are thus available in every kitchen for the Chinese New Year.

Lemon Verbena

Aloysia triphylla

Crush the dried leaves of lemon verbena, and a sweet lemon scent wafts into the air. It is the best of the lemon-scented herbs.

Historical Origins Introduced to European horticulture in 1784, lemon verbena differs from most herbs in that it does not come from the Mediterranean region. Rather, it is a deciduous tree originating from central South America—northern Chile, Argentina, and adjacent Peru. Today, it is commercially grown in France, Morocco, and Algeria to produce essential oil—the raw material used in perfumery.

Botanical Facts One of about 35 species in the genus *Aloysia* in the verbena family, lemon verbena is a tender deciduous shrub, often growing to 10 ft (3 m) or less in gardens in the US, though it can reach a height of more than 20 ft (6 m) in subtropical climes. The lance-shaped leaves, 3–4 in (8–10 cm) in length, exude a delicious lemony fragrance when crushed.

Culinary Fare Better known as a fragrance source rather than a culinary herb, it makes a fine herbal tea, traditionally believed to have a mild calming effect. It is also an ingredient in liqueurs. A sprinkling of the dried leaves is great in fruit salads, with melons, or added to flavored iced beverages. The dried or fresh leaves are equally useful as tea, food, flavoring, or fragrance for potpourris, soaps, and perfumes.

ABOVE: *Lemon verbena belongs in the verbena (Verbenaceae) family; all its species contain volatile oils in their leaves. The fragrances resemble citrus, lavender, camphor, and mint, and are utilized in perfumery.*

Dill

Anethum graveolens

Dill has been used since ancient times. Today, its distinctive flavor is familiar to all in the form of its best-known use—dill pickles, or pickled cucumbers.

Historical Origins Dill was described in the writings of the first-century Roman naturalist Dioscorides. The name derives from the old Norse *dilla*, meaning carminative. Alfric, onetime Archbishop of Canterbury, used the name in the tenth century, and dill, or *till*, was used in Germany the following century. Dill is found in Swiss Neolithic lakeshore settlements, attesting to its ancient use, cultivated since at least 400 BCE.

Botanical Facts An annual member of the parsley family, dill is one of the 10 members of the genus *Anethum*, originating from the Mediterranean basin and western Asia. Growing from 1 to 2 ft (0.3–0.6 m) in height, the finely divided, feathery leaves are threadlike in appearance. The yellow umbrella-shaped flowerheads feature dozens of tiny yellow flowers and are up to 8 in (20 cm) across. The flattened seeds have prominent ribs.

Culinary Fare Dill is possibly among the most versatile of culinary herbs, whether using the fresh leaves, seeds (fruits), or the dried leaves, known as "dill weed." Dill leaves, when harvested fresh before flowering, complement cucumbers, potatoes, carrots, and fresh or smoked salmon. Dried dill weed (leaves) can be used similarly. The pungent, pleasantly aromatic seeds are excellent with cabbage dishes, stews, soups, vinegars, and bread.

ABOVE: *Finely chopping dill leaves releases this herb's pungent fresh aroma, which teams up delightfully with carrots, along with a touch of honey and butter.*

Angelica

Angelica archangelica

You will not find angelica leaves, roots, and seeds in the supermarket herb section. It is more likely to be encountered as a flavoring in liqueurs such as Chartreuse.

Historical Origins Enchanting in lore, myth, as well as in the garden, the "root of the Holy Ghost" earned its angelic title as a guard against witches and a remedy against poisons and the plague. In his *Herball* (1633), John Gerard tells us that a piece of the root held in the mouth, or chewed, will drive away pestilential air. A crown of angelica leaves placed on the head inspired poets through its unique fragrance. Angelica is a plant of the European far north, first cultivated in English gardens around 1568.

Botanical Facts Angelica is a bold stout biennial or short-lived perennial to 7 ft (2 m) high. The broad aromatic leaves appear the first year, then a thick hollow stem an inch (25 mm) or more in diameter branches toward the sky with globular seed heads the size of baseballs. The genus *Angelica* is in the parsley family, and includes over 100 species. Angelica is from northern Europe—Norway, Sweden, Austria, the Alps, and Lapland.

Culinary Fare The large hollow stems can be cooked in sugar syrup, then dried and enjoyed as a sweetmeat. Stripped of their outer rind, the stems make a unique taste treat. Laplanders believe the herb strengthens life, and chew the dried root as a tonic.

Chervil

Anthriscus cerefolium

RIGHT: *Easily mistaken for parsley, chervil requires somewhat different growing conditions. Disliking direct sunlight and humidity, it is best grown on a shady windowsill or deck. It needs regular watering.*

Once considered a weedy annual to which very little attention was paid, in recent times chervil has become elevated to a higher culinary purpose, earning it the name "gourmet parsley."

Historical Origins Chervil has long been known in English gardens as an annual that sprang up in hedges and waste places. Its origins were unknown until the late nineteenth century, when it was found to be native to southeastern Russia and the Caucasus south to the northern mountains of modern-day Iran. Unknown to the Ancient Greeks, chervil first appears in the writings of Roman authors of the Christian era.

Botanical Facts Chervil is an annual herb with leaves similar to those of parsley, growing to a height of 2 ft (0.6 m) on wiry stems. The tiny loose clusters of white flowers turn into small seeds (technically fruits) about ⅛ in (3 mm) long. Today, it is found throughout much of Europe and has naturalized in the northeastern regions of the US. Chervil is easily grown from seed.

Culinary Fare An ingredient in *fines-herbes*, chervil is often combined with parsley and chives, as well as basil or tarragon. The delicate anise-scented leaves are an excellent addition to omelets and soups, and tomato and fish dishes. Finely-chopped, a sprinkling of the fresh leaves adds a lively hint of anise flavor to salads. Chervil is added at the later stages of cooking to preserve its flavor. It is best used as a fresh herb, tending to lose much of its flavor when dried.

French Tarragon

Artemisia dracunculus

Remarkably, French tarragon is a seed-sterile selection, propagated only by cuttings. It has been known in France for at least 800 years. Tarragon is one of the few *Artemisias* without a harsh bitter flavor.

Historical Origins As a culinary herb in the nineteenth century, tarragon was best known as a flavoring for pickles, salads, and medicated vinegars. Then came its use in a style of French cooking, which catapulted a once lowly weed to the darling of *haute cuisine*. The name tarragon and the Latin specific name *dracunculus* are both derived from the same Greek word root, meaning "dragon." The name's origins have become lost in history, possibly referring to the serpentlike appearance of the roots.

Botanical Facts Tarragon is a widespread species of Europe, Asia, and North America. If you buy seeds, you are getting so-called "Russian" tarragon, which has a harsher, more balsamic odor than French tarragon. True French tarragon (*Artemisia dracunculus*) can only be propagated by cuttings. There are more than 350 species of *Artemisia*, which is in the daisy (Asteraceae) family.

Culinary Fare Turn a fine fowl into a delectable delight with a generous sprinkling of dried tarragon leaves in butter, or fresh tarragon sprigs rubbed over the skin. Few herbs improve the flavor of fish like tarragon does. Judicious use in any white sauce will add a memorable flavor. The fresh leaves, finely chopped, are also tasty in salads.

ABOVE: *Best used fresh, French tarragon can also be frozen or dried. As its strong flavor can easily overpower other herbs, use it sparingly or on its own.*

Pot Marigold

Calendula officinalis

Sometimes known as poor-man's saffron or Egyptian saffron, pot marigold is valued for its yellow-orange blossoms. It is often used as a saffron substitute.

Historical Origins The genus name *Calendula* is from "calends," referring to the fact that in warmer climates, calendula can be in bloom every month of the year. It was formerly known in England as "golds," "gouldes," or "Marygold," honoring the Virgin Mary. The founding father of taxonomy, Carl Linnaeus, observed that the flowers open at 9 in the morning and close at 3 in the afternoon. Shakespeare was more poetic, describing "The marygold, that goes to bed with the sun, and with him rises weeping."

Botanical Facts *Calendula* includes a dozen or so species of marigold native to the Balkans and western Asia. They have been cultivated in European gardens for centuries and grow naturally in the vineyards of France and cornfields of Italy. Garden cultivars include some with double flowers but not all are edible, so it is best to use pot marigold.

Culinary Fare Medieval European markets, like modern herb markets in places such as Cairo, Egypt, featured barrels of dried marigold flowers, used as a base for soups and broths. In today's kitchens, fresh or dried flower petals are used as a yellow-orange coloring for white sauces, or a colorful garnish for butter, salads, and desserts. As the authors of old herbals put it, it is used "greatly to comfort the heart and the spirits." The leaves are edible; when chopped they make an excellent salad ingredient.

BELOW: *A bed of marigolds makes an attractive garden display. For culinary purposes, pot marigold is the preferred species.*

Cilantro, Coriander

Coriandrum sativum

BELOW: *Cilantro is among the most widely used herbs in the world, crossing every cultural barrier.*

Cilantro, coriander, and Chinese parsley refer to the fresh herb, while the dried seeds are known only as coriander.

Historical Origins Coriander has been used since ancient times. In India, it is mentioned in Sanskrit texts dating back nearly 7,000 years. In Ancient Egyptian tombs, like that of Tutankhamen, coriander was offered as food for the afterlife. In China, coriander first appears in an herbal published in 1061 during the Song Dynasty.

Botanical Facts Cilantro is an annual, easily grown from seed, believed to originate in southern Europe, North Africa, and West Asia. Like Italian parsley, the scalloped leaves are broad, shiny, and flat. The seeds are globular, and about ⅛ in (3 mm) in diameter.

Culinary Fare Cilantro is a traditional component of all Latin American cuisines. In Armenia, a small plate of fresh cilantro leaves is served as a side dish with nearly every meal. In Egypt, the restaurants serve cilantro as a garnish in the same way Westerners use parsley. In China, the finely chopped leaves are often sprinkled over fish or meat dishes. It is a particularly common fresh herb in Thai cuisine. Add a sprinkling of chopped fresh cilantro to enhance flavor in salads, beans, rice, omelets, soups, lamb—the possibilities are endless.

Coriander seed is generally used as a spice to enhance bean and meat dishes, breads, and desserts. It, too, is a versatile, underutilized flavoring agent.

Lemongrass

Cymbopogon citratus

RIGHT: *Lemongrass leaves are too tough to eat on their own. They need to be crushed or very finely chopped before use in cooking, or added in large pieces that are later removed, prior to serving.*

To the untrained eye, lemongrass looks like an ordinary overgrown clump of grass, but stroke the leaves and its lemony essential oil fills the air.

Historical Origins Commonly planted throughout the Old World tropics and subtropics, lemongrass probably originated in Malaysia, though its exact origin is uncertain. It has been grown for many centuries throughout Southeast Asia.

Botanical Facts Lemongrass is a large clumping grass, growing from 2 to 6 ft (0.6–1.8 m) in height, and taller in flower. The leaves, up to 1 in (25 mm) wide, have a prominent midrib.

When crushed, the leaves release a distinctly lemon scent with a somewhat earthy undertone, almost suggestive of citronella, to which this species is closely related. In cooler climates it can be grown as a container plant. Cut back in winter and keep in a cool dark place, then move to a warm sunny location for use during summer months.

Culinary Fare Lemongrass leaves make an excellent herbal tea. The fleshy heart of the young stems, stripped of the green leaves and cut up like green onions, is an excellent flavorful addition to a vegetarian stir-fry, or a side dish or garnish with rice. In Asia, lemongrass leaves are often used to flavor fish or are added to spicy sauces and curries. The crushed leaves can be used as a flavoring in soups or meat dishes.

Fennel

Foeniculum vulgare

The seeds of fennel have long been considered a very useful aromatic carminative to help relieve flatulence in children. In India, the seeds are often served after a spicy meal to calm the palate and rejuvenate digestion.

Historical Origins The Romans valued fennel for its aromatic seeds and succulent edible stalks. Charlemagne (742–814), perhaps the greatest of medieval rulers, stimulated fennel's spread from the Mediterranean region into central Europe by cultivating it on his imperial farms.

Botanical Facts Depending on the cultivar, fennel is an annual, biennial, or short-lived perennial. It grows to 6 ft (1.8 m) or more in height and has smooth, striated branching stems with a mild anise or "licorice" flavor. The tiny yellow flowers are borne on flat umbrella-shaped tops. The seeds (technically fruits) are up to ¼ in (6 mm) long, releasing a prominent anise fragrance when crushed.

Culinary Fare Fennel refers to fennel seed, leaves, and the succulent, swollen, celerylike stalks of dulce, Florence, or finnochio fennel. The seeds are often used to flavor baked goods such as cookies, cakes, and breads. Bruised in a mortar and pestle, they make a pleasant herbal tea. Fennel can be used raw in salads, sauteéd, or roasted. The thickened stems of finnochio fennel make a wonderful vegetable. The fresh or dried leaves, finely chopped, are an excellent complement to fish.

ABOVE: *Although the leaves of fennel bear a resemblance to those of dill, fennel leaves appear to be less delicate and have a stronger licoricelike aroma.*

Licorice

Glycyrrhiza glabra

The flavor we generally associate with licorice as a candy is actually the flavor of anise or fennel. The licorice root itself, sometimes called licorice sticks, has a very sweet earthy flavor.

Historical Origins Well-known to ancient Roman writers, the name licorice is a corruption of a medieval form of the Latin name *glycyrrhiza*. Written as *glyliquiricia*, it eventually evolved to become the English word licorice. Cultivation in northern and central Europe did not occur before the thirteenth century, but by the sixteenth century it was extensively cultivated in Germany. Licorice has long been famous as a remedy for coughs.

Botanical Facts Licorice refers to the rhizome or underground stem, chewed for its sweet root but famous for turning the teeth black if taken up as a persistent habit. Licorice is a member of the pea family. European licorice (*G. glabra*) originates from southern Europe and adjacent West Asia. The genus *Glycyrrhiza* contains about 20 species, most producing long, ropy, underground stems—the licorice root of commerce. Most species are concentrated in the Asian steppes of northwestern China. A closely related species, Chinese licorice (*G. uralensis*), is known as *gan cao*, meaning "sweet herb."

Culinary Fare Licorice is rarely used as an herb in cooking. Rather, licorice extract is used as a flavoring for a wide variety of candies and baked goods, which are mostly manufactured by confectioners. Some licorice manufacturers in Europe have been operating their businesses for centuries.

Bay

Laurus nobilis

RIGHT: *Bay leaves can be used fresh, but they do mature with age and after drying they impart more flavor. Fresh leaves make an attractive garnish.*

Also known as bay laurel, sweet bay, Grecian bay, laurel, or bay tree, the stately bay features in legends, myth, and honor. Its leaves are found in most kitchens.

Historical Origins Few aromatic trees have as high a place in history as the bay tree. A crown of the berry-laden leaves marked military, scholarly, or literary achievement. In medieval and later centuries, bay leaves were strewn about the floors to improve a home's overall fragrance, and quince paste was stored between layers of bay leaves.

Botanical Facts Valued for its sweet aromatic leaves, the "noble laurel" *Laurus nobilis* is one of only two members of this genus in the laurel family. Hailing from Asia Minor, this evergreen tree spread throughout southern Europe and North Africa several millennia ago. In subtropical regions, it is a tree that will attain a height of 50 ft (15 m), but grown as a container plant in cooler climates, it is generally regarded as a shrub, reaching from 3 to 6 ft (0.9–1.8 m) in height. The oblong to lance-shaped, glossy, aromatic leaves are familiar in nearly every kitchen. The oval, dark purple berries, seen on mature trees, are about the size of a small olive.

Culinary Fare Unlike most herbs, which are sold in "cut and sifted" form, bay leaves are nearly always available as whole dried leaves. They are a component of the classic *bouquet garni*. Whole bay leaves are added to stews, soup bases, and cream sauces. Add a leaf or two to impart a delicate aromatic flavor to poultry, fish, lamb, pork, veal, meatloaf, and game dishes. Bay leaf adds a subtle piquancy to root vegetables, such as carrots and onions. Although some believe it is lucky to find a bay leaf in their meal, others have actually choked to death on them. Therefore, it is advisable to remove leaves before serving.

From Poet Laureate to Baccalaureate

Fleeing the embraces of the Greco-Roman god Apollo, Daphne, daughter of Peneus, reached the banks of the stream from which she originated. Here she called on the river god for protection, and was changed into a bay tree. In the great Roman age, the bay became the emblem of victory and clemency. Roman soldiers carried a sprig of the herb in triumphal parades after a victory, while generals were crowned in a wreath of bay leaves. The tradition continued in later centuries as the writer Chaucer mentions that a crown of bay was bestowed on the Knights of the Round Table. Bay or laurel branch wreaths crowned the heads of distinguished poets in the Middle Ages, hence the name "poet laureate." Scholars of medicine, divinity, or law in European universities of bygone ages were crowned in berry-laden branches of laurel. From the Latin *baccae* and *laurae*, or "laurel berries," we have the *baccalaureate*, from which is derived the title of *bachelor*.

A fifteenth-century painting by Italian brothers Antonio and Piero Pollaiuolo depicts Daphne metamorphosing into a bay tree while fleeing Apollo's amorous advances.

Lovage

Levisticum officinale

Sweet lovage is a plant whose leaves are suitable for sparing use in the kitchen, while its root and seeds were once common to the apothecary.

Historical Origins The genus name *Levisticum* comes from the Latin *levo*, meaning to assuage, in reference to the once-common use of the seeds to relieve flatulence, a fact perhaps best not shared with dinner guests.

Botanical Facts Lovage is a stout hardy perennial, generally growing to a height of 4–6 ft (1.2–1.8 m), though it can reach far greater heights in rich garden soil. This member of the parsley family produces a tough hollow stem and flat divided leaves somewhat like those of celery. It originates in the eastern Mediterranean region and is naturalized throughout much of Europe and occasionally in eastern North America.

Culinary Fare The celerylike leaves of lovage have a pungent spicy flavor that can only be described as the "flavor of lovage." It is not an herb for the fainthearted, but a unique ingredient for the more adventurous cook. The fresh or dried leaves can be used sparingly in salads or soups, or else delicately sprinkled on potatoes, carrots, cabbage, and tomato dishes. The hollow stout stems were once the garnish from which a Bloody Mary was sipped, now universally replaced by a celery stick. Lovage is also occasionally used for flavoring vinegars.

Lemon Balm

Melissa officinalis

Melissa is the Greek word for bee, given to this herb because honeybees love its sweet blossoms. Lemon balm tea sweetened with sugar is one of the most pleasant of herbal teas.

Historical Origins Long appreciated in the Mediterranean basin and in North Africa, this herb was considered useful to strengthen the nerves and promote cheerfulness. It was thought to improve memory and decrease stress and anxiety. A casual look at the leaves suggests it is a mint, but lightly crushing the leaves releases a pleasant, almost oily lemony balm, an enjoyable surprise to the nose and palate.

Botanical Facts Lemon balm is one of 5 members of the genus *Melissa* in the mint family. It is native to sandy soils and scrubland in southern Europe. The opposite scalloped-edged citrus-scented leaves are about 2 in (5 cm) long and 1 in (25 mm) across. The tiny white flowers of this very easy-to-grow perennial are seldom noticed. It is commonly naturalized throughout Europe as well as much of North America.

Culinary Fare Lemon balm is not to be found in the supermarket spice section, primarily because once the leaves are dried, they lose much of their pleasant lemon scent. Therefore, only fresh leaves, or freshly dried leaves, are suitable for the kitchen. A perfect garnish for iced-tea, a sprig of the leaves, or a single leaf, can be floated atop a beverage.

ABOVE: *As lemon balm needs to be used fresh and is easy to grow, a pot or garden plant makes it readily available.*

Mint

Mentha x piperita Peppermint *Mentha x rotundifolia* Apple Mint *Mentha spicata* Spearmint *Mentha suaveolens* Pineapple Mint *Mentha x villosa* Salad Mint, Hairy Mint

Mints are known for their crisp clean flavors and are used in cooking, tisanes, culinary fare, and herbal medicines. The genus *Mentha* in the mint family contains upwards of 24 true species, but since they hybridize readily there are at least 2,000, if not more, named variations.

Historical Origins The name *Mentha* comes from the Ancient Greek legend of the nymph Minthes, daughter of Cocytus, who was changed into a mint by the goddess Persephone in a jealous fit. In remembrance of this, young Greek girls used to weave sprigs of mint into their bridal wreaths. The Greek name translates as "sweet odor."

Nearly every medieval book on herbs included mints. Convent gardens dating to the ninth century all grew mints. In herbals before the year 1700, almost all writers spoke of the garden mint as spearmint.

Peppermint is not mentioned in any work prior to 1696, when the English botanist John Ray first described it. This is because peppermint is actually a hybrid and did not evolve under cultivation until sometime in the mid- to late-seventeenth century. However, by the 1720s it became the mint of choice for medicinal use, given its stronger action to calm the stomach. Commercial cultivation of peppermint began in the 1750s in Mitcham, Surrey, a county in southeastern England.

Botanical Facts Peppermint is a mint particularly high in menthol, which gives this mint its distinctive flavor. It has sterile flowers in terminal flowerheads or elongated spikes. The leaves are egg-shaped, usually nearly hairless, with purple to reddish stems.

Spearmint, while originally a native of Europe, is naturalized around the world. This mint has smooth or slightly hairy stems and leaves, though the leaves are nearly stalkless.

Apple mint is quite hairy and the leaves, when crushed, have an applelike scent, as the name implies. Its leaves have mostly parallel sides, with a broadly rounded base.

The sweet-scented pineapple mint is a robust plant with a distinct fruity odor that

BELOW: *When it comes to mints, pick a flavor! From left to right: apple mint, pineapple mint, red raripila mint, basil mint (2 sprigs), black peppermint, and Moroccan spearmint (2 sprigs).*

Culinary Fare The highly variable essential oils give mint distinctive flavors and fragrances, such as those of pineapple mint, apple mint, and even chocolate mint. Mints tend to be underrated as culinary herbs.

LEFT: *Mints can be used fresh or dried but are at their best when picked fresh before flowering, as afterwards they tend to become harsher or more bitter.*

is reminiscent of pineapple. Leaves are long, up to 1½ in (35 mm) in length. Salad mint has the distinct clean odor of spearmint, with oval or nearly circular wrinkled hairy leaves.

Mints are particularly easy herbs to grow. They generally grow in relatively rich moist soils. In their native or naturalized haunts in Europe and North America they are often found flourishing near a stream. When grown in a home garden, mint needs to be planted in a confined spot because its underground stems readily invade nearby areas and will take over suitable habitats.

Mints, particularly peppermint and spearmint, make some of the best herbal teas, served iced or hot. Add sprigs of leaves to any iced tea. Fresh peppermint or spearmint makes an excellent "sun tea." For sun mint tea, take a full handful of the leaves, twisting or bruising them to release the essential oils, then place these in a gallon jar in the sun for about 30 minutes, then ice as a refreshing summertime beverage.

Fresh or dried mint, mint sauce, or mint jelly always goes well with lamb. Try using mint instead of basil in a favorite pesto recipe. It is particularly tasty in fresh Vietnamese spring rolls. An often forgotten ingredient in salads, finely chopped mint leaves of any flavor add a bright high note.

Mints are also a great addition to desserts, complementing a wide range of chocolate desserts; either incorporated into the dish as a flavoring or used as a garnish. Choose pineapple or apple mint for a particularly tangy sorbet. Mint's fame in every category of candymaking is also well known.

Arise and Shine—Time for a Julep!

A favorite alcoholic beverage of the old South in the US is the mint julep. In Kentucky it was made by placing a bunch of peppermint, top-down, in sweetened bourbon with water, taking care not to bruise the stems to impart bitterness into the drink. A julep is a mixed drink flavored with mint, a word that has been in use for over 500 years. Writers on early travels to the Americas described the practice, like this observation from a 1787 issue of the *American Museum*, "An ordinary Virginian rises about six o'clock. He drinks a julap made of rum, water, and sugar, but very strong." John Davis, in his *Travels to America* (1803), observed, "The first thing he did on getting out of bed was to call for a *Julep*. A dram of spirituous liquor that has mint steeped in it, taken by Virginians of a morning." In time, juleps lost this morning association; a writer traveling through Cincinnati in 1838 suggested drinking a mint julep before breakfast, during hailstorms, at dinner, and at night.

Sweet Cicely

Myrrhis odorata

In older works it was called "sweet chervil," "sweet-scented chervil," "great chervil," or "garden myrrh." Historically called *Scandix odorata* and *Osmorrhiza odorata* by botanists, sweet cicely is largely relegated to obscurity, except by the most adventurous herb gardeners and cooks.

Historical Origins Use of sweet cicely in cooking declined in the late nineteenth century. In the early twentieth century, its North American use was limited almost exclusively to immigrant populations.

Charles Bryant, in *Flora Dietetica, or History of Esculent Plants, Both Domestic and Foreign* (1783), sums up the use of the herb for the last 300 years: "[The flowers] are succeeded by long, angular deep-furrowed seeds, which when chewed, have a sweet, aromatic, flavor like Anise-Seeds. The leaves have nearly the same flavor, and are employed in the kitchen as those of the *cerefolium* [chervil]. The green seed chopped small and mixed with Lettuce or other cold sallads, give them an agreeable taste, and render them warm and comfortable to the stomach."

Botanical Facts Sweet cicely is a graceful plant, with finely divided fernlike, downy, gray-green leaves, with a slight anise fragrance. It grows 2–3 ft (0.6–0.9 m) high. In summer it produces umbels of white flowers that may stretch to 3–4 ft (0.9–1.2 m) in height. Found throughout much of Europe in open woods along stream banks, it is an excellent plant to naturalize along the edge of a garden where there is dappled shade and moist soil.

The long, black, shiny seeds are up to 1½ in (35 mm) long, and retain their viability only when fresh. Once a year old, they will no longer germinate. Native to the Alps, Pyrenees, and mountains of the western Balkan Peninsula, this herb is widely naturalized throughout the rest of Europe and cultivated elsewhere.

Culinary Fare In the modern kitchen it is a forgotten herb that requires creativity to weave into salads, soups, desserts, breads, and to use as a decorative garnish. The fernlike leaves can be finely chopped and sprinkled atop salads, especially fruit salads with apples, for a mild anise flavor. Whole leaflets can be used as a decorative center or edge for salads. The sweet stems make excellent swizzle sticks in cocktails.

Pick the delicate white flowerheads and pluck the tiny flowers, spreading them like lace over a cake, salad, or soup.

The white, carrotlike roots have a flavor that is similar to the leaves and, sliced thinly, make an excellent addition to soups.

BELOW: *Sweet cicely is an excellent bee plant. Beekeepers once rubbed a handful of leaves on new beehives to attract bees.*

Basil

Ocimum basilicum Basil *Ocimum citriodorum* Lemon Basil
Ocimum americanum Spice Basil *Ocimum sanctum* Sacred Basil, Holy Basil

Often associated with Italian culinary fare, basil is actually a native of the Old World tropics, but has been cultivated elsewhere for thousands of years. Lemon basil is a hybrid that has evolved only in the last century. India's sacred basil is an ancient, highly venerated herb.

Historical Origins In Italy, basil was a plant that was associated with courting. Young girls wore a sprig of the leaves on their waist or bosom as a symbol of availability. Married women wore basil in their hair to mark their position in life. In Tuscany, when courting, young women wore a sprig behind the ear, earning this herb the name *amorino*, or cupid.

Sacred basil, known in India as *tulasi*, has many parallels to the veneration of basil in Europe. Ancient texts were devoted to this plant, beloved by the gods and those who worshipped them. Sacred basil protected the body in both life and death; its supposed ability to enhance fertility earned it the name "giver of children." An immortal plant of the gods, it is worshipped in hymns and embodies perfection itself—the mystery of the creator is in the mystery of *tulasi*.

Botanical Facts Basil loves hot weather, exploding in growth during summer months when temperatures stay above 60°F (15.5°C) at night. It will succumb to the first frost.

Common basil is an erect annual, growing to about 3 ft (0.9 m) high. Lemon basil is a hybrid between basil (*Ocimum basilicum*) and spice basil (*O. americanum*) characterized by a pleasing citrus scent. This dwarf basil is an annual about 1 ft (0.3 m) high. Sacred basil is a native of Malaysia and India. It grows 1–2 ft (0.3–0.6 m) tall and has many branches. Basil experts believe this herb is seldom seen in US horticulture.

Culinary Fare Basil, especially fresh basil, has become one of the most popular culinary herbs. It is often associated with Italian cooking for tomato sauces and pesto.

ABOVE: *Basil varies from pale to dark green through to dark red and purple. An excellent container plant, pinch back regularly to encourage bushy growth.*

Pesto is made by pounding fresh basil leaves with garlic, pine nuts, olive oil, and grated parmesan; a blender or food processor does the job easily. Typically associated with Genoa and the Ligurian region of Italy, it is used as a pasta sauce. Pistou, a similar puree from Mediterranean France, is a traditional accompaniment to a rich vegetable soup. A few basil or lemon basil leaves make an excellent addition to green salads.

In Thai cuisine, sacred basil flavors stir-fries of chicken, added toward the end of cooking so its fragrance is retained. The small black seeds of purple basil, when soaked in water, swell and become jellylike, and are used in Southeast Asian drinks and desserts.

Marjoram

✳ *Origanum majorana* Sweet Marjoram
✳ *Origanum onites* Pot Marjoram

RIGHT: *Marjoram makes an excellent container herb. It thrives in full sun but needs some shade during summer; it needs watering often to avoid drying out.*

Unless pot marjoram is obtained from a speciality herb nursery, sweet or common marjoram (*Origanum majorana*) is likely to be the marjoram encountered at the local nursery or as a dried herb. Marjoram has a sweet pleasant fragrance and is sometimes used in perfumes for its warm-spice and woody notes. Pot marjoram (*O. onites*) is not quite so gracefully scented.

Historical Origins Introduced from Portugal to England in 1573, sweet marjoram is often associated with the English cottage garden. Pot marjoram has been cultivated in English gardens since the middle of the eighteenth century. Believed to be the *amaracus* of writers such as the Greek Theophrastus and the Roman Pliny, that name derived from a Greek myth in which a youth named Amarakos, employed by the king of Cyprus, was carrying a case of perfumes to the court. He dropped it, and fell unconscious, either from embarrassment of his carelessness or the overwhelming fragrance of the perfumes. The gods then changed him into a sweet herb, *amarakos*—the Greek name for pot marjoram.

BELOW: *Marjoram features in* Tacuinum Sanitatis, *a c.1385 Latin handbook on health. It was translated from the eleventh-century Arabic work* Taqwim al-Sihha *by Ibn Butlan.*

Botanical Facts Marjoram, better known as sweet marjoram, is a tender perennial often treated as an annual in US gardens, growing to a height of 18 in (45 cm). It has small, oval, mostly toothless leaves, and white flowers in a compact knotlike flowerhead. It originates from North Africa and Southwest Asia and is naturalized in southern Europe. Sweet marjoram has been cultivated for hundreds of years. Pot marjoram is a tender and finicky perennial with a hairy stem, growing to over 1 ft (0.3 m) in height, blooming from July through November. In southern Europe, it is a dwarf shrub with woody stems at the base. The flowers are in tight ball-like flowerheads. It is native to the Mediterranean region, particularly Crete and Greece, and the southern Adriatic coast.

Culinary Fare Sweet marjoram lends its delicious aroma to omelets, stuffings, soups, and tomato dishes where oregano may be deemed too harsh. In England, it is a necessary ingredient in veal potpie. Where a good beef stock is required for a stew or soup, sweet marjoram lends a delectable piquancy. A sprinkling of dried sweet marjoram or a small amount of fresh leaf, finely chopped, is an excellent complement to omelets or scrambled eggs, meatloaf, meatballs, soups, poultry, salads, and sauces. Dried sweet marjoram is widely available wherever dried herbs are sold.

Pot marjoram is most widely used in Turkey, where its strong oregano flavor makes it a popular herb to flavor lamb and bean dishes. With boiled onions or shallots, pot marjoram extends a warm flavor dimension to otherwise predictable fare.

Oregano

Origanum vulgare

Say spaghetti sauce or pizza and the culinary mind inevitably turns to oregano. In commercial flavoring, oregano surprisingly refers more to a flavor than an individual plant. Plants with an oregano flavor have an essential oil with a compound called carvacrol as their main flavor component.

Historical Origins Throughout most of history, oregano was more likely an herb of the apothecary's shelves rather than the shelves of the kitchen cabinet. A tea of oregano was considered useful as a mild digestive stimulant and an aid in relieving flatulence. Until recently, sweet marjoram was considered the finer choice for the kitchen, but the popularity of pizza over the last half-century or so and its association with the flavor of oregano has greatly increased this herb's use.

The genus name *Origanum* is said to originate from two Greek words, one that means mountain and the other meaning joy; hence it was the "delight of the mountains." John Lightfoot, author of the *Flora Scotica* (1789), observed mountain peasants in Sweden adding oregano leaves to their ale to give it a more intoxicating quality and prevent the ale from turning sour. The state name Oregon is said to express Spanish explorers' impression of a small, oreganolike, aromatic herb growing near the state's coast.

Botanical Facts Oregano is the best known of the more than 36 species of the genus *Origanum*, mostly associated with Mediterranean climates. Native from the Mediterranean region to central Europe, oregano is by far the most widespread species in the genus. It is naturalized in the northeastern United States, occasionally being found around the sites of very old gardens or farms. A perennial, reaching 2–3 ft (0.6–0.9 m) in height, it has branching, soft, hairy stems. Leaves are mostly oval and can be smooth or hairy; the white to purple-red flowers bloom from summer to autumn.

Culinary Fare The herb is given surprisingly little attention in many classic herb books published before the 1940s. Oregano's popularity in the US began in the 1940s, following the return of US servicemen from Italy after World War II. They had become accustomed to eating pizza that was flavored with oregano. This herb is most widely used to flavor all manner of Italian tomato sauces. It is excellent with many beef dishes, beans, deviled eggs, omelets, and rice dishes. The list is as endless as the imagination! Oregano is almost always used in cooked fare, as it has a somewhat biting quality when fresh. It is best used sparingly, as the strong flavor can overwhelm any dish that uses it.

BELOW: *Oregano is a hardy plant needing little maintenance. Though not fussy, it prefers a sunny spot. Its insect-repelling qualities make it an excellent companion plant in a vegetable garden.*

Rosemary

✳ *Rosmarinus officinalis*

Rosemary is an herb whose position is established in the annals of literature, medicine, and cuisine. Its virtues were often extolled by Ancient Greek, Roman, Arab, and European herbalists. The species name "*officinalis*" means it was once the "official" rosemary of the apothecary's shop. Today it is the official rosemary of the kitchen and the culinary artist.

Historical Origins The genus name *Rosmarinus* is derived from the Latin *ros*, meaning dew, and the Latin for seacoast, *marinus*. Often found on rocky outcrops along the Mediterranean coast, rosemary actually means "dew of the sea." Though a plant of southern Europe, it was cultivated in Britain prior to the Norman Conquest in 1066, having been taken there by the Romans. In the eighth century, Charlemagne ordered rosemary be grown on his farms. In thirteenth-century Spain, it was listed among the aromatic plants traded by Arab physicians.

Rosemary was once highly valued for its medicinal properties. The Roman physician Galen recommended an infusion of rosemary to cure jaundice. Nineteenth-century herbalist Nicholas Culpeper wrote that rosemary flowers not only "comfort the heart" but also "expel the contagion of the pestilence."

BELOW: *Tying a bunch of fresh rosemary leaves to retrieve from pureed soups allows the leaves to flavor without adding texture.*

Most famous for its association with comforting the brain and strengthening the memory, rosemary became a symbol of the fidelity of lovers, worn in the hair or strewn on the ground at weddings. Hungary Water, believed to be a secret recipe delivered to a Hungarian queen by a hermit, was for many centuries touted as a panacea. According to a manuscript of 1235, Hungary Water is a distillation of rosemary, lavender, and myrtle.

Botanical Facts Rosemary is native to the Mediterranean region of North Africa and southern Europe, including Portugal, Spain, southern France, Italy, Greece, and the Mediterranean islands. This densely leaved evergreen shrub, prostrate to erect, and over 6 ft (1.8 m) tall in habit, has linear, near stemless leaves. They are thickened around the edges, and when stroked emit a lovely, strong, almost piney scent. Studied closely, the relatively small white-to-purple flowers appear almost like a small orchid.

Rosemary's native haunts are coastal regions of the western European and western North African coasts. Here it is often found sprawling over rocks, basking in the afternoon sun, or glistening in foggy mists in the evenings and mornings. With a very similar Mediterranean climate, the coast of

LEFT: *Rosemary likes a hot dry position and will grow in any type of soil as long as it is well-drained–it does not tolerate wet winter conditions. Prune severely after flowering to maintain a compact habit.*

California, from San Francisco south to San Diego, in the United States, provides the perfect habitat for rosemary, where it is often planted as an evergreen ornamental groundcover and herb in many gardens.

Since rosemary is so widely cultivated as an ornamental, numerous cultivars have emerged in the US horticultural trade. There are over 70 cultivars described. Some, such as 'Albus' sport white flowers. Others, such as 'Aureus' feature green foliage with yellow striations. Many cultivars boast beautiful blue-violet flowers. 'Cascade' and 'Alida Hyde' are among the prostrate cultivars available. Selections such as 'Arp', named for Arp, Texas, are known as hardy rosemary, surviving winter temperatures from 0°F (–18°C) or warmer. Cultivars for the garden should be selected on height, flower color, and fragrance. Stroke the leaves before buying a plant at a nursery and pick a cultivar that is pleasing to the nose.

Culinary Fare Rosemary is one of the most versatile herbs in the kitchen. In the fifteenth century, rosemary was commonly used as a condiment for salted meats, undoubtedly not only adding flavor, but helping to preserve the meat as well. Rosemary extracts are used commercially in the modern food industry as a preservative due to the herb's antioxidant activity.

Fresh or dried rosemary goes well with any red meat, particularly lamb and also for beef, such as steaks, filet mignon, roast, even hamburgers. Fresh leaves can be bruised and rubbed on the meat before cooking or barbecuing. In closed gas barbecues, add generous sprigs of rosemary to infuse meat with its delicious flavor.

Finely cut, rosemary is an excellent flavoring for potato dishes and other root vegetables such as carrots and onions. It is a delectable addition to tomato-based soups, stews, and sauces. Added to spaghetti sauce, it brings out the flavor of other ingredients.

"There's Rosemary, That's for Remembrance"

Shakespeare's Ophelia utters these words in *Hamlet*, reflecting the long-held belief in rosemary as an emblem of remembrance. In his collection of sonnets, *A Handful of Pleasant Delights* (1584), Clement Robinson penned the more memorable line "Rosmarie is for remembrance" from the lines of "A Nosegay, Always Sweet for Lovers ..." Nicholas Culpeper repeated the long-held belief in his 1653 herbal that "It helpeth a weak memory and quickens the senses."

There may be more reasons than ever to remember to use rosemary in cooking. Researchers have studied the antioxidant activity of rosemary. The damage caused to cells by highly reactive oxygen forms can lead to such problems as coronary arteriosclerosis and diabetes, both associated with reduced brain function in the aging process. Rosemary has been shown to have antioxidant potential to protect tissues and cells against various oxidative stresses, which may also have the unintended consequence of improving brain function.

Ophelia offers rosemary to her brother Laertes to remind him to remember their father.

Parsley

Petroselinum crispum

ABOVE: *There are many cultivars of garden parsley, with a wide variety of growth habits, leaf forms, and flavor strength. The main types are Italian parsley (left) and curled parsley (right). Each has distinctive applications.*

Parsley, the most heavily consumed fresh herb in the United States, is one herb that is familiar to all. For many decades it has been the primary herb used as a garnish in restaurants. Often it is the only thing left on the plate at the end of the meal, when it may actually have been the most nutritious item.

Historical Origins Parsley has been cultivated for many centuries and is now naturalized throughout much of Europe. Thought to have hailed from southeastern Europe or western Asia, its exact origins are now obscured in history. Parsley's genus name *Petroselinum* comes from 2 Greek words—*Petros* meaning rock, from its propensity for rocky cliffs and old stonewalls; and *selenium*, an ancient name for celery —so one can think of it as "rock celery." An old English superstition suggests that bad luck will prevail if one transplants parsley, so be sure to plant it in its final place in the garden. In Ancient Greece, parsley was once planted on graves, and the phrase "to be in need of parsley," meant death was imminent.

Botanical Facts A biennial growing up to 2 ft (0.6 m) in height when in flower the second year, it can reach about 1 ft (0.3 m) in the first year before flowering. Parsley is generally treated as an annual, providing tasty fresh leaves only in the first year. There are 3 basic types: curled parsley has crisped or wrinkled leaves, and is the most familiar parsley of commerce; Italian parsley is flat-leaved (its leaves are not crisped); and 'Tuberosum', seldom grown today, is produced for its large carrot- or turniplike roots.

Culinary Fare Fresh young parsley leaves, whole or chopped, are an excellent addition to any salad. Parsley salads, such as the Middle Eastern tabouleh, are not only delicious, they are also extremely nutritious.

The fresh leaves are an admirable adornment for any meal. Its use as a garnish arises from a centuries-old belief that at the end of the meal, chewing a few fresh parsley leaves freshens the breath. It was even believed that chewing the leaves would make the odor of garlic imperceptible. Parsley is also an outstanding flavoring for tomato dishes, baked potatoes, various fish dishes, egg dishes, and white sauces. To make a delicious herb butter, take two tablespoons of finely chopped, fresh parsley leaves and knead them into a quarter of a cup of softened butter.

Not only tasty, parsley is high in vitamins A and C, fiber, potassium, magnesium, and iron. Surprisingly, the leaves also contain a significant amount of protein.

Common Sorrel

Rumex acetosa

Common sorrel is sometimes called sour dock. The leaves were once considered a cure for scurvy, unsurprising given their high level of vitamin C.

Historical Origins Thought to originate in Asia, common sorrel is a perennial found throughout almost all of northern Asia and Europe. It is naturalized in North America, where it is regarded as a weed or, shall we say, a "forgotten" herb. Chinese names for the plant are *niu-er-da-huang*, translating into "ox-ear rhubarb," and *jin-bu-huan* meaning "no-exchange-for-gold," perhaps denoting its common status.

Botanical Facts In Europe it is a common plant of meadows and grassy pastures. In North America it is found in fields around old homesteads. The long-stalked lower leaves are oblong, somewhat arrow-shaped, and blunt at the summit, with 2 teeth at the leaf base. Upper leaves are without stalks, hugging the stem, and are narrower. It produces panicles of tiny red-brown flowers in spring.

Culinary Fare The tangy acidic flavor of common sorrel is best appreciated in soups and sauces, especially when paired with salmon. Young leaves can be added sparingly to salads to impart a somewhat acidic spinachlike dimension. Once the plant begins to flower and develop red veins in the leaves, it is considered less edible.

ABOVE: *Sorrel's slightly succulent, vivid green leaves can be used in salads or cooked. Young leaves harvested before flowering are used as a potherb in China.*

Rue

Ruta graveolens

Rue is a great plant for the herb garden, given its beauty earned it the name "herb of grace." It was used as a condiment in the ancient world, but now has been replaced by more palatable condiments, such as black pepper.

Historical Origins Rue is a plant always approached with caution. Since the writings of the first-century Greek naturalist and Roman physician Dioscorides, folk tradition has warned that internal consumption can be dangerous during pregnancy. Rue, like poison ivy, is also known to cause contact dermatitis from handling the fresh leaves or the plant. However, poisoning by ingestion of rue is primarily of historical interest.

Botanical Facts Rue hails from the Balkan Peninsula and Crimea, but has been naturalized for many centuries throughout much of Europe, escaping from gardens. A casual glance might not suggest it, but rue belongs to the same botanical family as citrus fruits—the rue family. It is a graceful perennial, with blue-green, smooth, rounded, lacy leaflets. Rue can grow up to 3 ft (0.9 m) tall, but usually only attains a height of 2 ft (0.6 m).

Culinary Fare Rue's culinary contribution is limited by both flavor and potential toxicity. The leaves are among the most bitter of herbs and have been used as a flavoring ingredient in pickles. A minute pinch of the fresh herb can be used as a flavoring on fish or perhaps in a salad dressing.

ABOVE: *Rue is easily grown in any well-drained soil, ideally in full sun. Once established, plants can be trimmed to shape, but hard pruning is seldom necessary. It can be propagated from cuttings or seed.*

Sage

Salvia officinalis Sage *Salvia elegans (syn. Salvia rutilans)* Pineapple Sage *Salvia sclarea* Clary Sage

ABOVE: *Common garden sage is easy to grow, liking sunny well-drained soils. It reaches 2 ft (0.6 m) high. It dislikes the cold; in cooler climates, it is best grown in a container and brought indoors in chilly weather.*

Sage, represented by the genus *Salvia* in the mint family, is one of the most diverse groups of aromatic plants, containing over 900 species. It is known for its variety of fragrances and flavors, along with its beautiful flowers, which are produced in summer.

Historical Origins The name *Salvia* is derived from the Latin word *salvare*, meaning to heal or save, which is sometimes translated as "good health." The species name of common sage "*officinalis*" means it was at one time the official sage of the apothecary.

The species name "*elegans*" for pineapple sage reflects this herb's elegant beauty.

Clary sage was once considered useful for removing objects from the eye; its species name "*sclarea*" means to clarify. If soaked in water, the seeds become mucilaginous and were used to treat inflamed eyes.

Botanical Facts Common garden sage, cultivated for centuries, is native to southeastern Europe and Asia Minor. Over half of the world's sage supply is still harvested from the wild in the mountains of Albania, Montenegro, and Croatia. It grows to about 2½ ft (0.75 m) high, and is characterized by long, narrow, white-woolly, grayish green leaves, with a surface reminiscent of reptile skin. The white to dusky mauve flowers are a chief ornamental feature. Pineapple sage has oval pointed leaves with a distinct pineapple fragrance and 2-in (5-cm) long brilliant scarlet flowers. The fuzzy, deeply veined, broad, oval leaves of clary sage, with its spikes of beautiful light violet flowers, make it an excellent biennial for the herb garden.

Culinary Fare In British cooking, sage is best known as the primary seasoning and flavoring ingredient in poultry stuffing and sausages, but in France and Italy it is often paired with pork. It is an important flavor component of condiments. Use sparingly, as it has a very strong, sometimes bitter, flavor, making it a vital ingredient in the manufacture of liqueurs and bitters. Pineapple sage makes a pleasant herbal tea, and it is an excellent addition to cold drinks, such as an iced green tea. The beautiful red flowers, plucked individually from the flower spikes, make an attractive garnish for fruit salads.

Clear Choice for the "Foolish Drunke"

Clary sage, sometimes called "muscatel sage," is primarily used as a flavoring for alcoholic beverages, adding a muscatel note to wines, vermouth, and liqueurs. The designation "muscatel sage" arose in Germany when wine merchants infused clary sage with elder flowers, then added the mixture to cheap wine to give it the flavor of muscatel wine. Matthias de Lobel in his 1576 *Plantarum Seu Stirpium Historia*, reveals, "Some brewers of Ale and Beere doe put it into their drinke to make it more heady, fit to please drunkards, who thereby, according to their several dispositions, become either dead drunke, or foolish drunke, or madde drunke."

Salad Burnet

Sanguisorba minor

The name salad burnet might imply that the leaves could be freely fluffed as a salad base. However, the small, scallop-edged leaves, with a discernible flavor at first, soon leave a slightly bitter aftertaste on the palate.

Historical Origins The genus name *Sanguisorba* comes from Latin words meaning "blood" and "drink-up," in reference to the herb's traditional property as a styptic to stop bleeding. Colonial soldiers in the American Revolutionary War (1775–1783) are said to have drunk salad burnet tea before engaging in a skirmish to prevent copious bleeding, should they be wounded.

Botanical Facts Only those possessing the technical skills of a trained botanist can look at salad burnet's tiny green flowers in their tightly crowded compact flowerheads and realize that this plant is actually a member of the rose family. This herbaceous perennial grows to 18 in (45 cm) tall, and is native to dry chalky grasslands and rocky ground from much of southern, western, and central Europe.

Culinary Fare Salad burnet is almost always used fresh, and sparingly at that. Like many rose family members, the leaves are high in tannins and produce an astringent or puckering sensation. The leaves are also used in cooling drinks. The tender young leaves of spring are the tastiest, giving salads a slightly cucumberlike flavor.

RIGHT: *The attractive, rounded, pink flowerheads of salad burnet appear in early to mid-summer. It grows best in a sunny position and likes a little moisture in the soil.*

Savory

Satureja hortensis Summer Savory *Satureja montana* Winter Savory

Winter savory is generally plucked fresh from the herb garden or dried for future use. It has a sharper, more biting flavor than summer savory.

Historical Origins The genus name *Satureja* is derived from an Arabic word *za'atar*, used to describe all herbs with a scent like oregano. The species name of summer savory (*hortensis*) means "of the garden." Winter savory's specific epithet *montana* denotes "of the mountains." Used medicinally, they were considered warming to the stomach, and, as Charles Bryant put it in his 1783 *Flora Dietetica*, " … good against crudities of the stomach."

Botanical Facts The genus *Satureja* includes about 30 species, most of which are dwarf shrubs hailing from the Mediterranean region. Summer savory is a bushy annual crowded with short narrow leaves about 1 in (25 mm) long and light blue flowers. Native to southern Europe, it is easily grown from seed. Winter savory is a perennial that commonly grows in rocky limestone outcrops throughout southern Europe.

Culinary Fare Tomato-based soups and sauces can benefit from a gentle sprinkling of dried savory. Summer savory is often used to flavor bean dishes. Cabbage's relatively bland flavor can be greatly improved with the judicious use of a little summer savory. Both types have been used as flavoring in sausages and stuffings, as well as vermouths and bitters.

BELOW: *Summer savory is similar to winter savory but it is the more widely known of these herbs. Enjoyed in Ancient Rome, it is the most commonly used herb in Bulgarian cuisine.*

Thyme

Thymus x citriodora Lemon Thyme *Thymus vulgaris* Garden Thyme
Thymus serpyllum Wild Thyme, Creeping Thyme

RIGHT: *Most thymes are small herbs and shrubs, woody at least at the base. They make ideal container plants. Allow them to dry out in between waterings.*

The genus *Thymus* in the mint family contains upwards of 350 species. Most of these are associated with the Mediterranean region, where nearly 70 species are known. The bulk of thyme species are spread throughout dry rocky soils in western Asia and Europe. At least 200 selections have entered horticulture.

Historical Origins Thyme is a favorite plant of bees, and classical writers such as Ovid, Virgil, and Pliny extolled the virtues of thyme honey, which has a delicious, light, aromatic fragrance. "To smell of thyme" is an old phrase applied as praise and an expression of admirable style and activity, both considered virtues. Such symbolic virtue is depicted in chivalrous embroidered scarves as a bee hovering over a thyme flower.

Botanical Facts
Native to the western Mediterranean, extending to southern Italy, garden thyme is a small shrub about 8 in (20 cm) high with smooth, strongly aromatic oval leaves and white to pale purple flowers. Wild thyme is found from northeastern France and Austria to northern Ukraine.

This low, creeping, woody plant, with roots at the nodes, produces tight heads of purple flowers. The leaves are slightly hairy at the base. Lemon thyme is a branching perennial hybrid, growing up to 1 ft (0.3 m) high, with pale lilac flowers. Thyme's exact origin is uncertain, though it does appear in horticultural works by the seventeenth century.

Culinary Fare Garden thyme is used to flavor meat dishes, sausages, poultry, and soups and stocks, and is an ingredient of the classic bouquet garni. It is almost always used in the dried form, but fresh thyme from the garden can be used as well. Lemon thyme, as the name implies, has a delicious lemony fragrance and can be used in meat and poultry dishes, as well as desserts and cold beverages. Wild thyme is seldom seen in the herb trade, so it is generally collected from the garden, if available, and used in a similar manner to garden thyme.

Thyme for Medicine

Thyme was commonly cultivated in England by the sixteenth century, but its reputation was that of a medicinal plant, primarily to treat lung conditions, rather than as a delightful flavoring in food. Today, best known as a culinary herb, it has once again emerged as a medicinal plant. In Germany, thyme is approved for therapeutic use, labeled for "symptoms of bronchitis and whooping cough, and catarrhs of the upper airways." About a teaspoonful of the dried leaves are steeped in hot water, and the tea sipped for medicinal benefit. Various proprietary phytomedicinal thyme products are also available in Europe. The chief constituent of oil of thyme—thymol—is a powerful antiseptic, widely used before the advent of antibiotics. Many consumers are unwittingly familiar with the fragrance and flavor of thymol as a chief ingredient of those mouthwashes with a "medicinal taste."

Thymol Toilett was a French thyme-based disinfectant used in the nineteenth century.

Fenugreek

Trigonella foenum-graecum

Cultivated since the time of the Ancient Assyrians, and found in the tomb of Tutankhamen, fenugreek seeds are an ancient herb that is native to southern Europe and western Asia.

Historical Origins Fenugreek is thought to have been cultivated as animal forage long before recorded history. The Egyptians made an incense, *kuphi*, with fenugreek leaves as a key ingredient for ceremonies associated with fumigation and embalming rites. Introduced to China in the Song Dynasty, fenugreek has been grown there for over 1,000 years. A famous Roman writer on agriculture in the second century BCE recommended fenugreek seed as a fodder for oxen. But don't tell the meat inspector you have fenugreek-fed beef. An inspector's handbook from 1907 suggests while fresh fenugreek fodder will rapidly fatten animals, a single feeding makes the meat take on a disagreeable taste and the odor of hog dung.

Botanical Facts Depending on the botanical authority, the genus *Trigonella* in the pea family contains upwards of 100 species. Fenugreek was cultivated in central Europe by Charlemagne as early as CE 812. It is an annual, preferring warm climates, growing to about 2 ft (0.6 m) tall, with smooth pealike leaves. The small white flowers produce a sickle-shaped pod about 3–4 in (8–10 cm) long, with 10 to 20 hard brown-yellow seeds, with a mucilaginous pea- or beanlike taste and a slight vanilla note.

Culinary Fare Fenugreek is these days primarily used in commercial flavoring more than as a culinary herb in the kitchen. It has a strong, somewhat bitter—and to some, unpleasant—flavor that relegated it to becoming a minor ingredient of curries during the nineteenth century. In India, the protein-rich leaves are available in markets when young, with only 2 true leaves on the stem. Bunches of the herb are boiled then fried with butter, but possess a taste that is "strongly bitter, and disagreeable to those who have not become accustomed to it," according to one nineteenth-century British observer. The leaves, as a potherb, are also used to make an imitation maple syrup flavor.

Fenugreek seeds contain coumarins, usually associated with a vanilla flavor, and are used in India for flavoring soups. They also impart body to soup, due to their mucilaginous nature. The seeds are also used in curries. In cultures of North Africa, the seeds have been ground and used to stretch flour. In Switzerland, the seeds are used commercially as a flavor for cheeses. Although not a common ingredient in the supermarket, fenugreek seeds are widely used as a flavoring in packaged foods.

BELOW: *The flavor of fenugreek seeds is variously described as reminiscent of burnt sugar or maple syrup. The bitterness of the raw seeds is toned down by roasting then grinding.*

Spices

Introduction to Spices

Although we take for granted the easy availability of spices, they were once highly prized, expensive commodities, their acquisition shrouded in secrecy. Like the spices themselves, their history is colorful, romantic, and slightly dangerous. Empires and fortunes were made and lost in their pursuit. Once attained, the taste for spice became an intoxicating need that ultimately shaped the world's history.

The word spice originated from the Latin word *species*, which came to mean a valuable product with little volume. A spice is defined as an edible product obtained from the fruiting body of a plant used to add aroma and flavor to food, though spices are often also used for medicinal purposes. Their flavor comes from the essential oils that are present in the spice.

ABOVE: *As spices lose their flavor and color over time, it is important to store them correctly. They will last longer when kept in airtight containers away from light and humidity. A spice rack, while attractive, is not ideal, especially if it is exposed to light.*

CLASSIFICATION OF SPICES

True spice classifications are difficult to establish because absolute boundaries between the various groups are nearly impossible to set. For that reason, spices are usually classified by the particular part of the plant that is used:

Flowers Cloves and saffron.
Seeds Nutmeg, mace, anise, cumin, sesame, mustard, caraway, poppy, mahlab, monkey pepper.
Roots, root stalks, and rhizomes (Rhizomes are the fleshy stems.) Galangal, ginger, turmeric, sassafras, and wasabi.
Leaves and stems Curry leaf, sassafras.
Fruit Cardamom, allspice, pepper, Sichuan pepper, vanilla, tamarind, star anise, sumac.
Bark Cinnamon.

GODS, CONQUERORS, EMPIRES, AND EXPLORERS

Early civilizations refer to spices in their lore, writing, and artwork. An ancient Assyrian myth claims gods consumed sesame wine the night before they created Earth. Cloves were found at a 2400 BCE Sumerian site and 1,000 years later, Egyptians used spices for medicine and embalming. By 1800 BCE, Arabic traders had established caravan routes through India and China. Arabic traders dominated the spice trade by 500 BCE, a position they maintained until the early first century.

After Alexander the Great conquered Egypt in 331 BCE, Alexandria was established as a key port for the Arabian spice trade. When it came under Roman rule in 80 BCE, Rome broke the Arab trade stronghold with the discovery of the East Indian trade winds. The monsoon cycle caused winds to reverse midyear—this allowed the Romans to sail to India quickly and safely. By the end of the precommon era, their trade with India flourished and the Arab monopoly was crippled.

By the tenth century, Muslim forces conquered trading posts and took control of the spice route, bringing down the "Islamic Curtain," ending the westward flow of spices. Eventually, the potential profits from the trading ports of Venice and Genoa proved too tempting for the Muslims and spices returned to Europe.

From 1097 to 1490, the Crusades introduced more spices to Europe and demand exploded. Private merchants such as Niccolo Polo and his son Marco set out from Venice to bring spices and other riches back from the Far East. The plague is thought to have

LEFT: *The interior of a nineteenth-century Parisian spice shop, as depicted in a French engraving. Most spices at this time came from India, Indonesia, and Malaysia.*

The bold aromas and vibrant flavors and colors of spices will always titillate the senses and offer up an exotic pathway to culinary mysteries from far-away lands, brought to our table.

accompanied the returning merchants. High mortality rates soon closed the overland routes to trade. Venice became the clearing-house of all spices entering Europe. In 1453, excessive tariffs incited the Portuguese and Spanish search for a bigger piece of the pie.

In the late 1400s, explorers Christopher Columbus and Vasco da Gama independently set sail in search of a direct route to the land of spices. Both explorers, and those who followed, paved the way for more exploration, European dominance, and colonialism.

Holland influenced the spice trade by supplying ships and crews to the Portuguese. By the late 1500s, the Dutch controlled the spice trade. The French cashed-in on profits by stealing spices from the Dutch and cultivating them on French islands in the Indian Ocean. In the 1600s, after failing to find their own spice route to the east, Britain chartered the British East India Company to gain control of India. Britain overpowered the Dutch, and by 1799, the Dutch lost all spice trading centers. In colonial America, merchants entered the spice race in the late 1600s. By the late 1700s, the American spice trade was ablaze. The United States remains the world's biggest spice buyer.

The spice trade has experienced high and low points with countless horrific atrocities resulting from the desire for spice.

STORAGE AND PREPARATION

The first foray into the world of seasoning was probably accidental. Food stored in bark and leaves, or porridges laced with seeds or nuts, produced distinctive flavors. Medicinal foods often included many popular culinary herbs and spices and served to introduce the unique flavors of spices into the diet. Natural curiosity and a developing palate gave rise to the quest for taste.

With few exceptions, spices are dried and can be used whole or ground into a powder. Dried spices are best stored in a cool, dark, airtight environment. Spices that are the plant's rhizomes, which are its fleshy roots, should be stored in the fridge. Whole spices last longer than ground spices. Dry spices release their flavor if they are lightly toasted before grinding, or heated in oil, just prior to cooking.

ABOVE: *Wooden spice cannisters from England, dated to around the early eighteenth century, were the most common form of spice storage at this time. They usually contained airtight tins inside.*

"... merchants stood as a driving force behind the heroes of the age of discovery; this first heroic impulse to conquer the world emanated from very mortal forces—in the beginning, there was spice."

Stefan Zweig (1881–1942), *Der Mann und seine Tat* (*Magellan. The Man and His Feat*), **p. 20, trans. by Marion Sonnenfeld, S. Fischer Verlag (1983).**

Galangal

Alpinia galanga

RIGHT: *The edible part of galangal is the rhizome or fleshy root stem of the plant. It is quite fibrous and needs to be very finely chopped or grated before adding to the pan.*

Galangal is a tropical plant in the ginger family that is most often associated with Asian and Indian cuisine.

Historical Origins *Alpinia galanga* originates in China but is also found in parts of India, East Asia, and Southeast Asia. Its original name, greater galingal, came from an Arabic word based on the Chinese word for ginger. Arab or Greek physicians probably introduced it to Europe. Galangal is used to prevent nausea, as a stimulant, an antibacterial agent, and a body deodorizer. Nineteenth-century pharmacists mention two ninth-century Arabs writing of the medicinal benefits of galangal and also reported Welsh physicians using it in the early thirteenth century.

Botanical Facts Galangal plants reach 6–7 ft (1.8–2 m) and have narrow upright leaves. Petite flowers are greenish white with deep red veins, and form red berries. Fleshy stems are brown to orange with dark brown rings and pink-tipped eyes. Galangal is propagated by planting small fleshy stems in rich moist soil in the shade. Plants require tropical temperatures. It is ready for harvest when the leaves have yellowed or died.

Culinary Fare Galangal tastes similar to ginger, but with a bit more peppery heat. It can be added chopped or grated to impart a pungent tang to Asian-inspired dishes. The leaves and young shoots are also edible. The dried form is sometimes called laos powder.

Chinese Ginger, Fingerroot

Boesenbergia rotunda

Also called fingerroot or Chinese keys, Chinese ginger, from the ginger family, is grown for its flavorful, fingerlike fleshy stems. It is most often associated with Asian cuisine.

Historical Origins *Boesenbergia rotunda* originates in southern China and Southeast Asia. Despite the name, Chinese ginger is not actually eaten in China. Its preferred use is medicinal—to treat stomach ailments, as a tonic after childbirth, and for infections.

Botanical Facts Plants can grow to just over 2 ft (0.6 m) tall. They have broad, leathery leaves supported on sturdy stems. Pale pink flowers with dark pink or purple ruffled edges are borne just above the ground and are often hidden by the foliage. The edible fleshy stems are tan to yellow and are tightly bunched, like long slender fingers. Chinese ginger is propagated by root division and planted in rich moist soil in a semishaded spot. Plants require semitropical temperatures and high humidity. It is ready to harvest when the leaves yellow and die back.

Culinary Fare Chinese ginger has a lemony gingerlike taste and aroma. It is frequently used grated or chopped in Thai and Indonesian cuisine as a flavoring agent. It can also be pickled and served as a condiment, or it can be cooked whole and served as a vegetable.

Black Mustard

Brassica nigra

Black mustard is an annual weedy plant in the cabbage family, grown for its tiny black seeds. In many cultures, the name mustard usually refers to the condiment rather than the seed. Unlike other species of mustard, the leaves of black mustard are not consumed. They are somewhat toxic to mammals, which keeps most animals at bay. Two other members of the mustard family are grown for their seeds—*Brassica juncea*, brown mustard, and *Sinapis alba*, white or yellow mustard. Mustard is associated with Mediterranean, Indian, and Asian cuisine.

Historical Origins Black mustard shares many characteristics with its close relative, white or yellow mustard, and is believed to have originated in the southern Mediterranean region of Europe. Brown mustard originates in Asia.

Mustard seeds were found in many prehistoric and ancient sites, suggesting it was probably one of the first cultivated plants. Theory has it that the biblical parable of the mustard seed referred to black mustard. Romans brought black mustard seeds to Britain and the plants naturalized in the wild. Black mustard was used extensively in European mustards until the introduction of mechanical harvesting. Since the 1950s, brown mustard seeds replaced black mustard seeds for many applications. Medicinally, mustard has long been used to relieve respiratory illness and muscle pain.

Botanical Facts Plants are very tall, often reaching up to 8 ft (2.4 m) in height. Leaves are large and slightly fuzzy with powdery fleshy stems. Small yellow flowers are borne on long stems producing seeds that separate from their pods when ripe. Since the seeds drop easily, mechanical harvesting is not successful with this species. Black mustard is propagated by seed in rich moist soil. The plants grow best in cool climates; in warmer climates, daytime wilting is not unusual. They easily self-seed and grow wild.

ABOVE: *Illustration of a black mustard plant from* The English Botany; or Coloured Figures of British Plants *written by Sir James Edward Smith, illustrated by James Sowerby, and published in 36 volumes over 24 years, between 1790 and 1814.*

Culinary Fare Black mustard is much stronger than the other mustard species. For this reason its widespread culinary use has waned, though it is still widely used in Indian cuisine. It is more commonly used as an oil-producing seed. Mustard oil is first heated and cooled before use as a flavoring agent. This process helps mellow the taste and bring out its flavor, similar to the toasting technique used with dried spices.

Seeds can be used whole or crushed and added to marinades, pickling spices, soups, stews, and vegetables. Adding the seeds near the end of the cooking process helps preserve the flavor. Seeds can also be heated in oil or ghee (clarified butter) and "popped." This cooking method brings out a nutty flavor that compliments Indian cuisine. If the seeds are ground, an acidic product (often wine or verjuice) is added to the paste to stabilize it.

LEFT: *The hard round seeds of black mustard can be black or brown to reddish brown. Seeds can be ground and their pungency and aroma is released after steeping in cold water for 10–15 minutes.*

Black Cumin

Bunium persicum

Black cumin, also called Kashmiri cumin, kala jeera, or black caraway, is a member of the parsley family. It is grown for its pungent seeds and mild roots and is most often associated with Asian and Indian cuisine. *Bunium persicum* should not be confused with *Nigella sativa*, also called black cumin.

Historical Origins Black cumin is native to central Asia and northern India. The seeds are widely used in these regions as an aromatic spice and also for medicinal purposes, to treat digestive disorders, colds, and fevers.

Botanical Facts Black cumin is a bushy, herbaceous plant with frilly leaves and a rounded taproot. It produces an umbrellalike flowerhead, each with a small white flower. Seeds are dark brown to black, ribbed, and have pointed ends. The seed heads shatter easily, so harvesting requires great care. Propagation takes place by seed or tuberous root division. These plants prefer cold winters, warm growing seasons, and rich well-drained soil. Early rain is essential to promote flower formation but drier conditions are best once seeding begins.

Culinary Fare Once toasted, the seeds have a delightful, nutty, earthy flavor that compliments Asian and Indian cuisine. They can be ground and used to season soups, stews, vegetables, and couscous. Seeds are often added to yogurt, chutney, and many spice blends.

Caraway

Carum carvi

ABOVE: *Caraway seeds are used to add a distinctive flavor to rye bread. This spice is most characteristic of Scandinavian cuisine and it is popular in other parts of Europe.*

Caraway is a biennial plant in the parsley family grown for its flavorful fruit, which also serves as its seed. Its unique licorice and parsley flavor is synonymous with northern European cuisine. Medicinally, the seed is utilized as a digestive aid.

Historical Origins Caraway is native to western Asia and the Mediterranean region. Some believe it is the oldest cultivated spice in Europe. The first recorded use of its highly pungent seeds was in an Ancient Egyptian medical document written in 1552 BCE, though seed traces have been found at archaeological sites dating back to 3000 BCE.

Botanical Facts Caraway has an upright growth habit with graceful lacy leaves and a long thin taproot. Plants attain their full height of 2 ft (0.6 m) in their second year. Delicate flower umbels show at the end of the second season. Caraway fruit is pale tan, ⅛ in (3 mm) long, and crescent-shaped. Propagation takes place by seed in rich well-drained soil, moderate temperatures, and a sunny location. To produce flowers early, sow seeds in the autumn, just after harvest. Flowerheads and fruit should set during the following summer.

Culinary Fare Caraway seed is used whole to flavor soups, vegetables, meats, stews, and sweets. It also flavors cheese, liquor, and liqueur. Roots can be cooked and served like carrots or parsnips and the leaves can be used as a garnish.

Cinnamon

Cinnamomum verum

An evergreen tree in the laurel family, cinnamon is grown for its aromatic bark. *Cinnamomum verum* is often confused with *Cinnamomum aromaticum* (cassia), which has a milder taste and is the more common type of "cinnamon" available.

Historical Origins *Cinnamomum verum* originated in Sri Lanka and southern India The pungent bark has been used in Egypt since 2000 BCE, making it one of the first known spices. It was burned as incense, used medicinally and in the embalming process, and to flavor food and drink. At the time, cinnamon was more highly valued than gold. The Romans believed cinnamon was sacred and the emperor Nero was said to have burned a year's supply to honor his dead wife at her funeral in 65 BCE. Cinnamon leaf wreaths often decorated Roman temples. Once Europeans developed a taste for cinnamon, the demand for this spice was one of the primary motives for exploration.

Botanical Facts The cinnamon tree can reach heights of up to 56 ft (17 m), but when grown for its bark, it is usually kept much smaller. Leaves are dark green on top and lighter green underneath. Plants produce small greenish white flowers with an unpleasant smell. The bark and the leaves both smell of cinnamon. Propagation takes place by seed or by root division under tropical conditions. Two-year-old trees are drastically cut back, compelling the tree to produce a dozen shoots the following year. The bark is stripped from the shoots and dried as cinnamon.

Culinary Fare Cinnamon has a sweet hot flavor that pairs well with sweet and savory dishes alike. While characteristic of Syrian and Lebanese cuisines, it is also associated with Indian, African, Asian, and Mediterranean dishes. It can be ground and added to soups, stews, spice blends, baked goods, and other desserts. It can be used whole or crushed in beverages, marinades, and in pickling.

ABOVE: *Cinnamon can be bought as whole sticks of bark or as a ground spice. A stick can be added to a curry or floated in a hot punch. It adds a delightful warm earthy note to a mug of hot chocolate.*

Beware the Flying Eye-gougers

By the fifth century BCE, Arabs kept the cinnamon harvest proprietary by creating horrific tales of peril faced by those collecting this valuable spice. According to a tale written by Greek author Herodotus, men disguised themselves in oxen skins and sought out the precious cinnamon trees that grew in the center of a lake. The trees were guarded by screeching batlike creatures, intent on gouging out the eyes of all who dared approach this revered lake. Great birds gathered sticks from the cinnamon trees to make nests secured to the sides of unapproachable steep cliffs. Cinnamon gatherers offered the carcasses of beasts of burden to tempt the birds. The birds carried the carcasses off to their nests and the weight caused the nests to break off and fall to the ground. The fearless Arabs ran over to snatch up their prize.

A marble bust of the Greek historian and writer, Herodotus (c. 485–425 BCE).

Saffron

Crocus sativa

RIGHT: *The saffron threads are the long, deep orange stigma of a crocus flower. The spice adds a rich deep yellow tinge to any food prepared with it.*

Saffron is the orange stigma of a crocus flower. It is often said to be the most expensive spice by weight because it takes about 75,000 blossoms or 225,000 stigmas to produce a single pound (0.5 kg) of the actual spice. Saffron is most often associated with Mediterranean cuisine.

Historical Origins Saffron is native to southwest Asia and was first cultivated near Greece. Its use as a spice and also as a medicine has been documented for the past 4,000 years. In the Mediterranean region, saffron was used in ancient times in perfumes, dyes, toiletries, and as divine offerings. Alexander the Great used saffron in rice as well as in his bath to help heal battle wounds. His troops brought the practice of bathing with saffron back to Greece. Cleopatra added saffron to her bathwater as an aphrodisiac. After the fall of the Roman Empire, European saffron cultivation declined.

Centuries later, Moors reintroduced the spice when their civilization spread. The Black Death in the fourteenth century increased the medicinal demand for saffron. Saffron became such a valued product that those found guilty of saffron adulteration were heavily fined, imprisoned, or even executed.

North America was originally introduced to the spice when thousands fled northern Europe to avoid religious persecution. They settled in eastern Pennsylvania and began growing the saffron crocus. Cultivation continues there to this day.

Botanical Facts The saffron crocus grows from a perennial corm. Leaves are 16 in (40 cm) tall, slender, and upright. They emerge from the ground in early autumn with flowers arriving shortly afterward. Light to dark purple flowers are borne on 12-in (30-cm) stems. Each flower produces 3 stigmas. Stigmas must be harvested by hand. Propagation takes place by planting corms in rich well-drained soil. Saffron crocus thrives in a warm environment with generous spring rain and a relatively dry autumn.

Culinary Fare The vivid golden hue of saffron is used to flavor and color foods such as grains, soups, stews, confections, liquors, and sauces. The flavor of saffron is described as sweet, earthy, and slightly bitter. Medicinally, saffron is said to help speed healing, improve cold symptoms, and it also yields antioxidants.

The Saffron War of 1374

The Black Death crisis in Europe caused the demand for medicinal saffron to skyrocket and local supplies were soon depleted. The lingering hostilities from the Crusades made Arab sources unavailable, so imports from Greece were the only ones available to supply central and northern Europe. Saffron sales made the merchants wealthy and powerful, and social unrest resulting from the plague caused the decline of the noble classes. As a last-ditch effort to regain lost prominence, a group of nobles hijacked a huge saffron shipment bound for Basel, Switzerland. This triggered the 14-week saffron war. It raged until Leopold of Austria negotiated a deal with the Bishop of Basel to return the saffron to its rightful owners.

Protective gear worn by a doctor during the plague included a beaklike mask stuffed with medicinal spices.

Cumin

Cuminum cyminum

Cumin is an annual herbaceous plant in the parsley family grown for its seedlike fruit. It is an essential component in curry powder, chili powder, and garam masala. Cumin characterizes Middle Eastern and Turkish food, but is also found in African, Asian, and Mexican cuisines.

Historical Origins *Cuminum cyminum* originated in the eastern Mediterranean region but it has been naturalized across Asia and North Africa. Cumin seeds have been found in Old Kingdom Egyptian tombs. Romans and Greeks used cumin medicinally and cosmetically; drinking cumin tea was thought to produce a pale complexion. Greeks kept cumin seeds at the dining table and used it to season their food, and Moroccans continue this practice. During the Middle Ages, cumin was believed to keep chickens from wandering away. Cumin was also associated with happiness and it may have been used to shower on newly married couples much the way rice is used today. Spanish colonists introduced cumin to the Americas and it has since become an integral part of many New World cuisines. Medicinally, cumin has been used to treat stomach ailments, nausea, and as a diuretic.

Botanical Facts Cumin plants are tall and slender, growing to a height of 1–2 ft (0.3–0.6 m). Plants have thin, fernlike, dark green leaves on delicate stems. Tiny pink or white flowers form on umbels just above the leaves. Fruits serve as seeds, forming from the flowers, and can weigh down the thin stems. Planting close together offers added support to the heavy seed heads. Seeds are ⅛–¼ in (3–6 mm) long, crescent-shaped, ridged, and slightly hairy. Color ranges from light to medium brown.

Propagation takes place by seed in well-drained rich soil in warm moist climates. Plants will not tolerate long periods of hot dry weather. It is grown as a winter crop in many areas. Seeds are ready to harvest when seed heads begin to shrivel and dry. The seed heads shatter easily, so cumin is often harvested by hand. The seeds mature in around 3–4 months.

Culinary Fare Cumin has an earthy, sour, unpleasant smell before toasting. Heating it brings out a pleasant nuttiness and reduces the odor. Seeds are normally ground after toasting and added to spice blends or used to season soups, stews, meats, vegetables, and grains. Whole, they can be added to cheeses, drinks, liqueur, marinades, brines, and pickling solutions. The strong taste necessitates a light hand when adding to cooking as it easily overpowers delicate foods.

BELOW: *Used extensively around the world, cumin is an essential ingredient in curry powder. The formula varies but often includes coriander, turmeric, and black pepper; some also add cloves, cinnamon, cardamom, or fennel.*

Turmeric

Curcuma longa

Turmeric is an herbaceous perennial in the ginger family grown for its fleshy stem. It is often used as a cheap alternative to saffron, although the only similarity is its color. Turmeric is associated with Indian, Asian, and African cuisines.

Historical Origins *Curcumba longa* originated in India or Southeast Asia. It has been used as a coloring and flavoring agent for fabric and food for at least the past 4,000 years. Its first use by the Vedic culture in India was as a culinary spice and for religious ceremonies. In 1280, Marco Polo commented on the resemblance between turmeric and saffron. In medieval Europe, turmeric was called "Indian saffron" and has been used as an inexpensive substitute ever since. It is one of the cheapest spices, while the spice it mimics is the most expensive.

Indonesians used to dye their skin with turmeric as part of the wedding ritual. Turmeric is still used in Hindu religious rituals and as a dye for holy robes because it is a natural product that is not synthesized. Medicinally, it is used as a digestive aid, stimulant, and antiseptic.

Botanical Facts Turmeric plants can reach heights of 3–4 ft (0.9–1.2 m). The dark green leaves are large and broad, with upright sturdy stems. Pale yellowish white, yellow-tipped flowers are spectacular and showy, borne in the center of the plant. Because of its natural beauty, turmeric is often grown as an ornamental. Propagation takes place by planting small fleshy stems in rich well-drained soil under tropical conditions. They are planted in autumn and harvested in the next season when the leaves begin to yellow and die back. The growing season is 9–10 months long. If not harvested, clumps need to be divided every 3–4 years. The fleshy stems are boiled or steamed, peeled, and dried before being ground into a fine powder.

Culinary Fare Turmeric is almost always used dry. Fresh forms are more common in areas where it is grown. It can be used to color and flavor dishes, pickles, and condiments. Dried turmeric makes a nice addition to soups, stews, curries, meats, vegetables, grains, and mustard. It is also added to many spice blends.

BELOW: *Turmeric is best known for its rich, dark yellow color and the combined flavors of ginger and pepper. Its name is derived from a Latin term that refers to the mineral-like color of the root.*

Cardamom

Elettaria cardamomum

An herbaceous perennial in the ginger family, cardamom is grown for its highly aromatic seeds. It is listed as the second or third most expensive spice behind saffron and sometimes vanilla beans. Its flavor is described as warm and gingery with a hint of pine and citrus. It is also called green cardamom or true cardamom. It is most often associated with Indian, Asian, African, and Scandinavian cuisine.

Historical Origins Cardamom first originated in Sri Lanka and southern India, and is now primarily cultivated in India and Guatemala. Cardamom pods have been traded in India and Sri Lanka for over 1,000 years. Romans and Greeks were purchasing the spice for seasoning, perfume, and medicine as far back as 400 BCE, and possibly much earlier. The Ancient Egyptians chewed pods as a dental cleanser. Cleopatra is said to have been so entranced with the scent of cardamom she burned the spice lavishly in her palace when Marc Antony came to call. Vikings discovered cardamom in Constantinople and introduced it to Scandinavia. In Sweden, it has become more popular than cinnamon. Medicinally, cardamom is used as a digestive aid and to treat a variety of infections such as colds.

Botanical Facts Cardamom is a tall, bushy plant that can reach heights of 6–16 ft (1.8–4.8 m). Its leaves are 1–2 ft (0.3–0.6 m) long, dark green on top and pale green underneath, slender, and sword-shaped. Green flowers with white and purple veins are borne on trailing leafy stalks forming at the base of the plant. Flower spikes often reach lengths of over 3 ft (0.9 m). Seeds are encased in a small, green, paperlike pod, each pod containing 10–20 seeds. Pods are hand-gathered just before they ripen to prevent splitting. They are dried in the sun or an oven. Propagation takes place by planting fleshy stems in moist fertile soil. The native habitat for cardamom is tropical rain forest, so plants require warm frost-free temperatures, plenty of moisture, and filtered sunlight. It can take up to 4 years for plants to mature and produce seeds.

Culinary Fare Seed pods can be left whole or split before cooking if they are to be taken out of the dish before eating. Pods do not offer much flavor and can impart a bitter taste if left in food. More often, seeds are removed from the pods and bruised or crushed and fried before adding other ingredients to the cooking pan. Cardamom is used in pulses (legumes), curries, grains, pickles, desserts, beverages, and baked goods.

ABOVE: *A few cardamom pods simmered in a brew of tea with milk and sugar makes a fragrant cup, a brew that is often drunk in Afghani homes.*

RIGHT: *Cleopatra was renowned for using exotic scents to attract suitors. Fond of cardamom, she used it to perfume her palace.*

Star Anise

Illicium verum

Star anise is a slow-growing tropical evergreen tree in the magnolia family, grown for its uniquely shaped seed pods. The flavor is powerful and warm and tastes strongly of licorice; it is similar in taste to anise (*Pimpinella anisum*). Star anise is an essential addition to Chinese five-spice powder, garam masala, and the classic Vietnamese noodle soup, *pho*. It is mostly associated with Asian and Indian cuisine.

Historical Origins *Illicium verum* is not known in the wild but it is thought to be native to southwestern China. It is grown extensively in southern China and parts of Southeast Asia. It has long been used in Asian cuisine and culture. Followers of Buddha introduced star anise to Japan, and it was planted near Buddhist temples. It was first brought to Europe in the seventeenth century.

Medicinally, it is used as a stimulant, a digestive aid, a diuretic, and a breath freshener. Shikimic acid extracted from the seeds of star anise was used to develop Tamiflu, a promising drug used to lessen the severity of bird flu. In 2005, there was a shortage of star anise because of these findings. Fortunately, drug manufacturers have since developed new ways to produce the required acid.

Botanical Facts This small- to medium-sized shrubby tree grows 8–20 ft (2.4–6 m) tall. Its leathery leaves are shiny, dark green, and elongate oval. When crushed, leaves give off a strong anise scent. Flowers are greenish yellow and daisylike with a darker yellow center. The unique decorative seed pods form as the flower petals drop. Pods are about 1 in (25 mm) in diameter, rust colored, and shaped like a star with each point taking on the shape of a boat with a single seed as its passenger. Seeds are small, dark brown, and oily. Pods are harvested before they open and are always dried before use.

Propagation takes place by seed in moist, rich, well-drained soil in a sheltered position. Plants can withstand some cold weather, but will suffer or die if exposed to extended periods of frigid temperatures.

Culinary Fare Pods can be ground, but whole pods are frequently used as a "bouquet" or mixed with other whole spices to flavor soups, stews, and broths. Sometimes the seed pods are strained out but they are often left in as a decorative touch. Star anise is a classic flavoring for Asian-inspired dishes using pork and duck. It is an essential ingredient in Chinese red cooking—a braising technique using soy sauce. Star anise is often used in the West as a cheaper substitute for anise in baked goods and liquor production.

BELOW: *Star anise has such an attractive seed pod it makes an ideal garnish as well as a pungent flavor addition to a variety of meals.*

Aromatic Ginger

Kaempferia galanga

Aromatic ginger is a tropical plant in the ginger family, grown for its flavorful fleshy stems. It is also called sand ginger and resurrection lily. The genus *Kaempferia* is named for the German botanist Engelbert Kaempfer (1651–1716). Although aromatic ginger shares many similarities with lesser galangal (*Alpinia officinarum*), and they are often confusingly called by the same name, the two should not be interchanged. Aromatic ginger is most often associated with Thai and Chinese cuisine.

Historical Origins *Kaempferia galanga* originates in southern India but it is now grown primarily in Southeast Asia and in China, where it is used more as a medicinal spice. These days it is rarely used in Indian cuisine. Historical reference to the use of the 3 different galangals—*Alpinia galanga*, *A. officinarum*, and *Kaempferia galanga*—is somewhat difficult to ascertain because of the similarity in descriptions and purpose of all 3 plants. Arabs were known to dose their horses with galangal, possibly aromatic ginger, because of the stimulant and medicinal quality of the fleshy stems. Eurasians used it in tea infusions. It was also used powdered as snuff, a perfume, and in brewing. At one time, aromatic ginger was widely used in Europe as a flavoring spice and as a medicine, but its popularity has diminished over the years and is barely used there today. Medicinally, aromatic ginger is burned and inhaled to increase energy and improve mood, and used raw or as an infusion to produce mild hallucinogenic effects or treat infections.

Botanical Facts Aromatic ginger is an almost prostrate plant with thick, broad, 3–6-in (8–15-cm) long, green, patterned leaves. Fragrant, white, orchidlike flowers are borne in the center of the leaf whorls. Flowers are extremely short-lived, often wilting within a few hours of blooming. The flowering period is about 2 months.

Propagation takes place by planting small fleshy stems in rich, moist, well-drained soil. Plants need tropical conditions and a frost-free climate. Rain during the growth period is very important, but drier conditions are required later in the season. The leaves die back in late autumn. The fleshy stems should remain in the ground until harvest the following spring; these are reddish brown on the outside and have white flesh. If they are dried, little or no heat should be used. Too much heat can cause the flavor to diminish.

Culinary Fare Aromatic ginger is used more often fresh than dried, much the same way as traditional ginger. It takes excessive fishiness out of seafood and pairs well with chicken, beef, and vegetable curries. It adds a delectable aromatic flavor to soups, stews, sauces, marinades, and pickles.

LEFT: *Aromatic ginger has much longer, thinner fleshy stems than traditional ginger. While both are used in a similar manner, their flavors differ distinctively.*

Curry Leaf

Murraya koenigii

BELOW: *Curry leaves must be used immediately after drying as they lose their characteristic flavor very quickly. They are mostly used fresh.*

This small, fast-growing, deciduous tree is grown for its highly scented leaves and sometimes its edible fruit. Though technically an herb, the leaves are used as a spice. The plant's genus and species were named for botanists Johann Andreas Murray (1740–1791) and Johann Gerhard Koenig (1728–1785). Curry leaf should not be confused with *Helichrysum italicum*, the curry plant. Curry leaf has a taste and smell reminiscent of anise and tangerines, while curry plant smells like Madras curry powder, and is used more for medicinal purposes. Curry leaf is associated with Indian and Asian cuisine and has many different common regional names.

Historical Origins *Murraya koenigii* is native to India and is now found throughout the Indian subcontinent, except at higher elevations of the Himalayas, and in Australia. South Asian migrants took curry leaf with them, spreading its use to Malaysia, Africa, and the Pacific Islands. Its use in India is documented as early as the first century CE. Spanish explorers most likely introduced it to Europe while the British probably acquired it from the Spanish or through the British East India Company (1600–1858).

Medicinally, curry leaf infusions can be used to treat nausea, skin disorders, fever, and infections. A compound found in curry leaf is attracting the interest of Western doctors because of its ability to slow the enzymatic breakdown of starch to simple sugar, a potential boon to diabetics.

Botanical Facts Curry trees can grow to a height of 12–18 ft (3.5–5.5 m) with a trunk up to 16 in (40 cm) in diameter. One- to 2-in (25–50-mm) leaflets grow opposite one another on a long slender stem. Fragrant small white flowers form during late spring to early summer. Flowers give way to shiny black berries that, while edible, contain quite poisonous seeds.

Propagation can take place by stem cuttings or by seed, though germination times can be lengthy. However, stem cuttings can produce rooted plants far more quickly.

Curry trees require a tropical or subtropical environment with rich well-drained soil and plenty of sun. Hot dry conditions will produce the most flavorful leaves. The leaves can be harvested at any time but young tender specimens are preferred.

Culinary Fare The delicate flavor can be lost after drying, so leaves are seldom dried unless this is done just prior to cooking. However, the leaves may be left on the stem and frozen for later use.

Like other spices, curry leaf is often fried in ghee (clarified butter) before other ingredients are added to the dish. Leaves can be used to flavor soups, stews, curries, legumes, grains, vegetables, chutney, pickles, seafood, and meat. Their soft texture allows the cook to leave them whole in most dishes.

Nutmeg and Mace

Myristica fragrans

Nutmeg and mace are both produced from the fruit of a large tropical evergreen tree. Nutmeg is not a nut but a seed, and mace is a lacy net covering the outside of the nutmeg seed.

Historical Origins *Myristica fragrans* originated in the Moluccas (Spice Islands) in Indonesia and is cultivated in the West Indies. The earliest reference to nutmeg in antiquity was in Constantinople in 900 CE. By the twelfth century, both nutmeg and mace had famously found their way to Europe. Nutmeg was not only a spice; its magical properties were exploited in amulets for protection, to attract admirers, and improve virility. Control of the lucrative nutmeg trade was squabbled over for centuries. The Arabs, Portuguese, Dutch, and British all claimed sole rights.

Botanical Facts Nutmeg trees have a round growth habit, reaching heights of 40–50 ft (12–15 m). Thick leaves are shiny green on top, light green underneath, and measure about 4 in (10 cm) in length. Small, pale yellow, richly scented flowers grow in bunches among the leaves. A fleshy coating, the pericarp, envelops the nutmeg seed. Propagation takes place by seed in rich volcanic soil and humid tropical conditions. Both male and female trees, planted at a ratio of 10:1, are necessary for proper fertilization.

ABOVE: *Flowering of* Myristica fragrans *will take 5 years and drupe formation, 15 years. The red netlike covering is the mace and the seed within is the nutmeg.*

Culinary Fare Nutmeg and mace both have an aromatic nutty scent and warm sweet bite. Experts disagree as to which spice has the stronger flavor.

Both spices are largely associated with Indian and Middle Eastern cuisine, but their popularity has spread worldwide. They are dried and sold whole or ground—though whole blades of mace are rare. Both spices are used in desserts, sauces, vegetables, stews, curries, and in spice blends.

Peter Piper Picked a Peck of Pickled Nutmeg

The Dutch tried everything to maintain their dominance over the nutmeg trade in the Moluccas. Pulling up unguarded trees, soaking nutmeg in lime to render seeds sterile, burning excess harvest, and executing thieves failed to address two important details—pigeons and French tenacity. Pigeons absconded with nutmeg seeds and arbitrarily dropped them on other hospitable islands. In 1770, French botanist Pierre Poivre broke the Dutch monopoly by smuggling plants from the Moluccas to Isle de France (now Mauritius), hence the allegorical tongue-twister. (His first name is Peter in English, and Poivre is French for pepper, which in Latin is written as piper.) Poivre made away with a peck—a quarter of a bushel—of nutmeg, which was given the familiar term "pepper" because it was the most widely recognized spice of the day.

Poppy Seed

Papaver somniferum

Poppy seed is the flavorful seed from an annual flowering plant also grown for opium. Once refined, opium produces narcotics such as morphine, codeine, and heroin. The plant is also called opium poppy. The seed has a pleasant nutty taste, and contains none of the dangerous opiate compounds. Poppy seed is most often associated with Middle Eastern, Asian, and northern European cuisine.

Historical Origins *Papaver somniferum* originated in the eastern Mediterranean region or West Asia. It is now grown all over the Middle East and Asia. Growth, use, and distribution of poppy fostered an air of mystery and tyranny throughout the ages. It is hard to say what came first, opium use or the culinary use of poppy seeds, but an educated guess would have to be opium use.

Opium use predates recorded history, so the poppy may have been one of the earliest plants cultivated in Europe during the Neolithic era. They may also have been cultivated for opium in Mesopotamia around 3400 BCE. Opium poppy was used medicinally by Ancient Egyptians and by virtually all civilizations since. Seeds were used as a flavoring for breads and medieval Europeans mixed seeds with honey as a spread. Poppies bloomed on the Flanders battlefields of World War I, and because of this red paper poppies adorn lapels on remembrance days. Medicinally, seed infusions are used to help toothaches. Syrup made from the flowers can be used to soothe coughs.

Botanical Facts Poppy plants have a wide range of characteristics and can grow to a height of 1–4 ft (0.3–1.2 m), depending on the cultivar. Large, grayish green, fuzzy leaves form a slightly upright rosette. Striking multihued flowers with dramatic black-and-gold centers are borne singly on tall bristly stems. Seed pods swell as petals drop, with each pod producing hundreds of seeds.

Propagation takes place by seed in rich well-drained soil with plenty of sun and moderate temperatures. Seed should be sprinkled over the soil surface and lightly watered in. Seeds are planted in early spring and pods are harvested as they ripen. Harvest pods after petals fall but before the pods dry and split.

Culinary Fare Poppy seeds add flavor, color, and crunch as a garnish on breads and other baked goods. Toasting seeds first brings out their nutty taste. Seeds can be soaked, ground, and cooked into fillings for pastries and cookies. Seeds fried in butter or ghee can be added to pasta, rice, or vegetables.

BELOW: *The capsules left after the flower petals have dropped contain the poppy seeds. Because the poppies that produce the edible seeds can also yield opium, in many places it is illegal to grow them.*

Allspice

Pimenta dioica

Allspice is the dried fruit from a tropical evergreen tree in the myrtle family. Its genus name comes from sixteenth-century Spanish explorers who confused allspice berries with peppercorns. The common name comes from its aroma, which smells like a combination of cloves, cinnamon, nutmeg, and ginger.

Historical Origins *Pimenta dioica* originates in the Caribbean island of Jamaica and now grows in the tropical rain forests of South and Central America, which makes it the only spice grown solely in the Western Hemisphere. Spanish explorers brought allspice back to Spain and it found its way to other parts of Europe and Britain.

Botanical Facts Allspice trees have smooth gray bark and will reach heights of 20–40 ft (6–12 m). Leaves are thick, glossy, dark green, and measure up to 6 in (15 cm) long. Small, fragrant, white flowers are borne in bunches within the leaves. Green berries follow the flowers, turning purple when ripe. Propagation takes place by seed in moist well-drained soil in warm tropical climates. Both male and female trees are needed for fertilization.

Culinary Fare Allspice is most often associated with Caribbean and northern European cuisine. It is used whole in marinades, curries, soups, braised dishes, pickled dishes, and spice blends. Ground allspice is used in baked goods, desserts, and with grains. The bulk of allspice grown in the world is used commercially in condiments such as ketchup.

Medicinally, allspice is used as a digestive aid, and to warm and ease sore muscles.

ABOVE: *Fruit from allspice takes time—it will develop after 5 years and fully mature at 20 years. The berries are usually hand-picked, sweated, then dried.*

Anise

Pimpinella anisum

Anise is a tender annual plant in the parsley family grown for its flavorful seeds, though its roots are edible as well. It tastes a lot like licorice.

Historical Origins *Pimpinella anisum* originated in the Middle East and is now grown in India, southeastern Europe, and North Africa. Medicinally, anise infusions can treat digestive problems, coughs, and infections. It has been used for culinary and medicinal applications since 1500 BCE in Egypt, if not earlier.

Romans ate anise cakes after heavy meals to settle their stomachs. In 1305, the English listed it as a taxable drug and collected fees that helped pay for repairs and maintenance of London Bridge. It was said anise could drive off the evil eye.

Botanical Facts Anise plants grow to 2–3 ft (0.6–0.9 m) high with a long central taproot. Bottom leaves are dark green and parsleylike; upper leaves are feathery. Umbrella-shaped flowerheads are borne on long graceful stems. Flowers are small, white, and resemble Queen Anne's lace. Propagation takes place by seed in rich well-drained soil in full sun.

Culinary Fare Anise is most often associated with Middle Eastern and European cuisine. It is used to flavor baked goods, liqueurs, soups, curries, seafood, vegetables, and confections. Traditional Italian biscotti are made with anise for a hint of aniseed flavor.

BELOW: *Anise needs 120 frost-free days to set seed. Seed heads are picked when ripe and dried before using.*

Pepper

Piper cubeba Cubeb *Piper longum* Long Pepper
Piper nigrum Black Pepper, Green Pepper, "True" Pink Pepper, White Pepper

Pepper is a woody perennial vine that is grown for its spicy flavorful fruit known as peppercorns. *Piper nigrum* produces black, green, "true" pink or red, and white pepper. *P. cubeba* and *P. longum* are different species with similar flavor profiles.

Historical Origins *P. cubeba*, which is sometimes called Java pepper or tailed pepper, originated in Indonesia. Although popular in common-era Europe, it has since fallen from favor. *P. longum*, or long pepper, originated in India and at the foot of the Himalayas. Its use may have predated *P. nigrum*, which also originated in India and is now grown in many other areas including Brazil, Indonesia, and Malaysia.

ABOVE: *Black pepper* (Piper nigrum) *from* Florindie ou Histoire Physico-économique des Végétaux de la Torride *(Plants of the Torrid Regions), 1789, by G. S. Delahaye.*

Pepper has served as a spice and medicine since prehistoric times. It was first noted in India over 4,000 years ago.

Ancient Egypt's King Rameses II was found with peppercorns lodged in his nose as part of the mummification ritual. Although it was obviously present in Ancient Egypt, how it got there from India remains a mystery. In the fourth century BCE, the Greeks used *P. nigrum* and *P. longum* interchangeably.

According to some, China may have been using *P. nigrum* during the second century BCE. By 30 BCE, Rome's conquest of Egypt provided an unfettered sailing route from Rome to the coast of India, so pepper of all types became valued imported commodities.

In postclassical Europe, *P. longum* was the more expensive of the three. *P. cubeba* took a greater role as a medicine although it did have culinary uses. *P. nigrum* was the primary culinary spice often mentioned in writings and cookbooks. All forms of pepper were so valuable they were commonly used as currency. In Rome's final years, Alrac the Visigoth and Atilla the Hun both demanded a huge ransom of peppercorns to help ensure the city's safety. Needless to say, their bribes did not work.

Pepper was just as highly valued in medieval Europe and its exorbitant price convinced explorers to cut out the middleman and find direct trade routes to the land of spices. The flow of pepper had its high and low points throughout the centuries. By the eighteenth century, it was more available to people in all stations of life.

Botanical Facts *P. cubeba*, *P. longum*, and *P. nigrum* tend to have the same growth habit. Pepper vines are usually grown on stakes or other trees for support. The pepper leaves are green, lightly ridged, and oval to heart-shaped with pointed tips. Small flowers are borne on nodding spikes that lengthen as the flowers fade and the fruit matures. With the exception of *P. longum*, berries form singly in long slender bunches along the flower spike. *P. cubeba* produces a fruit with a bit of stem or a tail still attached. The berries from *P. longum* fuse together to form a single elongated rod, which looks something like a catkin. Once dried, the fruits of *P. longum* resemble slender miniature pinecones.

Propagation takes place by rooted stem cuttings in rich, moist, well-drained soil. Stem cuttings are tied to trees or supports. Tropical conditions and a steady water supply are necessary for a good harvest. Plants first bear fruit in their fourth or fifth year and will continue through until their seventh year. New cuttings are introduced regularly to produce a constant supply of berries. One single stem of a pepper vine will produce 20–30 fruiting spikes.

ABOVE: *Of the 3 species listed, the only pepper in full production today is* Piper nigrum. *It accounts for a fifth of the world's spice trade and is the most widely traded spice.*

Culinary Fare Pepper ultimately became the most coveted spice of all. It defined the history of the spice trade and is the "king of spices." The flavor of all 3 pepper species listed here is described as warm, sharp, hot, and sometimes even floral.

Ground pepper is a universal seasoning for virtually all savory dishes. Peppercorns are usually dried and ground before use. As pepper loses its punch within hours of grinding, it is best used freshly ground. Wooden grinders keep out heat, humidity, and light. Whole peppercorns are used in marinades, pickling spices, and spice blends. The complex Moroccan spice mix *ras el hanout* includes 3 peppers—black, long, and cubeb—along with nutmeg, cinnamon, cardamom, cloves, and other spices. Some sweet baked goods, such as spice cookies or breads, can benefit from the bite pepper offers. Medicinally, cosmetically, and superstitiously, pepper has been used to aid digestive disorders, dye hair, sooth muscles, and exorcise demons.

So Many Peppers, so Little Time

This list may just solve the culinary mystery of which peppercorn to choose for any occasion or type of cuisine.

Black Peppercorns Mature, unripe berries, picked while green or yellow, and dried. They offer classic pepper taste for seasoning. Types: telicherry, malabar, sarawak, and lampong.

White Peppercorns Mature, ripe berries with the outer skin removed. They offer a milder taste, with a sharp, hot finish, and are good with light-colored products.

Green Peppercorns Immature green berries, picked small. They can be hot but are usually not as strong as black peppercorns. Brined: salty, resinous, good for sauces. Dried: can be ground, crushed, or rehydrated and used in the same manner as for brined.

"True" Pink or Red Peppercorns Mature, fully ripe berries offer the biting hot flavor of black pepper with the fresh notes of green pepper. Available as for green peppercorns and used in the same way.

Mahlab

Prunus mahaleb

Mahlab is an aromatic spice made from the seeds of a small deciduous black cherry tree, the St. Lucie cherry.

Historical Origins *Prunus mahaleb,* commonly known as the St. Lucie cherry, grows wild in Turkey and other areas of the Mediterranean. Mahlab was used for perfume in Turkey and the Middle East. Culinary use of the seed followed shortly afterward. Medicinally, the seed and bark are used as an anti-inflammatory.

Botanical Facts St. Lucie cherry is currently cultivated for its seeds in Turkey, Iran, and Syria. It is also widely grown as an ornamental in other areas. It can either be described as a tree or a large shrub. It grows to a height of 6–30 ft (1.8–9 m), often bushing out rather than growing up. Small green leaves are slightly ridged, and oval, with serrated edges and pointed tips. Propagation takes place by seed or rooted cutting in well-drained soil in temperate locations.

LEFT: *St. Lucie cherry has fragrant white flowers clustered in groups of 3–10. Each tiny flower produces a fruit with a single pit. It is often used as rootstock for sweet cherries.*

Culinary Fare The flavor is described as a cross between cherry and almond. Mahlab is associated with Middle Eastern and eastern Mediterranean cuisine. Buy whole seeds as the powder deteriorates rapidly. The seeds are cracked to remove the inner kernel and ground before using. It is used to flavor rice, breads, pastries, desserts, and confections.

Sumac

Rhus coriaria

Sumac is a bushy shrub related to poison ivy, grown for its tangy fruit and sometimes its bark. Fruit has a sour citrus taste. Sumac is largely associated with Middle Eastern cuisine.

Historical Origins Sumac grows wild throughout the Mediterranean and other warm regions of the Northern Hemisphere.

It is widely used in Lebanese cuisine. Native American tribes were known for making beverages from different species of sumac and used the bark as a dye. Medicinally, sumac acts as a diuretic and a digestive aid.

LEFT: *A good pinch of sumac in the dressing will enhance the flavor of salads. Sumac is also called Sicilian sumac or elm-leafed sumac.*

Botanical Facts Sumac bushes grow to a height of 10 ft (3 m), though other varieties are bigger. Up to 11 leaflets grow on both sides of a central stem. Stems produce a sticky resin when broken or cut. Leaves make a dramatic showing in autumn, changing from green to scarlet. White flowers give way to central upright clusters of dark red berries, enclosed in a fine fuzzy brown coat. Propagation takes place by seed in well-drained soil in temperate conditions. Poor soil seems to work best with sumac.

Culinary Fare Berries are dried and crushed or ground to produce a purplish red powder that can be sprinkled directly into food or reconstituted with water and used as a citrusy juice. Sumac is used to flavor meats, seafood, vegetables, soups, stews, yogurt, marinades, and dressings.

Sassafras

Sassafras albidum

Sassafras is a deciduous tree in the laurel family, and is often grown for its flavorful roots and useful leaves.

Historical Origins *Sassafras albidum* originates in the eastern areas of North America. Choctaw Native Americans first used dried ground sassafras leaves to flavor and thicken food. Other Native American tribes realized the medicinal and culinary benefits of the roots and leaves, passing on knowledge to settlers. It acts as a stimulant and an anti-inflammatory. News of the benefits of sassafras made it back to Europe along with the plant. The use of natural sassafras roots and leaves has since been banned because of the presence of a harmful chemical. Sassafras products available today have had the chemical removed.

Botanical Facts Sassafras trees grow to a height of 50–60 ft (15–18 m). They have an unusual leaf habit, having 3 distinct leaf shapes on the same tree. Tiny yellow flowers form in the spring followed by little black fruits. Propagation takes place by seed in well-drained neutral soil. Leaves and roots can be harvested at any time.

Culinary Fare Root bark is steeped to make tea and beverages, such as root beer, and to flavor confections. Leaves are dried and ground to make fine powder used primarily as a thickener. Young leaves can even be eaten as salad greens.

ABOVE: *Sassafras roots appear uninspiring but they pack a punch. The volatile oil they contain, called safrole, is banned in the United States, and only prepared extracts without this component can be safely consumed.*

Sesame

Sesamum orientale

Sesame is an annual flowering plant that is grown for its tasty little seeds. Both the white and black varieties can be used interchangeably. Sesame is associated with Middle Eastern and Asian cuisine.

Historical Origins *Sesamum orientale* is believed to have originated in Africa or India. Sesame is quite possibly the oldest spice known. It was the first cultivated for its oil. Egyptian tomb drawings dating back 4,000 years show bakers adding sesame seeds to bread. At the same time, the Chinese were producing soot for ink as a byproduct of burning sesame oil. Ancient Greeks and Romans ate sesame seeds plain or mixed into a paste with cumin. Sesame entered Europe by means of the spice trade. African slaves brought sesame to North America and "benne" seeds, their African name, became a popular ingredient in the south.

Botanical Facts Sesame plants grow 2–6 ft (0.6–1.8 m) tall and give off a disagreeable smell. Fuzzy leaves are slender and oval. Small white to purple flowers form fat seed capsules. Propagation takes place by seed in well-drained soil in hot climates. Plants require over 100 days to mature.

Culinary Fare Sesame seeds taste best if lightly toasted before using. Seeds can be added whole to savory dishes, breads, cakes, cookies, and confections or ground and added to spreads. Sesame oil is used for cooking, adding a hint of sesame flavor to stir-fries. It is also used in soaps and emollients.

Clove

Syzygium aromaticum

RIGHT: *A clove or two popped into mulled wine or hot punch will give it a flavor wallop; try adding a few to pickles. Remove whole cloves before serving dishes containing them. Though not pleasant to chew, biting on a whole clove can relieve toothache.*

Clove is an evergreen tree in the myrtle family grown for its highly aromatic unopened flower buds. The word clove is derived from the Latin word *clavus*, meaning nail. Cloves, along with nutmeg and pepper, were among the most sought after spices in ancient and modern times.

Historical Origins *Syzygium aromaticum* originated in the Moluccas (Indonesia's Spice Islands), where people planted a clove tree to commemorate the birth of each child. It is now grown in many different areas including Brazil, the West Indies, Mauritius, Madagascar, and Sri Lanka. Indonesia remains the world's largest producer of cloves.

Archaeologists have uncovered some clove remains at a site in Syria, which dates as far back as 2400 BCE. Cloves were prized in Ancient Rome and mentioned in Chinese literature in the fourth century BCE. In the second century BCE, it was written that those approaching the Chinese emperor must have cloves in their mouths to avoid offending him. Arab traders brought cloves to the Romans; Portuguese explorers introduced them to Europe, and their popularity and high prices remained throughout the nineteenth century until larger-scale cultivation made cloves more available and less expensive. Today, Indonesians, in particular, favor an aromatic clove cigarette called *kreteks*; so much so that imported cloves are required to meet the heavy demand.

ABOVE: *A clove tree, from* Florindie ou Histoire Physico-économique des Végétaux de la Torride *(Plants of the Torrid Regions), a 1789 watercolor by G. S. Delahaye.*

Botanical Facts Clove trees grow to a height of 30–40 ft (9–12 m). Glossy green leaves are very fragrant and oval in shape with a pointed tip. Immature flower buds are green at first, gradually turning bright red before they open. Buds are handpicked just as they turn pink. If the buds are not harvested, small red flowers form in triple clusters on branch ends. Propagation takes place by seed in well-drained rich soil in tropical climates. Trees benefit from a partially shaded location and cooler temperatures. They require regular rainfall with dry periods before and during harvest. Flowering begins when trees are 5 years old. A single tree can produce 40 lb (18 kg) of cloves.

Culinary Fare This tiny bud packs a powerful punch with its warm pungent flavor. Cloves are associated with Asian, Middle Eastern, and African cuisine. Cloves are dried and used whole to stud meats before roasting and in spice blends, sauces, pickles, marinades, and beverages. Ground cloves enhance desserts, baked goods, grains, stews, and soups. Medicinally, clove infusions and clove oil have antibacterial, antiviral, antifungal, antiseptic, and analgesic properties. Cloves also serve as a digestive aid and are a simple and quick breath freshener.

Tamarind

Tamarindus indica

ABOVE: *Tamarind pods can either be purchased whole or as a paste made from the pulp.*

Tamarind is a tropical tree in the legume family. It is primarily grown for the tart pulp found in its seed pods but it has many other useful applications. The tangy sweet flavor is associated with Asian and Latin American cuisine.

Historical Origins *Tamarindus indica* is native to tropical Africa and is sometimes falsely considered a native to India as well because of its naturalization there in prehistoric times. It is currently grown throughout the tropical and subtropical regions of the world. India is the main commercial producer of tamarind.

Ancient Egyptians were well acquainted with African tamarind. There are also quite a few references to tamarind trees in Hindu mythology. Arabs and Persians were most likely first exposed to Indian tamarind, also called *tamar hindi* or Indian date. Tamarind was used by Greeks as early as the fourth century BCE. Medicinally, tamarind is a digestive aid and soothes sore throats.

Botanical Facts Tamarind trees are quite long-lived and very large, sometimes achieving heights of 100 ft (30 m). They have very sturdy trunks, strong branches, and thick gray bark. The bright green leaves are about 4–6 in (10–15 cm) long, fernlike, and feathery. In tropical regions, trees are evergreen but they may drop their leaves if exposed to long periods of dry hot weather. The sweet-scented flowers are yellow with pink streaks and resemble small orchids. Abundant quantities of seed pods are found growing on new branches. Pods start off green and change to brown; they ripen and swell. When ripe, the leathery pods house seeds surrounded by a sticky sweet pulp—nature's way of attracting animals to spread the seeds.

Propagation takes place by seed or cuttings in moist rich soil in tropical or subtropical conditions. Trees can bear fruit as early as the fourth year. However, depending on the conditions, fruiting may require 10–14 years' growth. Trees will bear for 50–60 years before productivity declines. Pods are frequently left on the tree for 4–6 months to dry before harvest.

Culinary Fare Pulp is used to flavor drinks, curries, stews, chutneys, legumes, grains, sauces, meats, and seafood. When purchased whole, the seeds (which are covered in the pulp) can be rolled in sugar and sucked as a confection. Sometimes the ground seed is added to cakes.

Tamarind is used in many places—such as the Middle East, Mexico, and Egypt—to make delicious drinks. The concentrate is a vitally important ingredient in the cuisine of Aleppo in northern Syria, where it is used to flavor many savory dishes; it adds a sharp but sweet note to sour soups in Southeast Asia.

BELOW: *Although valued mostly as a food source, the tamarind tree is also used by South Asian craftsmen to produce furniture, tools, boats, and dyes.*

Vanilla

Vanilla planifolia

RIGHT: *Vanilla flowers only produce a seed pod—the source of the spice—after they have been pollinated. Without bees, they need to be pollinated by hand; a tricky task given the flowers bloom for just a single day at best.*

Vanilla is a tropical climbing orchid grown for its sweet aromatic seed pod or vanilla "bean." The name came from the Spanish for "little sheath," based on the bean's appearance. Along with saffron and cardamom, vanilla is one of the most expensive spices. The flavor is highly favored worldwide.

Historical Origins *Vanilla planifolia* originated in Mexico and is now grown in other tropical areas, including Madagascar, Indonesia, Uganda, and Tonga. Madagascar is said to produce the highest quality beans.

The Totonac people on the Gulf Coast of Mexico were probably the first to cultivate vanilla. They believed the vanilla orchid grew from the blood of a mythical princess, so the plant was considered a gift from the gods. The Aztecs from central Mexico were introduced to vanilla when they conquered the Tontonacs in the fifteenth century.

In the early sixteenth century, Spanish explorers carried vanilla back to Spain and eventually Europe, where it became very popular. Until the mid-nineteenth century, Mexico was the main source for vanilla. Plant growth in other areas proved unsuccessful.

BELOW: *Vanilla is the world's most labor-intensive crop, which explains its high cost. The United States is the largest consumer of vanilla in the world, with most of it going to the dairy industry.*

Got a Bee in your Bonnet?

In 1819, French entrepreneurs shipped vanilla cuttings to Reunion and Mauritius islands in the South Indian Ocean in order to enter the vanilla trade. They were able to grow flowering vines but were stymied by the plant's inability to set seed pods. In their native habitat, vanilla orchids are pollinated by bees that pierce the flower's membrane, ensuring fertilization. A young Mauritian slave discovered how to mimic the bee's actions by using a sharpened bamboo sliver to pollinate the flowers by hand. Shortly afterward, vanilla orchids were sent to other tropical islands with pollination instructions. By 1898, the islands where the French introduced the vanilla orchid were producing around 80 percent of the world's supply of vanilla.

Botanical Facts Vanilla orchids are spare vines cultivated on trees or poles for support. Fleshy leaves are sword-shaped and evenly spaced on alternate sides of the central vine. Flowers grow in bunches and usually bloom one at a time. Blossoms fade within one day. Long seed pods form when flowers drop. Propagation takes place from rooted plant cuttings in moist rich soil in a shaded location under tropical conditions. Flowers appear in the second or third year. Outside Mexico, flowers are hand-fertilized. Before harvest, pods must remain on the plant for at least 9 months after they form. Vanilla pods must be dried and fermented to produce the characteristic aroma and flavor.

Culinary Fare Beans are sold whole or made into an extract. Both forms of vanilla can flavor all types of desserts and baked goods. Beans will give a more intense luscious flavor to cooking. They can be infused in the milk used in sauces, ice creams, or custards, but it will color them. Keeping a bean in the sugar jar to flavor it for use in baking will avoid this discoloration. Though vanilla is associated with sweets, savory applications are emerging. It gives a rounded note to tea.

Wasabi

Wasabia japonica

Wasabi is a semi-aquatic perennial plant in the cabbage family grown for its pungent lower stem. True wasabi is expensive and uncommon in North America. Many cheaper products labeled as such are artificially colored horseradish. Wasabi has a short-lived, hot, biting flavor most directly associated with Japanese cuisine.

Historical Origins *Wasabi japonica* originated in the mountainous regions of Japan. One story tells of a villager who came across wasabi root growing in a stream at Mt. Wasabi-yama and took the plant home, successfully growing it. "Wasabi" has been found in Japanese dictionaries and botanical texts dating back to 794 CE. Traditional cultivation techniques in Japan, where wasabi is a highly prized crop, are well guarded. Today it is also cultivated in China, Taiwan, Australia, New Zealand, and North America. Demand for true wasabi far exceeds supply. Medicinally, wasabi has long been used to fight the effects of food poisoning, and as a detoxifying and antibacterial agent. New studies indicate it may be a powerful anti-carcinogenic.

Botanical Facts Wasabi grows to a height of about 2 ft (0.6 m) with a thick central stem. Large rounded leaves form at the end of 12–18-in (30–45-cm) stems. As the central stem grows, leaves drop from the bottom. Small white flowers form on stalks toward the center of the plant. Propagation takes place by seed or replanting side shoots forming off the main plant. Wasabi grows in freshwater streambeds with running cold water. Cool temperatures, high humidity, and shade are required for optimum growth. Plants can naturally take up to 3 years to mature and easily reseed themselves.

Culinary Fare Preparing wasabi is somewhat ritualistic. Purists recommend light peeling of damaged areas only before grating. The thick end of the wasabi should be grated first because it offers more flavor and heat. The preferred grating tool is made out of sharkskin rather than metal, as metal is said to react badly with wasabi. An acceptable alternative is a porcelain ginger grater. Once grated, it should be eaten within 30 minutes to avoid flavor loss. However, allowing the paste to sit for a few minutes before serving will improve the flavor.

Wasabi paste is usually served as a fresh condiment with an assortment of Asian-inspired foods. It is more commonly associated with Japanese sushi and sashimi. In modern cuisine, wasabi is finding its way into many unconventional dishes. Leaves can also be eaten raw, pickled, or fried like chips.

BELOW: *Beware the more common imitations. These are the 3 types of genuine wasabi: fresh wasabi root alongside wasabi paste and wasabi powder.*

Monkey Pepper

Xylopia aromatica

Monkey pepper is a tropical evergreen tree grown for its pepperlike seeds. The tree is also called a peppertree and the seeds can go by the name negro pepper.

Historical Origins *Xylopia aromatica* originated in the lowland forests of tropical Africa. Monkey pepper was used in Europe as a less expensive substitute for *Piper nigrum* (pepper) until the fourteenth century. Its popularity waned with the increasing availability and lower prices of *Piper nigrum*. It was used very little after that, usually in times of war when other spices were in short supply. Today, monkey pepper is rarely exported out of Africa.

Botanical Facts Monkey pepper trees will grow to a height of 50–100 ft (15–30 m) on a thick sturdy trunk. It has leathery green leaves that are oval with pointed tips. One to 2-in (25–50-mm) pods form like miniature bunches of bananas after the tree flowers. Each pod contains 5–8 seeds. Propagation takes place by seed in well-drained soil in a tropical climate. Trees can take many years to flower and produce seed.

Culinary Fare Seeds have a sharp spicy flavor with a slightly bitter aftertaste. Pods are always dried and sometimes smoked before grinding. Monkey pepper can be used in much the same way as *Piper nigrum*.

Sichuan Pepper

Zanthoxylum piperitum

RIGHT: *While it lacks pepper's pungent bite, Sichuan pepper's woody citrus tones blend agreeably with ginger and star anise. Used in Himalayan cuisine, it is a flavor in momo dumplings. Remove stems and thorns before use as they can pierce the palate.*

Sichuan pepper is a deciduous aromatic shrub or small tree grown primarily for its highly aromatic fruit.

Historical Origins *Zanthoxylum piperitum* originated in the Sichuan province of China. Its use is not very widespread in the West. From 1968 to 2005, the US Food and Drug Administration banned the import of Sichuan pepper into the United States because it was capable of carrying a dangerous bacterial disease with potential to harm citrus crops. The peppercorns must now be heat-treated prior to import. Medicinally, Sichuan pepper serves as an analgesic and a digestive aid.

Botanical Facts Sichuan pepper gets its name based on the appearance of the tree's fruit. Also called Szechwan pepper, it is not actually a member of any pepper family.

Sichuan pepper trees grow to a height of about 9 ft (2.7 m). Green shiny leaves are oval with pointed tips. Both the central stem and leaf branches have nasty looking thorns. Whitish pink flowers are borne in small clusters throughout the foliage. Red fruits form as flowers die off. Fruits are dried after harvest. Propagation takes place by seed in rich moist soil. The trees require little attention.

Culinary Fare The dried fruit has a warm pepperlike flavor. Sichuan fruits are brittle when dry. The tiny black seeds inside the fruit are discarded. Fruits are usually toasted and ground before use. Sichuan pepper is mostly associated with Chinese cuisine. It is added to meats, poultry, and seafood. The leaves have a lemony taste and are used as an herb in Japan.

Ginger

Zingiber officinale

Ginger is a tender creeping perennial that is grown for its aromatic and flavorful rhizomes, the fleshy stems of the plant. It is in the same family as turmeric.

Historical Origins Known only under cultivation, ginger's origins are difficult to pinpoint. It may have originated in Southeast Asia, though there are also references to its use in Ancient China. Ginger is mentioned in the Koran, Hindu literature, and by Confucius, so its use is documented back to 650 BCE.

Egyptians used ginger in cooking and as a medicine. It was highly prized in the Roman Empire and Greece, more as a medicine than a culinary spice. Roman soldiers probably brought ginger to Europe and England and it eventually became almost as popular as pepper. During the thirteenth and fourteenth centuries, it was one of the most commonly traded spices. Apparently, back then, a pound of ginger was worth the cost of a sheep. In the sixteenth century, the Spanish introduced ginger to the New World to begin cultivation and circumvent traders who monopolized the business in Europe. Jamaica, West Africa, and Brazil all became important suppliers.

Botanical Facts Ginger plants produce long, slender, lance-shaped leaves that grow to a height of 2–3 ft (0.6–0.9 m). Fleshy stems spread sideways, so plants grow out rather than up. Flowers form on long spikes that come up beside the plant or in the center of the leaves, depending on cultivar. The flower colors vary. Propagation takes place by planting fleshy stems in rich, moist, well-drained soil in a tropical or subtropical climate. Plants grow best in lightly shaded areas. Fleshy stems should begin to grow in about 10 days and are ready to harvest in 9–10 months.

Culinary Fare Most often associated with Asian cuisine, ginger can be used fresh or dried. Fresh is generally the preferred choice for soups, vegetables, stir-fries, curries, grains, marinades, chutneys, beverages, dressings, pickles, and also in some desserts and confections. Dried ginger is best for most baked goods. In Japan, finely sliced ginger is used in *amazu shoga*, a pink pickle, as a traditional accompaniment for sushi.

ABOVE: *Used as a sweet spice, ginger is perhaps best known for its use in gingerbread. In the story by the Brothers Grimm, the ravenous Hansel and Gretel were lured into a wicked witch's enticingly edible gingerbread house.*

Eat Ginger Every Day, Keep the Doctor Away

Ginger is one of the few spices to have virtually all of its medicinal claims verified. The use of ginger as a medicine predates its use as a culinary spice. It has been found to successfully treat motion sickness, postoperative nausea, bacterial dysentery, malaria, coughs, and migraines. Ginger extracts have been found to improve blood cholesterol levels, elevate low blood pressure, and prevent cancer in animals. Gingerols, the chemicals responsible for ginger's heat, are helpful in treating pain and fever, and its volatile oils may have a positive effect on cold and flu viruses.

Plants Used in Beverages

Introduction to Plants Used in Beverages

ABOVE: *From the Mayan through the Aztec era, cocoa beans became a major commodity, being traded and used for currency. These scenes from daily Aztec life show the growing, harvesting, and trading of cocoa beans.*

Water is life. Nothing can exist without a steady supply. Our bodies are made up of two-thirds water. Our earliest ancestors traveled in small nomadic bands of hunter-gatherers, always tied to, and constrained by, the need for fresh water. The desire to quench thirst is a physical imperative, and the desire to spice things up with a liquid that is flavored is probably as old as humankind. The sharing of delicious drinks with friends has always been, and remains today a universal symbol of friendship, hospitality, and trust. We still drink wine from a shared bottle, coffee or tea from a shared pot, or yerba maté from a shared straw, just as our ancestors did hundreds of years before us. The tradition of raising a glass to health, happiness, and success is liquid history, and has been a sign of social unity and cohesion for thousands of years.

ADDING FLAVOR WITH PLANTS

Flavored beverages have been made in many ways throughout the centuries: by steeping natural plants, herbs, and spices in water; by fermenting grapes, grains, and plants; and by distilling those fermented products. Each of these beverages and their consumption is part of the history of civilization. Though the popularity of a particular beverage may ebb and flow, beverages in general have been unifying forces for the people drinking them, for social purposes, for health, for their use in trade, and for their religious significance.

FERMENTED BEVERAGES

The origins of the fermented beverages beer and wine are lost in ancient history, their discovery so long ago that it is recorded only in legend. Fermentation is a simple chemical process where yeast consumes sugar and gives off alcohol and carbon dioxide. The ability of a beverage to induce altered consciousness must have seemed magical, and the earliest civilizations considered fermentation a gift from the gods. The earliest people fermented anything they thought might be suitable—honey, pomegranates, and grains for beer, and grapes for wine. The alcohol content in these beverages served a number of purposes. It helped extend food stores in days long before refrigeration, as alcohol acts as a preservative; it helped make water safe to drink, as alcohol kills harmful microorganisms; and in the form of fermented beverages, it provided a source of calories for hungry people.

The transition of people from small tribes of nomadic hunter–gatherers to stable communities of farmers began when cereal

RIGHT: *Yerba maté is a popular tealike beverage made from an infusion of the leaves of* Ilex paraguariensis *of South America. It is brewed in a cup or hollowed-out gourd and sipped through a metal straw.*

grains were deliberately cultivated. Archaeologists argue whether the advent of farming created a steady supply of beer, or whether it was the desire for a steady supply of beer that led to the beginning of the agricultural era. Because grain could be stored, it allowed people to stop roaming and to settle in one place. These early settlements soon became cities, and the solid production of grain, bread, and its liquid counterpart, beer, were the companions and building blocks of every early civilization, just as cereal was the basis of every economy.

> *"The morning cup of coffee has an exhilaration about it which the cheering influence of the afternoon or evening cup of tea cannot be expected to reproduce."*
>
> Oliver Wendell Holmes, Sr. (1809–1894), *Over the Teacups*

FROM GRAPES TO WINE

In addition to grains, grapes were fermented to produce wine. Remains of cultivated grapevines and pottery shards from storage vessels have been found from 9000 BCE. Wine is perhaps easier to make than beer, as the fermentable sugars are right in the grape, whereas in beer the starch in the grains has to be converted to sugar before fermentation can begin. Wine was the water that flowed through the veins of the ancient Greek and Roman civilizations. And wine came to be associated with religious rites in Christianity as a celebration of communion, although in Islam, wine—along with other forms of alcohol—was banned.

COFFEE AND TEA

Coffee, chocolate, tea, and tisanes are all beverages in which water is flavored with the essences of plants, their ground and roasted seeds, their leaves, or their flowers. In China and Japan, tea was associated with religion, and elaborate rituals sprung up around both the preparation and the drinking of tea. The Aztecs adored frothy hot chocolate. Coffee was popular in Arabia, perhaps as a substitute for wine. However, these beverages soon slipped the leash in their homelands and became worldwide favorites.

With the advent of mechanization, it was inevitable that someone would figure out how to keep the alcohol in a fermented beverage and discard the water, and by the mid-seventeenth century, people were distilling the materials that grew naturally in their homeland. Cold-climate regions distilled grain-based beers, warmer climates distilled fruit-based wines, and tropical climates distilled plant-based liquids.

A MIXTURE OF OLD AND NEW

As the demand for increasing flavor in beverages grows, more options are created to satisfy growing worldwide demand. Elixirs that once were used for health purposes have become sodas. Fruit juices and energy drinks are popular, and bottled mineral waters and flavored waters are a growing market.

Beverages available today are a blend of old and new. We still love beer, wine, coffee, and tea, even though they have been drunk for thousands of years. Water is life, but it is those flavored waters, the beers, wines, coffees, and teas that make life worth living.

BELOW: *Arabica coffee beans are cultivated commercially to produce coffees of distinction. Harvesting by hand is essential to ensure the quality of the beans.*

Anise Hyssop

Agastache anethiodora

RIGHT: *The purple flower spikes of the anise hyssop bloom throughout the summer. Its leaves are used for teas and flavorings. The plant can survive for up to 10 years.*

Hyssop is a herb that is best known for its anise-scented foliage. Few herbs are prettier than anise hyssop. It is native to North America, China, and Japan, where it grows in dry scrub and fields.

Historical Origins The name "hyssop" can be traced back throughout recorded history. In the New Testament, a sponge soaked with wine and placed on a hyssop branch was offered to Jesus of Nazareth on the cross before he died, and this plant may be similar to the hyssop we know today.

Botanical Facts Hyssop is an ornamental herb, about 2 ft (0.6 m) high with anise-scented leaves and beautiful purple flowers. There are 20 species of very aromatic perennials in the *Agastache* genus, a member of the mint (Lamiaceae) family. The leaf shapes, as well as the colors, of the flowers vary. Hyssop grows in a lush upright manner and the whole plant exudes aroma. Bees, butterflies, and hummingbirds are attracted to the plant, and wild birds eat its seeds. Both the leaves and flowers are edible.

Culinary Fare The flowers of the plant are edible and make attractive garnishes. Leaves are used fresh for salads, or to make tea that has a light licorice flavor. Teas made from the anise hyssop plant are supposed to be good for coughs.

Tequila

Agave tequilana

BELOW: *Tequila, which is named for a town in Mexico, is produced from the fermentation of the heart of the agave plant.*

Tequila is produced from only one of the many species of agave, the *Agave tequilana* (or blue agave), a plant more closely related to a lily than a cactus.

Historical Origins Native American peoples have long used agaves for food, fibers, soap, medicine, and beverages, and the plant was of such importance that it had mystical and spiritual significance for Native American tribes. Today, the Mexican government protects this national treasure, and the production of tequila is very strictly regulated.

Botanical Facts The name "agave" is taken from the Greek *agavos*, meaning stately or noble, a reference to the bearing of the plant. *Agave* is found from the southwest United States to Mexico, and the Caribbean, into Colombia and Venezuela. Most species take decades to bloom. There are a multitude of species, subspecies, and cultivars within the genus. Its common name in Mexico is maguey.

Culinary Fare The agave rosette bears fruits just once in its lifetime, each fruit taking between 8 and 12 years to mature. The fruit or heart of the agave (called the *piña*, which is Spanish for pineapple, as it resembles that fruit) is harvested and either steamed or baked to release the fermentable sugars, and then distilled into tequila. There are many styles of tequila, based on the amount of agave used in production, and the length of aging.

Wormwood

Artemisia absinthium

One of the bitter herbs mentioned in the Bible, wormwood—legend has it—sprang up in the trail left by the tail of the serpent as it slithered out of the Garden of Eden. The dark history of absinthe continued in nineteenth-century Europe as absinthe, a notorious drink once favored by artists and writers, was subsequently widely banned.

Historical Origins The earliest known description of wormwood is found on an Egyptian papyrus dated from 1600 BCE. The herb was used medicinally to rid the body of worms, as a tonic for stomach ailments, and to ease the pain of childbirth. Its medicinal role continued in the modern absinthe, created as a health tonic toward the end of the eighteenth century and drunk by the French Foreign Legion to ward off malaria. Commercial absinthe was first produced in the 1790s by Dr. Pierre Ordinaire, whose formula was subsequently purchased by Henri-Louis Pernod.

Botanical Facts Native to the temperate regions of Europe, Asia, and North America, *Artemesia absitinthium* has naturalized throughout the world. Wormwood is an evergreen perennial with gray and white leaves. It bears clusters of pale yellow flowers from early summer to early winter.

Culinary Fare Both the leaves and the tops of the flowers of wormwood are used to make a medicinal tea. Its leaves are used as a flavoring in absinthe. However, absinthe contains thujone, a toxic chemical that was believed to cause hallucinations and psychosis. Today, with this theory disproved, absinthe is enjoying a resurgence of interest and many countries are allowing its production and sale.

Calamint

Calamintha grandiflora

Plants in the mint family have a long history of both medicinal and culinary use. The leaves of the calamint are harvested as the shrub begins flowering and are dried for later use as a flavoring in tea, or as a warm poultice for easing aches and pains.

Historical Origins The genus name *Calamintha* comes from the Greek words *kallis* meaning "beautiful" and *mintha*, meaning "mint." The name is appropriate as the plant is an attractive addition to gardens. The mint smell of the leaves has long been used to mask odors. This species is native to southern Europe and North Africa.

Botanical Facts The *Calamintha* genus of the mint family, Lamiaceae, is made up of 7 species of shrubby perennials found in the northern temperate zones. Calamint is a bushy plant with dark green, downy, toothed leaves, and pink, tubular-shaped flowers that are attractive to bees. Calamint leaves have moderately pungent, but pleasant, lightly minty-herbal aromas.

Culinary Fare Calamint has a long history of medicinal use for upset stomachs, but today is usually restricted to a simple infusion of the leaves, which produces a lightly aromatic tea smelling of sweet mint.

ABOVE: *Calamint blooms almost all summer long, bearing up to 5 flowers in one cluster. Its blossoms make a tasty addition to a summer salad.*

Tea

Camellia sinensis

ABOVE: *The Japanese tea ceremony is an intricate ritual based on 4 principles: harmony with people and nature, respect for others, purity of mind and heart, and tranquility of spirit. Each step in the ceremony has significance.*

Apart from water, tea is the most popular beverage on the planet, and the one that has been used longest. Drinking a cup of tea has evolved over the centuries into a pleasurable daily ritual in many countries around the world, a refreshing respite during a busy day. Tea contains caffeine, a mild stimulant that increases mental clarity. However, many of the negative effects associated with caffeine are thought to be mitigated by the other compounds such as polyphenols—powerful antioxidants—found in tea.

Tea and Buddha

One ancient Chinese tale about the invention of tea tells the story of Bodhidharma, the Indian saint who founded the Japanese school of Buddhism. In CE 520, he traveled from India to China to teach Buddhist doctrines. In order to set an example, Ta-Mo (the White Buddha) as he was called by the Chinese, vowed he would not sleep until his mission was accomplished. After 9 years of meditation, he was overcome by fatigue and fell asleep. In disgust, he cut off his eyelids, and threw them to the ground, where Buddha caused them to sprout seeds, which grew into the first tea plant. The shrub is a symbol of eternal wakefulness, whose leaves resemble the shape of an eye, and the beverage brewed from the leaves gives the gift of mental alertness.

Historical Origins Tea is thought to have originated in China, and fossil records show that the plant has existed for millions of years. Long before tea was consumed as a beverage it was a medicine and a source of food. Prehistoric peoples chewed the leaves for their stimulating effect, and rubbed the leaves on wounds—practices that continued for thousands of years. Tea leaves boiled and made into balls mixed with dried fish, oil, and salt were eaten as food.

The origins of tea are intertwined with myths and legends. Oral histories in China connect the drinking of tea with philosophy and religion. Tea played an enormous role in the religions of Taoism and Buddhism, as an invaluable aid to meditation. The shifting dynasties in China, with the subsequent rewriting of history with each new dynasty, make an accurate account of the development of tea drinking difficult. By the fourth century, tea drinking was so ingrained in the population, that tea had to be cultivated rather than grown wild.

The first book devoted to the subject of tea was from the Tang dynasty, around the eighth century. The Taoist poet Lu-yu wrote the "Cha-Sing" (or "Classic Art of Tea"). After his death the legend is that Lu-yu became the genie of tea called Chazu, and his effigy is still honored today by tea merchants around the world. During the Tang dynasty, the art of elegantly making a perfect cup of tea, as well as the ability to recognize type and quality, became hallmarks of sophistication among the Chinese people. Chinese knowledge of tea spread to Japan in the twelfth century when a Buddhist monk named Eisai cured the ailing shogun, Minamoto Sanetomo, with the help of some strong tea. The shogun was thankful for his recovery, and promoted the new drink, and its popularity spread from the royal court to every level of Japanese society. The climate of southern Japan is perfect for tea cultivation. Tea's adoption by the Western

LEFT: *Freshly picked green tea leaves are oxidized, crushed, and dried to produce the small, dark leaves used to create black or red tea.*

world was slow to begin with, but once tea took off, it swept through Europe and the Americas. In North America, struggles over tea taxation and control helped ignite the American Revolution.

Botanical Facts A member of the Theaceae family, the *Camellia* genus has nearly 300 species, all native to the coastal and mountain regions of East Asia. The tea plant is an evergreen shrub with thin, smooth-edged, tapering leaves. Although the plant can grow as high as 30 ft (9 m), commercial tea bushes are trimmed to a height of about 4 ft (1.2 m) to facilitate picking. Commercialized plants can have the leaves picked up to 30 times per year. The best tea plants are grown at higher altitudes where the cooler air encourages complexity of flavor to develop in the leaves.

Culinary Fare Tea is an infusion made by steeping dried tea leaves in boiling water for a few minutes. There are a number of styles of tea: white, green, oolong, and black. All are from the tea plant, but differ in the amount of oxidation the tea leaf is exposed to once the waxy outer surface is broken. There are hundreds of varieties of tea, each with its own aroma and flavor characteristics. Additional flavorings and seasonings may be added to the tea as well—herbs such as mint and chamomile, fruits, spices, and flowers. Different cultures prefer different styles of tea as well as different additions such as milk, sugar, and lemon. Tea has lately been an exciting addition to the modern bar scene, with tea being used to create liqueurs and to flavor drinks, including martinis.

BELOW: *Harvesting tea is a labor-intensive process. Only the bud and the top leaves are plucked to ensure teas of high quality.*

Chicory

Cichorium intybus

All parts of the chicory plant have been valued for centuries—the tender leaves eaten as salad or brewed into a medicinal tea—but the herb is probably best known for the use of its root. Whenever coffee has been unavailable or too expensive, people have turned to the ground and roasted chicory root as a widely available and less expensive (and also caffeine-free) alternative. The use of chicory in coffee was one of the triggers for the development of pure food legislation of the nineteenth century, in response to demand from consumers who wanted to know how much chicory was in the coffee they were purchasing.

Historical Origins The cultivation of chicory dates back at least 4,000 years. It was planted in Egypt and irrigated by the flooding of the Nile River. The Greeks and Romans used chicory as a vegetable and for salad. In Tudor England, Queen Elizabeth I drank chicory broth as a health tonic. Chicory seeds were planted throughout Europe and the Americas, and were especially suited to a range of climates in the United States. Thomas Jefferson imported chicory seeds from Italy, and in a letter to George Washington in 1795 he described the herb as "one of the greatest acquisitions a farmer can have." By the early 1800s, chicory was abundant, and a common household product in Europe and North America. In the late eighteenth century, coffee was very fashionable and very expensive in Europe. The Dutch began roasting and grinding the roots of the chicory plant for use as a coffee substitute. During World War II and the resulting shipping disruptions, most coffee consumed in Europe and North America was produced from chicory.

Botanical Facts Chicory is a tall, herbaceous perennial native to the Mediterranean Basin and Asia Minor. The version of chicory seen today is a cultivated form of the original wild chicory, with a long, thick root. The large-rooted type of chicory was developed in the Netherlands in the eighteenth century as horticulturalists attempted to find alternatives to coffee. Naturalized in North America and Australia, chicory grows wild today throughout much of Europe and the United States, and its slender stalks with blue flowers are a common sight along US highways.

Culinary Fare Ground chicory can be used as a beverage by itself, but it is also used to soften the harsher taste of coffee. When added to coffee beans, it adds aroma, color, and sweetness, offsetting the characteristic bitter taste of coffee. The chicory–coffee blend is sometimes called Creole coffee, and is still very popular in Louisiana. Chicory is also used in the food industry as a sweetener.

RIGHT: *The roots of chicory have commonly been used as a substitute for coffee for many years. Because it has a higher sweetening power than sucrose, chicory is also added to foods as a sweetener.*

Coconut

Cocos nucifera

Coconut is perhaps the world's most useful plant. Every part of it can be used. The interior of the nut is used for food, drink, medicines, and chemicals, and the exterior is used for thatching materials and timber. There is an Indonesian saying that there are as many uses for coconut as there are days in the year, and in many cultures the coconut is considered the tree of life.

Historical Origins Although there is some disagreement about the origins of the coconut, it is generally thought to be native to Malaysia and surrounding areas. It spread with travel by seafaring people, or simply by the coconuts falling and floating to new locales. In early Sanskrit records, which describe the use of the coconut in India, it is called "the tree that gives all that is necessary for living." Marco Polo mentions coconuts in his accounts of his thirteenth-century travels. In the late fifteenth century, Portuguese and Spanish explorers thought that the 3 little "eyes" in the coconut shell resembled a grinning monkey, and they named the coconut *coco* (which translates as "goblin"). The species name, *nucifera,* is Latin for "nut-bearing."

Botanical Facts Coconuts grow on palms that flourish on the seashores in the coastal regions of all tropical seas around the world. The coconut tolerates salty ocean-side conditions. It prefers sandy soil as its roots need air, which is provided by the ebb and flow of the tide as the water draws air in, and then expels it again. The roots of the coconut palm are shallow, but spread widely, so strong winds give the trees their characteristic lean. The coconut palm can grow to be 100 ft (30 m) high, and is topped by a crown of large leaves. Its drupes grow in bunches of about a dozen coconuts each, and each tree yields thousands of coconuts over an average 70-year lifespan.

Culinary Fare The coconut itself is technically a drupe, meaning a fruit with a hard stone in the center. The coconut as it is sold commercially is that hard stone, with the rest of the exterior cut away. The outside husk is very smooth and tough. It is colored greenish to reddish brown, and becomes a gray color as the fruit matures. Between this outer husk and the nut itself is a layer of loose, thick, brown fibers, which protects the pulp that adheres to the interior walls. The center of the coconut is filled with a pearly liquid known as coconut water, which has a refreshing, slightly sweet taste. Grating coconut and mixing it with milk or water makes coconut milk, which is a staple in cooking in many countries. In the Philippines, it is often used as a sauce to accompany desserts. Coconut is a prized addition to many alcoholic beverages, flavoring tropical drinks such as the popular pina colada. Coconut is also used to flavor spirits such as rum.

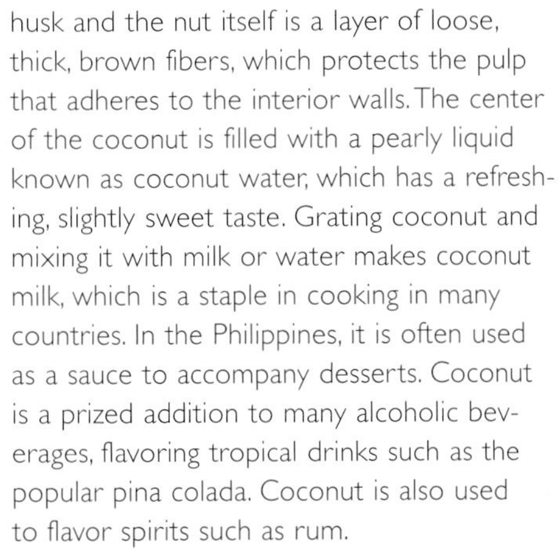

ABOVE: *Coconut water, straight from the nut, makes a sweet and refreshing drink.*

BELOW: *The leaves of the coconut palm are among the largest of any plant. Healthy palms can have up to 30 leaves, one of which dies off and is replaced by a new one every month.*

Coffee

✳ *Coffea arabica* Arabica Coffee ✳ *Coffea robusta* Robusta Coffee

RIGHT: *Coffee plants produce oval-shaped berries that become a red-purple color when ripe. Most berries contain 2 seeds, or beans, but sometimes they have only one.*

Coffee is one of the world's most popular beverages. For the most part, the plant is tended and harvested by hand, the tiny beans supporting an industry that employs about 30 million people. Coffee is drunk in every country of the world, and the rich traditions and interesting tales that are associated with this dark beverage are almost as intoxicating as the heady brew itself.

The Goatherd and the Coffee Bean

With the origins of coffee unrecorded in our early history, many legends have sprung up around its birth. One such legend tells of an Ethiopian goatherd who noticed his goats acting in an unusually frisky and frolicsome manner after eating the berries from a particular tree. The goatherd soon determined that it was the berries that caused the goats' eccentric behavior. After chewing a few himself and noting his increase in energy, he took them along to the local holy man, who experimented with the berries, finding a way to dry and roast them and then brew them into a beverage. The holy man discovered the ability of the beans to aid in staying awake during lengthy prayer sessions, and a coffee tradition was born.

Historical Origins Coffee is produced from an evergreen shrub thought to have originated in Ethiopia and nearby parts of tropical Africa. At one time, African tribes used mortars to crush the ripe berries, and mixed them with animal fat, shaped them into balls, and carried them into battle for energy. The earliest coffee drinks probably were not at all like they are today, but were most likely made with the juice of the fermented berries. Eventually, coffee made its way from Ethiopia to Arabia. Coffeehouses were serving customers in Mecca by the middle of the fifteenth century.

By the early seventeenth century, the coffee bean was just beginning to be known outside of Arabia, and there was a growing demand. The Arabs tried to keep a monopoly on coffee, by making it law that no bean could be exported without first being boiled or dried, so the seeds could not germinate. Of course, smuggled beans eventually made it out of the region and were successfully planted in other areas, particularly India and Sri Lanka. The enterprising Dutch established coffee plantations in Java, quickly overtaking the Arabs to become the leading exporter of coffee. (It is no coincidence that in the United States, "java" is slang for coffee.) Arabian coffee was often called Mocha after the port on the Red Sea from which it was shipped, and sometimes these two coffees were blended together, hence the name, Mocha Java. The Dutch dominance in coffee production ended by the mid-nineteenth century, due to plant disease and political disruption, but by this time, coffee plantations had been established in other parts of the world, notably in Brazil.

Botanical Facts This tropical African and Asian genus is in the madder family (Rubiaceae), and includes about 40 species of evergreen shrubs. The plants are quite pretty with lush, green foliage. They bear clusters of white, fragrant flowers on the leaf axils, which

LEFT: *The ripening berry of* Coffea arabica *turns from green to yellow, and finally to red. Berries take 7–9 months to ripen.*

are followed by colorful berries. These ripen to red-yellow or purple berries known as cherries, in the centers of which are usually 2 coffee beans. Depending on the species, coffee plants can reach a height of more than 20 ft (6 m), but they are usually trimmed in commercial planting to make picking easier. The plants are usually 4–6 years old before they begin producing beans, and may stay productive for over 30 years. The berries ripen at different times, so it is possible to have both ripe and unripe coffee cherries on the same plant. Because of this, hand harvesting is almost a necessity for good-quality beans.

To produce coffee beans, 2 species, *Coffea arabica* and *C. robusta*, are used to the exclusion of all others. Arabica coffee is named for the Arabs who were the first to cultivate it commercially. Beans from *C. arabica* contain less caffeine than beans from the *C. robusta* plant, and are considered to have more expressive, subtle aromas. About 70 percent of the beans grown for the coffee market are *C. arabica*. *C. robusta* makes up the remaining 30 percent. With their higher caffeine level and harsher, more bitter flavor, Robusta beans fetch a lower price than Arabica beans, and are often blended with them. In cultivation, *C. robusta* has two advantages over *C. arabica*: It grows well at lower altitudes and warmer temperatures, and tends to be more resistant to disease.

Culinary Fare To produce beans for coffee, the outer pulpy layer of the coffee cherry is removed, either by drying the beans in the sun then removing the pulp, or by removing the pulp immediately after picking. Some of the material from the pulp still clings to the bean. The beans are then placed in a water bath, where fermentation causes a chemical change that softens the clinging pulp enough so that it can be removed. Next, the beans are dried, usually in the sun, and continuously monitored and turned to prevent uneven drying. Depending on the weather, the beans need 1–2 weeks to reach the appropriate stage of dryness. The beans are then sorted by color, size, and weight, and the last layer of filmy parchment is removed. The beans are roasted before being ground.

BELOW: *After the coffee beans have been removed from their cherry, they are spread on a drying table to dry in the sun before being sorted and packaged.*

Coca Leaf

Erythroxylum coca

RIGHT: *The leaves of the coca plant have green, thin, tapering leaves. They are rich in vitamins, protein, and calcium. The plant itself can live and produce its red, oval fruits for up to 40 years.*

Coca has a long history of both legal and illegal use around the world. The coca leaf has been used for centuries as a natural stimulant, but the coca leaf is also the raw material used for the manufacture of the modern drug, cocaine.

Historical Origins Coca originated in the Andes Mountains of South America, and was revered by pre-Inca peoples as early as the sixth century. Pottery figures from the times show figures with bulging cheeks full of coca leaves. Mummies were buried with a store of coca leaves for the afterlife. The leaves were chewed as part of religious ceremonies. People in the region doing physical labor chewed the leaves for their stimulating and appetite-suppressing effects.

Botanical Facts Coca is a tropical shrub native to the Andes Mountains, but cultivated in Africa, Taiwan, and Southeast Asia. Coca thrives in hot, damp environments, but the better-quality leaves come from drier sites.

Culinary Fare Native South Americans chew the leaf, mixing it with lime or plant ashes to release the stimulating alkaloids. *Mate de coca*, or coca tea, is made from the leaves of the coca, and is a popular drink in many South American countries. The popular beverage Coca-Cola was named for its two original primary ingredients: the coca leaf and the kola nut.

Pina Palm

Euterpe oleracea

This small, slender palm is grown for its fruits and for the cluster of new leaves more commonly known as hearts of palm.

Historical Origins The palm trees in the Amazon region have given people in the area food, drink, clothing, materials for shelter, and leaves for thatching, weapons, and tools, for thousands of years.

Botanical Facts The genus of feather-leafed palms (family Arecaceae) consists of 7 species native to the lower Amazon region of Brazil. The pina palm is known as a clumping palm, with several tall, slender trunks and a drooping crown of feathery fronds. The palm produces small purple-black fruits in clusters.

Culinary Fare The fruit is edible, and the juice of the fruit is used in wines and liqueurs as well as a juice, juice blends, and smoothies. A local drink called *acai* is very popular in the Amazon region, particularly in Brazil, where it is widely available and inexpensive. There has been great interest in *acai* juice as a health tonic in the United States because of its high level of anthocyanins, the same phytochemicals that are found in red wine and blueberries.

RIGHT: *The pina palm is a fast-growing tree that bears both male and female flowers on the same tree. It is rarely grown outside the Amazon region.*

Hops

Humulus lupulus

Hops, a vining plant whose flowers have been routinely used since the eleventh century in the production of beer, is a member of the cannabis family. Hops gives beer its characteristic bitter taste, and acts as a preservative to help extend the shelf life of this low-alcohol beverage. Many different types of hops are grown around the world, and beer makers select the types of hops best suited to the style of their beer, much like chefs select spices for cooking.

Historical Origins In the first century CE, the Roman naturalist Pliny discussed hops in his study of natural history. The Romans said hops "grew wild among the willows like a wolf among sheep" and named it *lupulus,* meaning the good wolf. Hebrews used hops to ward off the plague. Folklore has it that hops turn green on Christmas Eve. Native Americans and American colonists used hops tea as a sleep aid and to improve digestion, or they made it into poultices to soothe toothaches or open wounds. The use of hops in beer originally met with some resistance in Europe, but gradually came into widespread use, and today, most beers use hops in their production.

Botanical Facts This genus contains 2 species, which are herbaceous, climbing, twining perennials from northern temperate regions. The plants are vigorous and can grow up to 70 ft (21 m) in a season. The hardy plants have dark, heart-shaped leaves and produce small, conelike flowers. The dried flowers of the female cones of *Humulus lupulus* contain bitter resins and essential oils that are an indispensable ingredient in beer. Cultivated hops are typically grown up strings in a hop garden. Well-known hop gardens are found in Poland, France, Germany, the Czech Republic, China, England, and the Pacific Northwest of the United States.

Culinary Fare Almost exclusively, hops are produced for use in the beer-brewing industry. The dried flowers impart a pleasantly bitter taste to beer that helps balance the sweetness of the malt. The degree of bitterness contributed by the hops is determined by the point in the brewing process at which they are added. The amount of bitterness is measured in International Bitterness Units (IBUs). The higher the number of IBUs in the beer, the more perceived bitterness. Hops also contribute citrus, floral, fruity, or herbal aromas, and brewers will often use a second addition of hops in order to create the desired aromas in the final beer. Importantly, hops also aid the activity of the yeasts used by brewers, as well as reduce the growth of less desirable organisms.

ABOVE: *As well as being an essential ingredient of beer, hops are used to improve the appetite and also in herbal preparations to help relieve anxiety and insomnia.*

Yerba Maté

Ilex paraguariensis

BELOW: *The freshly cut leaves of the yerba maté are shredded and brewed as a tea. It is used to boost the immune system and as an aid to digestion.*

Maté is a tealike beverage produced by steeping the dried leaves and twigs of the yerba maté in hot water. The tea is brewed in a cup or a hollowed-out gourd and shared with friends through a common straw. Argentineans living in other countries miss the ritual of sharing the maté with friends almost as much as they miss the drink itself.

Historical Origins The Guarani Indians used yerba maté as a basic food and currency. The Guaranis shared their love of the plant, and the tea brewed from its leaves, with the sixteenth-century Spanish conquistadores. In the mid-seventeenth century, Jesuit missionaries grew yerba maté on plantations.

Botanical Facts The genus *Ilex* belongs to the holly (Aquifoliaceae) family, and has more than 400 species. *Ilex* is native to sub-tropical South America in Argentina, Brazil, Uruguay, Paraguay, and Bolivia. Yerba is from the Spanish *hierba*, meaning herb, and maté means cup—so the name means "cup herb."

Culinary Fare Yerba maté's main use is as a tea. The tea is brewed by slowly filling cups or gourds packed with the leaves with almost boiling water. The infusion is sipped through a metal straw (called the *bombilla*). The tea can be drunk straight, or sweetened, and sometimes other herbs or milk are added. Yerba maté is so popular in Argentina, Paraguay, and Uruguay that a cup of maté with friends is a daily ritual.

Chilean Wine Palm

Jubaea chilensis

BELOW: *This young wine palm may grow to 80 ft (24 m) in height.*

The sap of this huge tree is fermented into an alcoholic beverage, called palm wine, hence the name of the tree. Over the years, heavy use of Chilean wine palms has drastically reduced their numbers, and local conservation groups have begun reforestation efforts.

Historical Origins The generic name *Jubaea* honors a North African king, Juba, who had an interest in plants. When the Spanish conquistadores first entered central Chile in the sixteenth century, they found these palms all over the the coastal ranges. The Spanish learned from the indigenous populations how to collect the palm sap as a sweetener, or to ferment it into an alcoholic beverage. The tree is now protected in Chile to ensure that it is not over-harvested.

Botanical Facts *Jubaea chilensis* is the only species of *Jubaea* in the palm family Arecaceae native to coastal areas of Chile. This species has been called the Incredible Hulk of the palm world because of the massive diameter of the trunk.

Culinary Fare To retrieve the sap, the palm tree is cut down, and the sap is allowed to drip out slowly over the course of 6–8 weeks. The sap can be fermented into palm wine, or else boiled down into a thick syrup called palm honey. The seeds are edible.

Juniper

Juniperus communis

The distinctive aromas and flavors from the leaves and berries of the juniper have been used throughout history with mentions as far back as Egyptian scrolls from 2800 BCE. Today, the best-known use of juniper is as the primary flavoring agent in the alcoholic beverage gin.

Historical Origins The bracing pine and menthol aromas of juniper have long been associated with purification. The berries were burned in the Middle Ages and beyond to clear the air of pestilence, and juniper-laced elixirs were used to ward off plague. Native Americans believed in juniper's cleansing and healing powers and used it to keep away infection, relieve arthritis, and cure wounds and illnesses. Juniper was used in the production of high-quality spirits, as well as to mask the aromas of cheaply made spirits during the US Prohibition era.

Bathtub Gin

Gin has been produced in the United States since colonial times, but during Prohibition, its popularity soared. For bootleggers making illegal alcohol on the sly, traditional brown whiskies that required long aging in oak were just not practicable to produce. Gin was fairly easy to make by mixing grain alcohol with juniper berry extract and other flavoring agents in a large, vatlike bathtub. The resulting brew could be sold right away without any need for aging. The pungent taste of the juniper berries masked some of the harshness of poor-quality spirit, and drinkers further masked the taste by adding soda or tonic to the drink, helping give rise to the advent of the mixed drink.

Botanical Facts Juniper is a prickly, evergreen shrub that grows throughout much of the Northern Hemisphere. It is a member of the large cypress family—the only one with edible fruits. The size and color of the shrubs vary according to where they are planted. A juniper tree has long, needle-shaped leaves, with berries ½–¾ in (12–18 mm) in diameter. The fruit of the juniper, often called a berry, is actually a cone with a pulpy resin. The berries are initially green, but they ripen to a bluish or purple color, a process that can take up to 3 years.

Culinary Fare In the production of gin, neutral spirits are flavored with juniper, and then enhanced with the addition of other botanicals, with most gin producers keeping their recipes a secret. Juniper is also used as a flavoring agent for some beers, schnapps, and the Scandinavian aquavits. Juniper is also a popular spice in European kitchens, where it is used to season meat and cabbage dishes, for pickling and as a marinade.

LEFT: *Juniper berries are usually dried before being sold commercially. When used in cooking, they are generally removed from the dish before serving.*

Chamomile

Matricaria recutita

BELOW: *Chamomile tea has an applelike taste. It is made by pouring freshly boiled water over loose, dried chamomile flowers, leaving it to steep, and then straining to serve.*

Chamomile is one of the most popular herbs in the world, beloved for centuries. Ancient Egyptians dedicated chamomile to their sun god, and prized the herb above all others for its healing properties. Today, chamomile is the best-selling herbal tea in the world, drunk for its soothing, relaxing qualities.

Historical Origins In documents as early as 1150 BCE, chamomile is mentioned as a beverage. The Egyptians not only drank chamomile, they used it to honor the gods, treat the ailing, and to embalm the dead. The Romans enjoyed drinking chamomile-flavored beverages, and the Anglo-Saxons regarded it as sacred. In the Middle Ages, it was widely used to ward off the plague and as an aromatic herb to mask the odor of unwashed bodies at gatherings and celebrations.

Botanical Facts There are both annual and perennial varieties of chamomile, with the annual version used for culinary purposes. Native to Europe and Asia, the bushy plant has delicate aromatic foliage and small daisylike flowers, with white turned-down leaves and a yellow center. The flowers of the plant are harvested and dried for culinary use. The leaves of the plant have a distinct aroma similar to that of apples.

Culinary Fare Chamomile is brewed into a tisane, an herb tea. It is quite popular, as chamomile tea is thought to taste less bitter than other herbal teas. The beverage is soothing, and encourages relaxation and sleep. With no caffeine, it is often drunk by pregnant women, nursing mothers, or given to restless children. Chamomile also seems to settle the stomach, and is used to combat morning sickness or upset stomachs.

Lemon Bergamot

Monarda citriodora

Plants in the *Monarda* family have long been valued for their ornamental, culinary, and medicinal uses. Lemon bergamot is a wildflower with fragrant purple to pink flowers and a delicate lemon fragrance. The plants are irresistible to butterflies, hummingbirds, and especially bees, hence the alternative common name, bee balm.

Historical Origins The genus *Monarda* is named for the Spanish physician Nicholas Monardes (1493–1588), who wrote about plants on his extensive travels, cataloging them for their medicinal uses. His genus is a member of the mint (Lamiaceae) family. The term *citiriodora* is probably derived from the Latin term *citrus* (or lemon tree) and *odora* (having the fragrance of). The name "bergamot" is probably derived from the similarity of this plant's odor to the aroma of bergamot oranges. Bergamot oil, which is extracted from the oranges, is used in the production of Earl Grey tea.

Botanical Facts Lemon bergamot is a hardy annual or short-lived perennial native to the western and midwestern United States. It contains the volatile oils citral and carvacrol, which are responsible for the plant's characteristic lemon scent.

Culinary Fare When the dried leaves of the plant are crushed, they release a lovely lemon scent, and are used to flavor teas, fruit dishes, and preserves.

Oswego Tea

Monarda didyma

Oswego tea is native to North America and was naturalized in Europe from seeds sent in the 1700s. The genus *Monarda* is named for the sixteenth-century Spanish physician and herbalist Nicolas Monardes, and is one of several plants that are members of the mint (Lamiaceae) family.

Historical Origins Oswego tea is named for the Oswego River in western New York where the Oswego tribe living in the area brewed tea from the dried leaves of the plant. The Native Americans shared this practice with the early colonial settlers. When black tea became scarce after the Boston Tea Party, oswego tea was a popular substitute.

Botanical Facts Oswego tea has beautiful scarlet flowers that are prized for their ability to attract butterflies and bees.

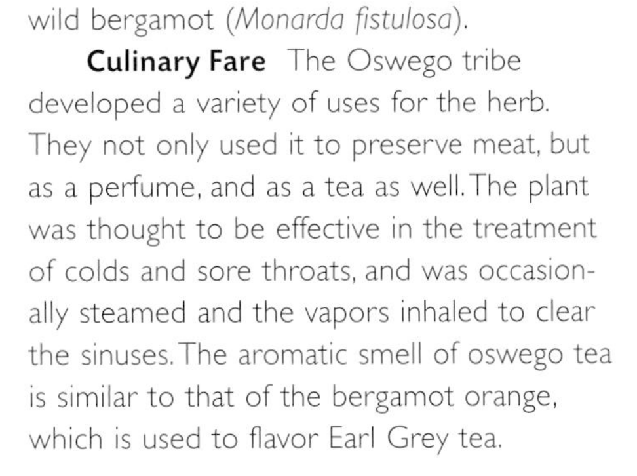

The entire bushy plant has a delicate citrus fragrance, but the flowers are thought to make the best tea. The plant is very similar to its cousin wild bergamot (*Monarda fistulosa*).

Culinary Fare The Oswego tribe developed a variety of uses for the herb. They not only used it to preserve meat, but as a perfume, and as a tea as well. The plant was thought to be effective in the treatment of colds and sore throats, and was occasionally steamed and the vapors inhaled to clear the sinuses. The aromatic smell of oswego tea is similar to that of the bergamot orange, which is used to flavor Earl Grey tea.

ABOVE: *Tea made from the oswego tea plant, also called bee balm, is believed to relieve abdominal pains. It is also used to treat nausea and vomiting.*

Wild Bergamot

Monarda fistulosa

Native Americans have long prized the beautiful leaves and flowers of the wild bergamot. The bushy plant was used as a type of medicine as well as a source of food and a beverage.

Historical Origins Native to North America, the plant was traditionally used for medicinal purposes. Native North American tribes used the leaves for a variety of purposes. The chewed leaves of the plant were placed in the nostrils to relieve headache, and the leaves and flowers were boiled together to create a beverage to relieve stomach pains, colds, and heart problems. Steamed leaves of the wild bergamot were used in sweat lodges.

Botanical Facts Like other members of the genus *Monarda*, this species has the alternative common name of bee balm because its flowers attract bees. The perennial is a member of the mint (Lamiaceae) family, and has a wide distribution, from Canada down to Mexico. It is very similar to the plant oswego tea (*Monarda didyma*). Wild bergamot's smooth-edged leaves are about 4 in (10 cm) long. The plant is showy and attractive in gardens, with bushy lavender-colored flowers.

Culinary Fare Both the leaves and the flowers of the plant are aromatic, smelling of citrus and delicate mint. The flowers are used to make fragrant herbal teas.

RIGHT: *Some Native American tribes added the leaves and flowers of wild bergamot for extra flavor when they cooked their meat.*

Catmint

Nepeta cataria

RIGHT: *The young leaves of catmint can be added raw to salads, while the older leaves are used to flavor meat dishes, or to make a minty herbal tea.*

The names catmint and catnip are used interchangeably. Catmint is so called because the minty aromas create a state of euphoria in cats once the aromatic leaves are rubbed (bees love it as well). People find the plant appealing too, as a popular herbal tea.

Historical Origins While no one is sure when the first cat discovered catmint, the plant has been used by their human friends for at least 2,000 years. The leaves have been brewed since Roman times into a pungent, mint-flavored tea that is used to ward off sore throats and colds.

Botanical Facts Catmint is a member of the mint (Lamiaceae) family, is native to temperate Eurasian and northern African areas, and is widely naturalized throughout Europe. It was brought to North America by the first settlers and is now common as a weed. The plant is a tall herb that resembles mint in appearance with grayish green leaves and mauve-blue to purple flowers.

Culinary Fare The aroma is unique—a powerful, minty, chemical smell. A tisane made from the dried leaves and flower heads has long been popular in Europe and the United States. Catmint was a more important herb in culinary uses in the past than it is today, although it is still used fairly commonly in Italy in salads, soups, and casseroles.

Guarana

Paullinia cupana

RIGHT: *The seeds from the guarana plant contain 3–4 times more caffeine than coffee beans. Guarana is mostly used as an energizer and a stimulant.*

Guarana has traditionally been a popular ingredient in Brazilian drinks and sodas. Interest in guarana has skyrocketed as the rest of the world reaches for the newly popular energy drinks.

Historical Origins The name of the genus honors botanist C. F. Paullini, a German botanist. Guarana is native to the rain forests of Venezuela and Brazil. It is thought to have been cultivated since pre-Columbian times. The plant plays an important role in the culture of the Guaranis, a tribe of Native South Americans.

Botanical Facts This climbing shrub features large, divided leaves and clusters of small, yellow flowers. It is prized for its yellowish to red fruit about the size of a chestnut. Each fruit contains 2–3 black-and-white seeds, which startlingly resemble human eyes, and contain large amounts of guaranine, a chemical that is essentially the same as caffeine.

Culinary Fare The seeds of the guarana are shelled, washed, roasted, and ground into powder or made into syrup. Guarana-laced juice drinks and sodas are popular in Brazil, and are thought to be a fatigue-fighting stimulant. In North America, guarana is becoming a popular ingredient in energy drinks and herbal teas that are promoted as increasing one's alertness and stamina.

Cocoa

Theobroma cacao

Extracted from the beans (seeds) of the cocoa tree, cocoa is the basic ingredient in the production of chocolate. It is hard to imagine a product more steeped in history, romance, and intrigue than chocolate—the taste that has fueled the palates and imaginations of centuries of people around the world, living up to its name "elixir of the gods."

Historical Origins Archaeologists believe that the Olmecs, an ancient people living in the area of what is now Mexico, ate the sweet, pulpy fruit of the cocoa plant and likely cultivated it in 1000 BCE. The Olmecs gave the tree its earliest name, *kakawa*. Clay chocolate-drinking vessels are found dating from CE 500, when it is thought that the Mayans were the first to make a drink from cocoa (roasting the beans and grinding them) and adding honey, chilies, and herbs.

Sixteenth-century Spanish chronicles of the conquest of Mexico by Hernán Cortés relate that when Moctezuma, emperor of the Aztecs, dined, he was served whipped, frothy chocolate, which he ate with a golden spoon from a golden goblet. Cortés brought chocolate and the knowledge of how to prepare it back to Spain in 1527. However, it took a few more decades—and the addition of sugar and vanilla to the beverage—for chocolate to become appreciated in Europe. Once it did take off, this new drink swept through fashionable Europe, where it was reputed to have aphrodisiac qualities. In 1847, the first chocolate bar was created, and chocolate makers have never looked back, developing ever more luscious creations.

Botanical Facts The *Theobroma* genus is from tropical America and is a member of the cacao (Sterculiaceae) family. A mature cocoa tree is about 25 ft (8 m) tall, and produces leaves, flowers, and fruits or pods year-round. The flowers arise directly from the leaf axils after the leaves have fallen, and are followed by large, fleshy fruits that contain many seeds (beans).

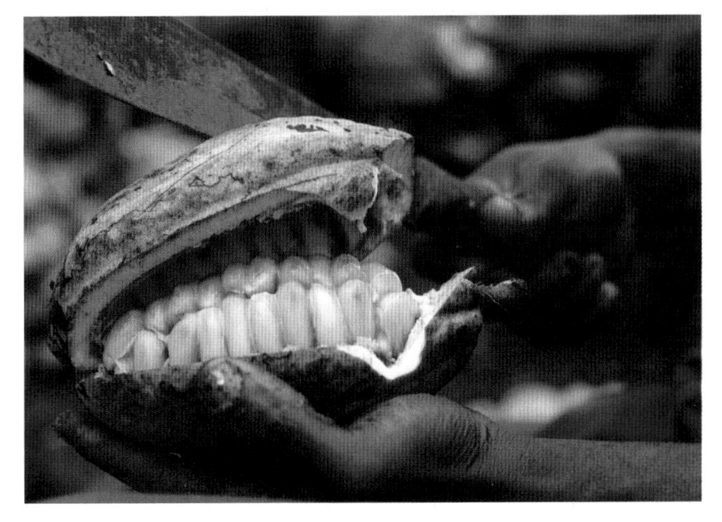

ABOVE: *The cocoa pod is cut open to obtain the beans (seeds) inside. These days, most cocoa beans come from the Ivory Coast and Ghana, which together account for more than half the world's production.*

Culinary Fare The beans are used in the manufacture of chocolate. The cocoa pod is filled with about 40 beans, which are protected by a sticky, sweet pulp. The beans and pulp are scooped out and allowed to ferment together for several days. This allows the flavors of the beans to emerge. The fermented beans are then dried in the sun, and roasted. Next, the roasted beans are heated and crushed between steel rollers to produce a dark, sticky paste, a little more than half of which is cocoa butter. Sugar and milk may be added, and different methods are used to create the final products. Like the Mayans before us, we love a soothing cup of cocoa. Cocoa is also a flavoring ingredient of the many chocolate-flavored liqueurs.

Columbus Discovers Chocolate

Christopher Columbus is thought to have been the first to bring cocoa beans to Europe. In 1502, he set off on his final voyage to discover India. Instead, the Italian explorer landed on the island of Guanaja, just off the coast of Honduras. Looking for riches, he was given a handful of cocoa beans by the local Mayan people. Columbus was surprised when, as his son Fernando reported later, a few of the beans dropped and the Mayans "scrambled for them as if an eye had fallen out of their heads." Columbus thought the beans were almonds, but seeing how the locals prized the shriveled-looking specimens, he took them back to Spain, where they were overlooked for the far more exciting riches aboard his galleon. It was just a few decades later that Cortés, who had discovered the Aztecs sipping a frothy chocolate drink, brought drinking chocolate from Mexico to Spain.

Plant Sugars and Other Products

Introduction to Plant Sugars and Other Products

Undeniably, the cultivation of plants marks an important turning point in the history of humankind. Humans needed to become settled before they could devote their time and labor to planting, nurturing, and harvesting crops. In the case of plant sugars and plant oils, that harvest was just the beginning of an equally strenuous extraction and refining process. The work for plant sugars often included pressing, boiling, and straining in order to extract and condense the plants' inherent sweetness. In order to extract oils, plants were often roasted, ground, and refined or strained in order to release the oil from the seed, stem, or bean. In both cases, societies had to be settled and had to have the human power needed in order to see each process through to the end. It follows that, as civilizations progressed, it was the developed countries that were able to enjoy the fruits, or in this case the sugars and oils, of their labors.

The production of both plant sugars and oils has been greatly aided by technology. Modern advances have offered humans an increasingly refined product and greater yield with correspondingly less labor. Like so many other foodstuffs, the cultivation of plant sugars and oils changed the world forever. Their cultivation has brought societies both great wealth and great despair.

BELOW: *The distinctive autumn foliage of the sugar maple sets it apart from other trees. Sugar maple trees are a major source of maple sugar used to make maple syrup.*

PLANT SWEETENERS

The sap of a tree, the stem of a grass, the leaves of a shrub, and the roasted husk of a bean are seemingly disparate items that all share one common trait: sweetness. The discovery of plant sugars came only after people were leading a settled life. Sweetness is, after all, not integral to our survival—it is a bonus, an extra, a luxury. In any culture, a community must be secure in its ability to survive before using its resources to add some sweetness to that survival. In effect, a discussion and examination of various plant sugars is also a discussion and examination of the progression of human beings from survival to security. However, with security came new struggles as communities began to fight between themselves for domination.

The history of plant sugars mirrors the history of power and domination throughout the world. The cultivation of sugar cane was propagated by the same societies that ruled the seas. Tracing the movement of sugar cane cultivation around the globe also traces the movement of slavery and the resulting horrors that came in its wake. Sugar's sweetness was not enjoyed by all.

In the modern world, plant sugars are still connected with politics and power. After too many years of freely imbibing in sugar cane, Western developed nations are looking for healthier solutions. Less refined sugarcane products and alternative sources such as sweetleaf are being re-examined with an eye toward reversing health trends such as obesity and diabetes. The stable existence that allowed humankind to cultivate and

LEFT: *In many countries, such as Reunion Island, the harvesting of sugar cane is done by hand. First the field is set on fire to rid it of the dead leaves and any snakes, but leaving the good stalks and roots unharmed. Then the harvester cuts the cane just above the ground.*

process plant sugars is now being blamed for the increase in unhealthy lifestyles. Best that we remember that sweetness is not integral to our daily survival; it is an extra something to be enjoyed sparingly.

PLANT OILS

Plant oils have a very long history of use, not only as foodstuffs but also as sources for heat, light, cosmetics, and preservatives. An individual country's use of an oil or fat is determined, to a great extent, by its environment or geography. For instance, in countries with wildly diverse weather and landscapes, animals of both land and sea provided a steady source of oils and fats. Europeans traditionally used both animal and marine fats because those sources were readily available to them. On the other hand, warmer countries in Africa, Asia, and parts of the Mediterranean had environments that were able to support an abundance of plant oil sources.

"Good oil, like good wine, is a gift from the gods. The grape and the olive are among the priceless benefactions of the soil, and were destined, each in its way, to promote the welfare of man."
George Ellwanger (1816–1906), *Pleasures of the Table*

The distinction between animal fats and plant oils is an important one. Animal fats are, for the most part, both plastic and saturated. A plastic fat is simply a fat that remains solid at room temperature. The more solid the fat, the higher the fat is in saturated fatty acids. Numerous studies have shown that saturated fatty acids raise blood cholesterol and increase one's risk of heart disease. Plant oils, however, are usually liquid at room temperature and, consequently, are also lower in saturated fats. It would seem logical that in the pursuit of better health, modern humans should increase their use of plant oils and decrease the use of saturated fats. If only it were that straightforward.

In the early twentieth century the process of hydrogenation was developed. On the most basic level, hydrogenation converts liquid fats into plastic fats. This process allows for plant oils to be used in products such as margarines and shortenings. A solid fat is, for the most part, easier to work with in the kitchen. Additionally, hydrogenation lengthens the life of the fat, increasing its stability against rancidity. The problem is, of course, that previously unsaturated and relatively healthy fats have been transformed into saturated and unhealthy fats. This has caused health problems around the globe.

Today, the move is away from unhealthy hydrogenated fats. As healthy alternatives are being sought, many plant oils are being used in their unsaturated and liquid state. When a solid fat is required, naturally saturated fats such as tropical oils are being used with greater frequency.

Plant oils are not plastic. They are liquids, and for many years they proved to be difficult to transport from one region to another. Most plant oils, olive oil being a notable exception, are processed in essentially the same manner.

ABOVE: *Extra-virgin olive oil is a healthy alternative to a plastic fat. To produce this oil, fresh olives are crushed into a paste from which the juice is extracted. No chemicals are used. This oil has a fruity aroma and contains antioxidants.*

Maple Syrup

Acer saccharum

Maple syrup is unique to North America, specifically northern New England and southern Canada. Its high price can be attributed to the limited area in which it is produced and to its short harvesting time.

Historical Origins Most historians agree that it was Native Americans who first produced maple syrup by boiling the sap of the sugar maple tree. French settlers arriving in Canada embraced the syrup-making process, as white sugar was not only expensive, but also extremely difficult to find.

Botanical Facts Temperature is vital to maple syrup production. For the sap to run, the days must be cold (but above freezing) and sunny, and there must be frost at night. Without these criteria, the sap will not flow from the tree's roots to its branches. Not surprisingly, the sugar season is short, often lasting as few as 4–6 weeks. Once the nights become warm and buds begin to form, the season is over. Collecting sap from the tree does not hurt it or hinder its growth in any way. On an average day, one tree will produce about 3 gal (12 L) of sap.

Culinary Fare Perhaps the best-known use for maple syrup is as an accompaniment to breakfast fare such as waffles, French toast, and pancakes. It is also used as a sweetener for glazed squash, sweet potatoes, and carrots. Maple-cured bacon and hams are popular throughout the southern United States, and many children enjoy maple creams and candies.

RIGHT: *To collect the sap from the sugar maple, a special tube called a spile is inserted into a hole that has been drilled into the tree's trunk. The sap flows through the tube and is collected by the bucket hanging from the other end of the spile.*

Welcome to the Sugarhouse!

Years ago, the sugarhouse was an integral part of traditional maple syrup production. Sugarhouses were found throughout the countryside as farmers sought to supplement their winter income. The steady white puffs of steam issuing from the chimneys of the sugarhouse were a sure sign that spring was just around the corner.

Decades ago the sap-collecting process was both tedious and labor-intensive. The transformation of the sap into a syrup required long boiling times. This meant that fires in the sugarhouse were burning around the clock for this period. The steam from the chimney often called together community members. Children and adults gathered to enjoy sausages boiled in sap and maple syrup candy made by pouring rings of hot maple syrup onto freshly fallen snow.

However, modern production methods have now sped up the syrup-production process, lessening some of the allure of the old-fashioned sugarhouse. No longer are the houses bursting with activity for the few magical weeks when the sap is running.

Sugar Beet

Beta vulgaris Crassa Group

Sugar beets contain 16–18 percent sucrose and are the source of white table sugar for much of Europe. Their origin as Europe's sweetener has its roots in politics and war.

Historical Origins Britain blockaded the transport of goods, including sugar, to France during the Napoleonic wars. The French then began to look elsewhere for a sweetener and found it in the sugar beet. Sugar beets do not need a tropical climate in order to grow (as does sugar cane), nor is the production process labor-intensive. These factors secured the beets' success in war-torn Europe, a success that continues to the modern day.

Botanical Facts The sugar beet is a biennial that stores its food in its roots. This store of food allows the beet to survive its first winter and gives it the energy to set seeds in the second year. In order to capture the roots' sweetness, harvesting must take place at the end of the first year's growing season. Once harvested, the beets are washed, sliced thinly, and flushed with water, which forces the sucrose from the plant cells. The resulting liquid is refined, reduced through evaporation, and then crystallized.

Culinary Fare Beet sugar is used in both the savory and sweet sides of the kitchen. Sugar provides baked goods with not only flavor but also color through caramelization. Its use adds tenderness and moistness to cakes.

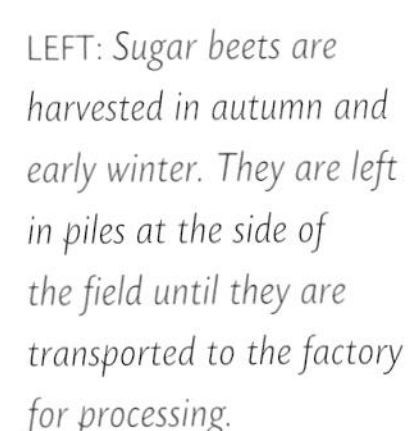

LEFT: *Sugar beets are harvested in autumn and early winter. They are left in piles at the side of the field until they are transported to the factory for processing.*

Jelly Palm

Butia capitata

Although native to central-southern Brazil and parts of Uruguay and Argentina, today the jelly palm can be seen throughout the world. It is the most commonly cultivated exotic palm in the southeastern United States. The jelly palm is grown as much for its ornamental qualities as for the fruit it produces.

Historical Origins Palm trees have long been an important symbol in history. Ancient Romans gave their champions palm fronds. Palm trees are mentioned in both the Bible and in Muslim holy books. To early Christians, the palm represented the victory of good over evil. The fruit of the jelly palm has long been used to make jellies and wine.

Botanical Facts The jelly palm can grow to more than 15 ft (4.5 m) in height, with 40–50 huge leaves. Each leaf can grow as large as 6 ft (1.8 m) in length. The trees are fairly adaptable, preferring sunny locations but needing only an occasional watering. They will grow in almost any kind of soil and are able to withstand high winds. The latter quality makes the trees popular with landscapers in hurricane zones.

Culinary Fare The color of the ripe jelly palm fruit can vary between orange and red, and its size can range from the size of a cherry to the size of an apricot. The fruits are picked as they ripen. The fruit's fibrous flesh surrounds a large central seed. Its flavor is distinctly fruity with pineapple, apricot, and vanilla notes. The jelly palm fruit is either eaten raw, or pureed, sweetened, and cooked to make jelly palm jelly.

ABOVE: *This young jelly palm will mature into a large tree which produces fruit that can be mashed, fermented, and aged to make jelly palm wine.*

Carob

Ceratonia siliqua

ABOVE: *The carob tree is native to the region fringing the Mediterranean and as far east as Arabia.*

Botanical Facts
The carob is part of the legume family and is an evergreen tree that can reach up to 55 ft (17 m) in height. It is cultivated for its seed pods, which contain seeds known as locust beans. The pods range from light to dark brown and are flat, somewhat curvy, and oblong shaped. Inside each pod are 10–13 hard seeds, which are used to produce locust bean gum. Carob powder is produced by roasting the pods before grinding them into a fine powder.

With its intrinsic sweetness and dark brown color, carob is often used as a chocolate substitute. Its lack of caffeine attracts consumers looking for a healthful chocolate alternative. Carob powder may look like cocoa powder, but it is not a cocoa product. As it is nearly 50 percent natural sugar, carob powder can also be used as a sugar replacement.

Historical Origins The carob tree grows in the Mediterranean and in times of famine, peasants there were able to eat the tree's pods in order to survive. Carob is sometimes referred to as "St. John's Bread" because it is believed to have sustained John the Baptist as he wandered the desert. In more prosperous times, the seeds inside of the carob pod were used to weigh diamonds and are the origin of the word "carat." These seeds were also used to make a gum, traces of which have been found in Egyptian tombs and are believed to have been part of the mummification process.

Culinary Fare Carob powder can be used as a chocolate substitute in baked goods, candies, and beverages. It is thought to be a healthy alternative to chocolate because it is non-fat and caffeine-free. Many of its health benefits are dramatically reduced, however, once it is combined with fat or sugar. While, visually, carob may resemble cocoa powder, its flavor does not. Carob has rich, flowery notes that, while enjoyable, may surprise someone expecting a large chocolate flavor.

Locust bean gum, made from carob seeds, is used as a commercial thickener in such products as ice creams, fruit fillings, and salad dressings. In addition to their use in locust bean gum, roasted carob seeds can be used as a coffee substitute, and in some countries they are simply mixed in with coffee beans as a matter of course. When the pods are coarsely ground and boiled in water, a thick, honeylike syrup similar in color to molasses is produced.

Palm Oil

Elaeis guineensis

Palm oil is obtained from the flesh of the oil palm's fruit. High in saturated fats, palm oil has seen a recent increase in popularity as consumers move away from the use of artificially hydrogenated fats and return to naturally saturated fats. It is also increasingly used as a biofuel, an inexpensive alternative to petroleum.

Historical Origins Palm oil has been discovered in Egyptian tombs dating from 3000 BCE. Large quantities of the oil were found, which suggest that it was probably used for culinary rather than cosmetic uses.

Botanical Facts A native of West Africa, the oil palm thrives on open, flat land. The fruit of the palm grows in densely packed bunches. A single bunch has up to 1,000 individual pieces of fruit, each about the size of a plum. A bunch of the fruit can weigh as much as 22 lb (10 kg). The introduction of labor-saving machinery, including hydraulic presses, bunch strippers, and clarification units, has greatly increased the yield of palm oil that can be produced from a single tree.

Culinary Fare In West Africa, some soups and baked dishes still incorporate unrefined palm oil. In the West, however, the oil is refined, deodorized, and bleached before being used in the kitchen. Because it is naturally hydrogenated, palm oil will remain solid at room temperature. It is able to withstand higher temperatures than many other oils, which makes it good for frying.

ABOVE: *The fruits of the oil palm are used to produce palm oil. The oil is suitable to use at high temperatures, and for that reason, is a popular ingredient in various baked goods.*

Soya Oil

Glycine max

Soy is part of the legume family with roots that go back all the way to the ancient Chinese world. Despite its long history, soy's popularity outside of the Asian world is fairly new. In recent years it has become one of the world's most widely used oils.

Historical Origins Soy was introduced to the West as a result of a 1915 boll weevil infestation that attacked cotton crops in the United States. The subsequent scarcity of cottonseed oil forced the states to look for alternative oil sources.

Botanical Facts Today there are more than 20,000 variations of the soybean plant. The size of the beans varies from ½–1½ in (12–35 mm) and their color from black to brown, green, or yellow. Soy has a short, 15-week, growing season. It can be grown quickly to meet the demands of the market and because it can be held and stored for long periods of time, any surplus will not spoil. Additionally, the beans can be shipped over long distances and harvested solely by mechanical means. These factors all work together to make soy an extremely profitable crop.

Culinary Fare Soybeans are the source of the first hydrogenated fat produced in the 1930s. Today, soybean oil is an integral part of products such as margarine, cooking oils, shortenings, and salad dressings. Its relatively bland taste makes soybean oil the perfect foil for use with other strongly flavored ingredients.

BELOW: *Soybeans are cleaned, dried, and hulled before being processed into soybean, or soya, oil. This oil is sold commercially as "vegetable oil."*

Sunflower Oil

Helianthus annuus

RIGHT: *The oil produced from sunflower seeds is high in polyunsaturated fats and low in trans fats. It is a popular choice in cooking because it keeps well, and can withstand extremely high temperatures.*

The use of the sunflower dates back to prehistoric times in North America. The seeds of the sunflower were collected for food, its petals used as a dye, and the long stems were used as a building material and sometimes dried and burned as fuel. Not surprisingly, the plant has long been cultivated.

Historical Origins In the sixteenth century, the sunflower was introduced to Europe where it was used as an ornamental plant. In the beginning of the eighteenth century, Peter the Great brought the plant to Russia. Because the plant was so new, it was not on the list of foodstuffs banned by the church during fast days. Initially its seeds were enjoyed both raw and roasted, and later the oil from the seeds was extracted and used as well. Russia soon became the largest grower of sunflowers in Europe.

Botanical Facts Appropriately, given its common name, the sunflower thrives in temperate, sunny climates. The sunflower's most striking physical characteristic is its large size. The flower can reach heights of 3–10 ft (0.9–3 m) and a single head can hold up to 8,000 seeds. The seed kernel is crushed and cooked before its oil is extracted using solvents. The seeds contain 20–32 percent oil, although new strains have been developed that contain more.

Culinary Fare Sunflower oil is used in margarine, shortening, cooking, and salad oils. It contains more vitamin E than other vegetable oils and is also low in saturated fats.

Rose Hip Syrup

Rosa canina

BELOW: *Each rose hip has up to 150 seeds contained inside the firm outer layer.*

Although a rose by any other name may smell as sweet, it is the fruit of the rose, not its petals, that is used in rose hip syrup.

Historical Origins The rose hip was first cultivated in ancient times in Persia (Iran). The deep red of the rose of the Mediterranean gave birth to the legend that the flowers spouted from the blood of Adonis.

Botanical Facts Rose hips are the fruits of *Rosa canina*. They are the ripe seed receptacles that remain long after the petals have fallen to the ground. The fruit is a bright red color and contains a large amount of citric and malic acids as well as sugars and a touch of vanillin. The sweetness of the rose hips is intensified after the first frost. One rose hip contains more vitamin C than an orange. When Britain experienced a shortage of citrus during World War II, the government urged everyone to use rose hip syrup to make up the deficit.

Culinary Fare Rose hip syrup is made by mincing the fruit and boiling it with water. The liquid is strained and cooked again with sugar. The resulting syrup can be enjoyed on its own or as an ingredient in jams, jellies, cordials, and desserts.

Olive Oil

Olea europaea

Currently about 90 percent of the world's olives are made into olive oil. Unlike other plant oils, olive oil is sold without being refined. It has a great depth of flavor that is not found in other oils.

Historical Origins Olives have been cultivated for thousands of years. Other sources of oil, such as various nuts and flax, do not produce as much yield as olives, so it was not long before olives were cultivated throughout the Mediterranean. While some skill is needed in order to grow olives, the oil itself is easy to produce, store, and ship. Production of olive oil soon brought great prosperity to various areas throughout the Mediterranean, especially Crete.

Botanical Facts Olive trees thrive in temperate climates with warm and dry summers. Not surprisingly, Spain, Italy, Greece, and Portugal are the world's largest olive oil producers. Although the trees are not immediately productive, they have extremely long life spans. Some Spanish olive trees are thought to be 1,000 years old.

Harvesting olives is labor-intensive. A tarp is spread underneath the tree, which is then lightly shaken to release the olives. Because the olives do not all ripen at the same time, this procedure must be repeated a number of times. Additionally, the olives must be processed within three days of being harvested, which adds to the expense.

Culinary Fare Most cooks are familiar with extra-virgin olive oil. It is an unrefined oil, distinguished from other plant oils by its color and deep flavor. Extra-virgin olive oil is made from the first cold pressing of the olives. Subsequent pressings require heat to release the oil and do not have the luxurious full body of extra-virgin olive oil. Because it is unrefined, the oil should be stored in dark bottles, away from light so that it does not go rancid.

A Country Divided

The cuisine of Italy is divided into northern and southern styles of cooking. The differences in the country's culinary traditions are the result of the differences in the country's geography. Northern Italian cuisine is based upon dairy products, so dishes there utilize butter, lard, and cream. The olive tree does not grow in northern Italy, and its absence is evident in the cuisine there.

In contrast, the olive tree thrives in the south of Italy. Southern Italian cuisine is resplendent with light dishes that showcase both the olive and its oil. Dishes are often flavored with just olive oil and a bit of garlic, even breads such as focaccia pay tribute to the olive. For the Italians, olive oil is more than a flavorful condiment; its presence, and absence, defines the country's very different culinary personalities.

RIGHT: *Olive trees need a long, hot growing season for the fruit to ripen. The fruit turns almost black when ripe.*

Sugar Cane

Saccharum officinarum

RIGHT: *To extract the cane juice, the cane is crushed between sets of huge rollers. The juice separates out from the fiber and is then cleaned before being further processed.*

BELOW: *One of the advantages of sugar cane is that it re-grows from the roots, so the plants last for a number of seasons.*

Sugar cane is the principal source of white granulated table sugar. Sugar is easy to store, lasts a long time, and can sweeten a product without drastically changing its flavor. These qualities have helped to make sugar cane the world's major sweetener. Once sugar cane was discovered, it quickly displaced other sweeteners, such as fruit syrups or tree saps.

Historical Origins Sugar cane does not have a particularly distinguished history. A lot of labor is required to transform the cane into sugar, so perhaps it is not surprising that as the demand for sugar grew so did the demand for cheap labor. Soon sugar plantations in the New World were importing slaves from Africa to harvest and process the cane. Only with slave labor could the sugar plantations make a profit. The New World had not only the subtropical to tropical climate needed to grow sugar cane, but also an abundance of land on which the cane could be planted, as well as woodlands to supply fuel for the sugar refining mills.

Botanical Facts Sugar cane is a perennial grass. It requires both a warm climate and a lot of rain in order to thrive. The cane can grow to be quite large. A stem of a mature plant can exceed 17 ft (5 m). Harvesting the cane can be difficult because its sucrose content is the greatest near ground level. The cane must be cut as close to the ground as possible without destroying its root system. Additionally, the sugar cane must reach the factory within 24 hours of being cut. Its leaves must be removed before it is milled or else they will absorb some of the stem's sucrose content.

Culinary Fare There are two steps to the sugar-production process. First, the cane is milled (crushed), which extracts the sucrose from its stems. That juice is then refined, resulting in a syrup that is heated until sugar crystals form. The crystallized mix is spun in a centrifuge, which, in turn, forces the crystals to separate from the molasses. The result is brown, raw sugar. Further refining produces white, granulated sugar.

Although white sugar is certainly sugar cane's best-known product, molasses and alcohol, namely rum, are other sugar cane products. In recent years, the consumption of white sugar has begun to level off in Western countries as health concerns turn consumers to the use of artificial sweeteners.

Sweetleaf

Stevia rebaudiana

A green shrub that is native to Paraguay, sweetleaf has 300 times the sweetening power of sugar cane and 30–45 times that of sucrose. Its native name in Paraguay is *ka'a he'e*, which means sweet herb.

Currently, sweetleaf is the number one sweetener in Japan. Other major consumers of sweetleaf include Brazil and the rest of South America, South Korea, China, and the Pacific Rim, including Australia. In Europe and North America, sweetleaf's categorization as a "food supplement" has effectively kept it out of the mainstream and placed it onto the shelves of health-food stores.

Historical Origins The indigenous people of Paraguay have used sweetleaf for hundreds of years. Europeans first learned of sweetleaf when the Spanish conquistadores of the sixteenth century sent word to Spain that the natives of South America were using a plant to sweeten their otherwise extremely bitter herbal tea, maté. The leaves were also chewed by themselves.

But the plant was only a curiosity until the early twentieth century, when Dr. Moises Santiago Bertoni of Asuncion, Paraguay, brought the plant to the attention of the Western world. After conducting research on it, he found that its sweetening powers were far superior to those of sugar.

The appearance of sweetleaf on the world scene is not without controversy. It is not approved for use as a sweetener in the United States or the European Union. Many believe that powerful artificial sweetener lobbies are behind the refusal of the Food and Drug Administration (FDA) to allow sweetleaf on the market as a sugar substitute.

Japan is one of the first countries to fully embrace the use of sweetleaf. Since 1977, the Japanese have replaced artificial sweeteners with sweetleaf. It is presently used in all manufactured soft drinks, and as table sugar.

Botanical Facts Sweetleaf is a perennial herbaceous shrub which grows in subtropical climates where it is always humid. It is a member of the sunflower family, and when it is properly cultivated it can grow to be more than 3 ft (0.9 m) in height. It has small green leaves, and the plant bears tiny white florets. The plant contains a chemical compound called stevioside, which gives sweetleaf its powerful sweetness. Stevioside is found in the liquid in the plant's leaves, which are dried and processed ready for use.

Culinary Fare Sweetleaf is praised as a sugar alternative because it has no calories or carbohydrates. It is suitable for those on diabetic diets, and does not promote tooth decay. The list of its uses is extensive. It has been used in beverages, desserts, candies, ice cream, chewing gum, and cereals. It can be used in soft baking, as in cookies, and although it does not caramelize in the same manner as granulated sugar, it does perform many other of sugar's functions in the bakery.

ABOVE: *As well as being used as a sweetener in processed vegetable, fish, and meat products, the dried leaves of sweetleaf are used to make a tea.*

Reference Section

Nutritional Tables

The following tables show the nutritional content of food plants in the Fruits, Vegetables, Nuts, and Grains chapters. Plants are listed in the order in which they appear in those chapters. Symbols represent percentage values for 13 selected vitamins and minerals important for good health, as well as fiber and protein levels (see the key at the base of each page).

For food labeling, the US Food and Drug Administration defines a "good" source for a specific nutrient as one that provides at least 20 percent of the Daily Value. Daily Values are reference points that help consumers make choices about foods and are derived from the Dietary Reference Intakes, recommended nutrient values for children and adults. It should be noted that most foods in these tables do not provide a "good" source of selected nutrients; they do, however, add "fine" contributions as a part of a total diet. As well, many foods provide contributing values of between 5 and 9 percent of the Daily Value.

The tables also show each food's antioxidant contribution and glycemic

Fruits

COMMON NAME	BOTANICAL NAME	VITAMIN A	THIAMIN (B1)
Calamondin, Limequat	*Citrofortunella microcarpa, Citrus × fortunella*		
Australian Finger Lime	*Citrus australasica*		
Limes	*Citrus aurantifolia, C. hystrix, C. latifolia*		
Oranges	*Citrus aurantium, C. aurantium* subsp. *bergamia, C. reticulata × C. sinensis, C. sinensis*		
Lemons	*Citrus ichangensis, C. limetta, C. limon, C. × meyeri, C. ponderosa*		
Kumquat	*Citrus japonica*	#	
Pummelo	*Citrus maxima*		
Citron	*Citrus medica*		
Grapefruit	*Citrus × paradisi*	#	
Clementine, Mandarin, Tangerine	*Citrus reticulata*		
Tangelo	*Citrus × tangelo*		
Avocado	*Persea americana*		
Date	*Phoenix dactylifera*		
Illawarra Plum, Plum Pine	*Podocarpus elatus*		
Apricot	*Prunus armeniaca*	•	
Cherries	*Prunus avium, P. cerasus, P. padus, P. serotina*	•	
Plums	*Prunus × domestica, P. institia, P. nigra, P. salicina, P. spinosa*		
Nectarine and Peach	*Prunus persica*	#	
Quandong	*Santalum acuminatum*		
Black Chokeberry	*Aronia melanocarpa*		
Quince	*Cydonia oblonga*		
Persimmons	*Diospyros kaki, D. virginiana*	•	
Loquat	*Eriobotrya japonica*	•	
Apple	*Malus × domestica*		
Medlar	*Mespilus germanica*	•	#
Sorb Apple	*Sorbus domestica*		
Pears	*Pyrus communis, P. pyrifolia*		

Key: • Good source: 20% and above of Daily Value ◊ Fine source: 10–19% of Daily Value # Contributing source: 5–9% of Daily Value

value. Antioxidants, which comprise vitamins, minerals, and phytochemicals, are beneficial to the body's defense mechanisms. They may help prevent or slow down the onset of some diseases. In the tables, antioxidants are rated high (H) or low (L).

The glycemic index (GI) classifies a carbohydrate food according to the glucose or sugar effect of that food on the body. Foods with a high GI greatly affect blood glucose levels. Conversely, foods with a low GI have a low effect on blood glucose levels. The values used to illustrate this effect are: above 70 = high (H); 55–70 = moderate (M); less than 55 = low (L).

It is important to note that there is no perfect food. Registered dietitians and nutritionists recommend eating a wide range of foods, consistent with the Food Pyramid food guidance system from the US Department of Agriculture. A half-cup food portion used for nutrient analysis for the tables is not meant to indicate an adequate dietary intake. When a specific food contributes some nutrient level, the goal is to eat these types of foods as part of a total diet rich in other plant foods that add some of the same and/or other nutrients.

Note also that there is no single source of nutritional information about plant foods. The tables have been compiled from the widest range possible of well-recognized authoritative sources. A small number of food plants that currently lack definitive nutritional data have been omitted from the tables.

(B3)	VITAMIN B6	FOLATE (B9)	VITAMIN C	VITAMIN D	VITAMIN E	FIBER	PROTEIN	CALCIUM	IRON	POTASSIUM	ZINC	ANTI-OXIDANTS	GLYCEMIC INDEX
			•									H	L
			•									H	L
			•									H	L
			•									H	L
			•									H	L
			•			•		#	#	#		H	L
			•							#		H	L
												N	H
			•									H	L
			•			◊						H	L
			•			◊						H	L
	◊	•			◊	•				◊		H	L
			•							#		L	L
												H	L
			◊			#				#		H	M
			•									H	L
			•							#		H	L
			#			#				#		H	L
		•	•			◊				◊		H	L
			•									H	L
			•			#				#		H	L
	#		◊			◊				#		H	L
	#									#		H	L
		#	•							#		H	L
	◊		•			◊				◊		H	L
												H	L
			#			◊						H	L

Portion size: 100 g portion or 1 medium fruit used for analysis. For reference: ½ cup = 113 g.

Fruits

COMMON NAME	BOTANICAL NAME	VITAMIN A	THIAMIN (B1)
Pineapple	*Ananas comosus*		
Cherimoya	*Annona cherimola*		
Guanabana, Soursop	*Annona muricata*		
Apple Annonas	*Annona reticulata, A. squamosa*		
Breadfruit	*Artocarpus altilis*		#
Jackfruit	*Artocarpus heterophyllus*	#	
Bilimbi	*Averrhoa bilimbi*		
Carambola, Star Fruit	*Averrhoa carambola*		
Akee	*Blighia sapida*		
Nance	*Byrsonima crassifolia*		
Papaya	*Carica papaya*	•	
Natal Plum	*Carissa macrocarpa*		
Casimiroa, White Sapote	*Casimiroa edulis*	#	
Caimito, Star Apple	*Chrysophyllum cainito*		
Seagrape	*Coccoloba uvifera*		
Elephant Apple	*Dillenia indica*		
Longan	*Dimocarpus longan*		
Ceylon Gooseberry	*Dovyalis hebecarpa*		
Durian	*Durio zibethinus*		•
Tropical Cherries	*Eugenia aggregata, E. uniflora*	◊	
Mangosteens	*Garcinia mangostana, G. prainiana*		
Dragonfruit, Pitaya	*Hylocereus undatus*		
Lychee	*Litchi chinensis*		
Acerola	*Malpighia glabra*	◊	
Mango	*Mangifera indica*	•	
Sapodilla	*Manilkara zapota*		
Brazilian Grape Tree	*Myrciaria cauliflora*		
Bananas	*Musa acuminata, M. balbisiana*		
Rambutan	*Nephelium lappaceum*		
Otaheite Gooseberry	*Phyllanthus acidus*		
Canistel, Egg Fruit	*Pouteria campechiana*	•	
Tropical Guavas	*Psidium guajava, P. littorale*	◊	
Tamarillo	*Solanum betaceum*	•	
Mombin	*Spondias mombin*		
Tropical Apples	*Syzygium jambos, S. malaccense*		
Tropical Almond	*Terminalia catappa*	◊	•
Babaco	*Vasconcellea × heilbornii*	#	
Chinese Jujube	*Ziziphus zizyphus*		
Pawpaw	*Asimina triloba*	•	
Figs	*Ficus carica*	•	
Strawberry	*Fragaria × ananassa*		

Key: • Good source: 20% and above of Daily Value ◊ Fine source: 10–19% of Daily Value # Contributing source: 5–9% of Daily Value

(B3)	VITAMIN B6	FOLATE (B9)	VITAMIN C	VITAMIN D	VITAMIN E	FIBER	PROTEIN	CALCIUM	IRON	POTASSIUM	ZINC	ANTI-OXIDANTS	GLYCEMIC INDEX
			•									H	M
	◊	#				◊				#		H	M
			•									H	L
			•									H	L
	#		•			•				•		H	M
	#		◊			#				#		H	M
			•									H	L
			•									H	L
		◊	•			◊				#		H	L
			•									H	L
		◊	•			#				#		H	M
			•						#	#		H	L
		#	•			◊			#	#		H	M
	◊	#	•			◊				•		H	L
												–	L
			•									–	L
			◊						#			–	L
							#		#			L	L
	◊	#	•			◊				◊		H	L
			•			#						H	L
												L	L
			•									H	L
			•									H	H
			•									H	L
	#		•		#	#				#		H	M
			•			•						H	L
												–	L
	◊	#	◊			◊				•		L	L
												–	L
												L	L
			•			•						H	–
			•			#						H	L
			•							#		H	L
			•									H	L
			•							#		H	L
		◊			•	•						H	L
			•							#		H	–
			•							◊		H	L
			•									L	M
	#					◊				#		H	M
		#	•			#						H	L

Portion size: 100 g portion or 1 medium fruit used for analysis. For reference: ½ cup = 113 g.

Fruits

COMMON NAME	BOTANICAL NAME	VITAMIN A	THIAMIN (B1)
Raisin Tree	*Hovenia dulcis*		#
Mulberries	*Morus alba, M. nigra*		
Pomegranate	*Punica granatum*		
Currants	*Ribes nigrum, R. odoratum, R. silvestre, R. uva-crispa*		#
Rubus Berries	*Rubus caesius, R. chamaemorus, R. fruticosus, R. ursinus* hybrid		
Raspberries	*Rubus idaeus, R. occidentalis*		
Loganberry	*Rubus × loganobaccus*		
Elderberry	*Sambucus nigra*	◊	
Chilean Guava	*Ugni molinae*	◊	
Blueberries	*Vaccinium ashei, V. corymbosum, V. lamarckii*		
Cranberries	*Vaccinium macrocarpon, V. oxycoccus*		
Bilberry, Whortleberry	*Vaccinium myrtillus*		
Lingonberry	*Vaccinium vitis-idaea*		
Kiwi Fruits	*Actinidia arguta, A. chinensis, A. deliciosa*		
Watermelon	*Citrullus lanatus*	◊	
Melons	*Cucumis melo var. cantalupensis, C. melo var. inodorus, C. melo var. reticulata*	•	•
Ceriman	*Monstera deliciosa*		#
Passionfruits	*Passiflora edulis, P. laurifolia, P. ligularis, P. mollissima, P. quadrangularis*		
Grapes	*Vitis labrusca, V. vinifera*		

Key: • Good source: 20% and above of Daily Value ◊ Fine source: 10–19% of Daily Value # Contributing source: 5–9% of Daily Value

Vegetables

COMMON NAME	BOTANICAL NAME	VITAMIN A	THIAMIN (B1)
Onions	*Allium cepa, A. fistulosum*		
Leek	*Allium porrum*	•	
Garlic	*Allium sativum, A. scordoprasum, A. ursinum*		◊
Celeriac	*Apium graveolens var. rapaceum*		
Horseradish	*Armoracia rusticana*		#
Beetroot	*Beta vulgaris*		
Kohlrabi	*Brassica oleracea*, Gongylodes Group		
Turnips	*Brassica napus var. napobrassica, B. napus var. rapifera*		
Rampion	*Campanula rapunculus*	•	
Achira	*Canna edulis*		#
Taro	*Colocasia esculenta*		
Carrot	*Daucus carota*	•	
Yams	*Dioscorea* species	•	#
Water Chestnut	*Eleocharis dulcis*		#
Jerusalem Artichoke	*Helianthus tuberosus*		
Sweet Potato	*Ipomoea batatas*	•	#
Cassava	*Manihot esculenta*		

Key: • Good source: 20% and above of Daily Value ◊ Fine source: 10–19% of Daily Value # Contributing source: 5–9% of Daily Value

(B3)	VITAMIN B6	FOLATE (B9)	VITAMIN C	VITAMIN D	VITAMIN E	FIBER	PROTEIN	CALCIUM	IRON	POTASSIUM	ZINC	ANTI-OXIDANTS	GLYCEMIC INDEX
	#					◊	#	#	◊	•		H	L
			•			#			◊	#		H	L
	#		◊							#		H	L
	#					◊	#	#	◊	•		H	L
		#	•		#	•				#		H	L
		#	•			•				#		H	L
			•			•						H	L
	◊		•		◊	•			#	#		H	L
			•			#						H	L
			◊			◊						H	L
			•		#	•						H	L
			◊			◊						H	L
		#	•		#	•				#		H	L
		◊	•		◊	◊				◊		H	L
			◊									L	H
										#		H	H
			•			•				◊		H	L
			◊			•			•	•		H	L
			•							#		H	L

Portion size: 100 g portion or 1 medium fruit used for analysis. For reference: ½ cup = 113 g.

(B3)	VITAMIN B6	FOLATE (B9)	VITAMIN C	VITAMIN D	VITAMIN E	FIBER	PROTEIN	CALCIUM	IRON	POTASSIUM	ZINC	ANTI-OXIDANTS	GLYCEMIC INDEX
	#	#	#									H	L
	◊	◊	•			#		#	◊	#		H	L
			•			#			◊	◊		H	L
	#		◊			#				#		L	H
	•		•			•		◊	#	◊		H	L
		•	#			◊			#	#		L	L
			•			◊				◊		H	L
			•			#				#		H	H
	◊	◊	•			#		#	◊	#		H	L
#	◊	•				#	#		◊	◊		L	L
	#				#	•				◊		L	L
	#		#		#	◊				#		H	L
#	◊		•			◊				◊		H	M
#	◊		#		#	◊				◊		L	L
		◊	◊			◊				#		L	L
#	◊		•			◊				◊		H	M
		#	•							#		H	H

Portion size: 100 g portion used for analysis. For reference: ½ cup = 113 g.

Vegetables

COMMON NAME	BOTANICAL NAME	VITAMIN A	THIAMIN (B1)
Lotus	*Nelumbo nucifera*		#
Oca	*Oxalis tuberosa*	•	#
Jicama	*Pachyrhizus erosus*		#
Parsnip	*Pastinaca sativa*		
Hamburg Parsley	*Petroselinum crispum* var. *tuberosum*	•	
Radish	*Raphanus sativus*		
Salsify	*Scorzonera hispanica, Tragopogon porrifolius*		
Skirret	*Sium sisarum*	•	
Yacon	*Smallanthus sonchifolius* (syn. *Polymnia sonchifolia*)		#
Potato	*Solanum tuberosum*		#
Celery	*Apium graveolens*		
Asparagus	*Asparagus officinalis*	◊	
Orache	*Atriplex hortensis*	•	
Winter Cress	*Barbarea vulgaris*	•	
Beets, Chard	*Beta vulgaris* var. *flavescens, B. vulgaris* var. *flavescens* subsp. *cicla*		
Sea Spinach	*Beta vulgaris* subsp. *maritima*	•	
Chinese Mustard, Gai Choy	*Brassica juncea* var. *rugosa*	•	
Kale	*Brassica oleracea*, Acephala Group	•	
Chinese Broccoli	*Brassica oleracea*, Alboglabra Group	◊	#
Cauliflower	*Brassica oleracea*, Botrytis Group		
Cabbage	*Brassica oleracea*, Capitata Group		
Broccoli	*Brassica oleracea*, Cymosa Group	•	#
Brussels Sprouts	*Brassica oleracea*, Gemmifera Group	◊	#
Asian Greens	*Brassica rapa* var. *chinensis, B. rapa* var. *nipposinica, B. rapa* var. *pekinensis, B. rapa* var. *rosularis*	•	
Gotu Kola	*Centella asiatica*	•	
Goosefoot	*Chenopodium album, C. bonus-henricus, C. giganteum*	•	
Endive	*Cichorium endivia*		
Sea Kale	*Crambe maritima*	•	
Mitsuba	*Cryptotaenia japonica*		
Globe Artichoke	*Cynara scolymus*		
Rocket	*Diplotaxis muralis, Eruca sativa, Hesperis matronalis*	•	
Lettuce	*Lactuca sativa, L. sativa* var. *augustana*	•	
Water Spinach	*Ipomoea aquatica*	•	
Rice Paddy Herb	*Limnophila aromatica*		
Alfalfa Sprouts	*Medicago sativa*		#
Miner's Lettuce	*Montia perfoliata*	•	
Watercress	*Nasturtium officinalis*	•	
Perilla	*Perilla frutescens*	•	
Japanese Radish Sprouts	*Raphanus sativus*		#
Rhubarb	*Rheum × cultorum*		
Marsh Samphire	*Salicornia europaea*	•	

Key: • Good source: 20% and above of Daily Value ◊ Fine source: 10–19% of Daily Value # Contributing source: 5–9% of Daily Value

(B3)	VITAMIN B6	FOLATE (B9)	VITAMIN C	VITAMIN D	VITAMIN E	FIBER	PROTEIN	CALCIUM	IRON	POTASSIUM	ZINC	ANTI-OXIDANTS	GLYCEMIC INDEX
	◊		•			◊				◊		H	L
	◊		•			◊				◊		H	M
	◊		•			•				◊		H	M
	#		•		#	◊				◊		H	H
		•	•						•			H	L
			•									H	L
	#		•		#	◊				◊		H	H
		•	•						•			H	L
	◊		•			•				◊		H	M
		◊	•			#				◊		H	H
	#		◊			#				#		L	L
	#	◊	◊			#			◊	#		H	L
			•		•	•		#	•			H	L
	#		•		#			◊		◊		H	L
		•	#			◊			#	#		H	L
					•				•			H	L
		◊	•			•		◊	#	◊		H	L
	#		•			#		#	#	#		H	L
	#	◊	•			◊	#	#				H	L
	#	◊	•			◊						H	L
		#	•			#		#		#		H	L
	#	◊	•			◊	#	#				H	L
	◊	•	•			◊	#			#		H	L
	#		•		#	#		#		#		H	L
			•			•	•	•				H	L
						◊						L	L
		#	#			◊						L	L
			•		#	#	#	◊	#	#		H	L
												L	L
		•										L	L
		•	•			#	#	◊	#	◊		H	L
		◊	#						#	#		L	L
					•				•			H	L
	#	#				#	#					L	L
		#	◊			#	#		#			L	L
		◊	#						#	#		L	L
	#		•		#			◊		◊		H	L
		◊	•			#	#	◊	◊	#		H	L
		#	◊			#	#		#			L	L
			◊			#		#		#		L	L
		•	•						•			H	L

Portion size: 100 g portion used for analysis. For reference: ½ cup = 113 g.

Vegetables

COMMON NAME	BOTANICAL NAME	VITAMIN A	THIAMIN (B1)
White Mustard Sprouts	*Sinapis alba*		#
Alexanders, Black Lovage	*Smyrnium olusatrum*	•	
Spinach	*Spinacia oleracea*	•	
Dandelion	*Taraxacum officinale*	•	◊
New Zealand Spinach	*Tetragonia tetragonoides*	•	
Corn Salad, Lamb's Lettuce	*Valerianella locusta*	•	
Standard Okra	*Abelmoschus esculentus*	#	#
Pigeon Pea	*Cajanus cajan*		•
Capers	*Capparis spinosa*		
Chilies and Peppers	*Capsicum annuum, C. chinense, C. frutescens*	◊	
Chickpea	*Cicer arietinum*		
Cucumbers	*Cucumis metuliferus, C. sativa*		
Pumpkins, Winter Squash	*Cucurbita maxima, C. moschata, C. pepo*	•	
Summer Squash, Zucchini	*Cucurbita pepo*	◊	
Hyacinth Bean	*Dolichos lablab*		◊
Calabash, Cucuzzi	*Lagenaria siceraria*		
Lentils	*Lens culinaris*		◊
Chinese Okra, Luffa	*Luffa acutangula*		
Tomato	*Lycopersicon esculentum*	◊	
Tepary Bean	*Phaseolus acutifolius*		◊
Adzuki Bean	*Phaseolus angularis*		#
Mung Bean	*Phaseolus aureus*		◊
Runner Bean	*Phaseolus coccineus*		◊
Lima Bean	*Phaseolus lunatus*		
Haricot Bean, Common Bean	*Phaseolus vulgaris*		◊
Peas	*Pisum sativum, P. sativum var. macrocarpon*	◊	◊
Tomatillo	*Physalis ixocarpa*		
Chayote, Choko	*Sechium edule*		
Aubergine, Eggplant	*Solanum melongena*		#
Asparagus Pea	*Tetragonolobus purpureus*		#
Broad Bean, Fava Bean	*Vicia faba*	#	#
Cowpea, Black-eyed Pea, Yard-long Bean	*Vigna unguiculata*	◊	◊

Key: • Good source: 20% and above of Daily Value ◊ Fine source: 10–19% of Daily Value # Contributing source: 5–9% of Daily Value

(B3)	VITAMIN B6	FOLATE (B9)	VITAMIN C	VITAMIN D	VITAMIN E	FIBER	PROTEIN	CALCIUM	IRON	POTASSIUM	ZINC	ANTI-OXIDANTS	GLYCEMIC INDEX
		#	◊			#	#		#			L	L
		•	•						•			H	L
		•	•			◊		#	◊			H	L
	◊		•			◊		◊	◊	◊		H	L
		•	•			◊		#	◊			H	L
		◊	#						#	#		L	L
	#	◊	•			#		#				H	L
		•	•			•	◊			◊		H	L
							#		#			L	L
	•	#	•			#			#	#		H	L
	#	•				•	◊	#	◊	#		H	L
			◊									L	L
	#	#	•			#		#		◊		H	L
	#	#	•							◊		H	L
	#	•	#			•	◊		◊	◊		H	L
		#	#			#				#		L	L
	#	•	#			•	◊		◊	◊		H	L
			•									H	L
			•			#				#		H	L
	#	•	#			•	◊		◊	◊		H	L
	#	•				•	◊		◊	◊		H	L
		•				•	◊		#	#		H	L
		•				•	◊		#	#		L	L
	#		◊			•	◊		◊	◊		H	L
		•				•	◊		#	#		L	L
	◊	◊	•			•	◊		#	#		H	L
			•			#				#		H	L
	#	#	◊			◊				#		L	L
						◊						L	L
		#	◊				◊	#	#	#		L	L
		◊	•				◊		#	#		H	L
	#	◊	•				#	#	#	◊		H	L

Portion size: 100 g portion used for analysis. For reference: ½ cup = 113 g.

Grains

COMMON NAME	BOTANICAL NAME	THIAMIN (B1)	RIBO-FLAVIN (B2)
Amaranth	*Amaranthus caudatus*		
Oats	*Avena sativa*	◊	
Quinoa	*Chenopodium quinoa*	#	#
Finger Millet	*Eleusine coracana*	◊	
Buckwheat	*Fagopyrum cymosum, F. esculentum, F. tartaricum*		
Barley	*Hordeum vulgare*	#	
Rice	*Oryza sativa*	#	
Proso Millet	*Panicum miliaceum*	◊	
Pearl Millet	*Pennisetum glaucum*	◊	
Rye	*Secale cereale*	#	#
Foxtail Millet	*Setaria italica*	◊	
Sorghum, Great Millet	*Sorghum bicolor*	◊	
Triticale	× *Triticosecale hybrids*	#	
Wheat	*Triticum aestivum, T. dicoccon, T. durum, T. monococcum, T. sativum, T. spelta*	#	
Maize, Sweet Corn	*Zea mays*	#	

Key: • Good source: 20% and above of Daily Value ◊ Fine source: 10–19% of Daily Value # Contributing source: 5–9% of Daily Value

Nuts

COMMON NAME	BOTANICAL NAME	THIAMIN (B1)	RIBO-FLAVIN (B2)
Candlenut	*Aleurites moluccana*	#	
Cashew	*Anacardium occidentale*	•	◊
Peanut	*Arachis hypogaea*	◊	#
Brazil Nut	*Bertholletia excelsa*	•	
Breadnut	*Brosimum alicastrum*		#
Canarium Nut	*Canarium indicum*		
Pecan	*Carya illinoinensis*	•	#
Shagbark Hickory	*Carya ovata*	•	#
Sawari Nut	*Caryocar nuciferum*	#	
Chestnut	*Castanea sativa*	◊	◊
Hazelnuts	*Corylus avellana, C. maxima*	•	◊
Walnuts	*Juglans cinerea, J. nigra, J. regia*	◊	#
Macadamia	*Macadamia integrifolia*	•	#
Pachira Chestnuts	*Pachira aquatica, P. insignis*	◊	◊
Pine Nuts	*Pinus koraiensis, P. pinea*	•	
Pistachio	*Pistachia vera*	•	#
Almond	*Prunus dulcis*	#	•

Key: • Good source: 20% and above of Daily Value ◊ Fine source: 10–19% of Daily Value # Contributing source: 5–9% of Daily Value

FOLATE (B9)	VITAMIN C	VITAMIN D	VITAMIN E	FIBER	PROTEIN	CALCIUM	IRON	POTASSIUM	ZINC	SELENIUM	ANTI-OXIDANTS	GLYCEMIC INDEX
				◊	◊	◊	◊	◊			H	H
			#	◊	#		#	#	#	#	H	M
◊				◊	◊	#	#	#	◊		H	L
									◊		H	M
				◊	#				#		H	L
				◊	#		#		#		H	M
				#	#				#		L	M
									◊		H	M
									◊		H	M
#							#				H	L
									◊		H	M
									◊		H	M
#					#				#		H	M
					#						H	L
#				◊	#				#		H	M

Portion size: 100 g portion used for analysis (except for rye, triticale, and wheat—1 slice bread). For reference: ½ cup = 113 g. Data for rice is for brown rice; for wheat, whole-wheat bread.

FOLATE (B9)	VITAMIN C	VITAMIN E	FIBER	PROTEIN	CALCIUM	IRON	MAG-NESIUM	POTASSIUM	ZINC	SELENIUM	ANTI-OXIDANTS	GLYCEMIC INDEX
		◊	◊	◊	◊	◊			◊		H	L
◊		#	◊	•	#	•		◊	•	#	H	L
			•	•	#	◊		◊	•	◊	H	L
#		•	•	•	◊	◊	◊	◊	•	•	H	L
◊	◊		◊	◊	#	◊	◊	•	◊		H	L
			◊	◊	◊						–	L
			•	◊	#	◊	◊	◊	•		H	L
			•	•	#	◊	◊	◊	•	◊	H	L
#				◊		#	◊			#	–	L
◊	•		◊	#		#		◊			H	L
•			•	•	◊	•		•	◊		H	L
			◊	◊	◊	◊	◊	◊	•		H	L
			•	◊	#	◊	◊	◊	#		H	L
◊	•		◊	#		#		◊			H	L
◊			◊	•		•		◊	•		L	L
		◊	•	•	◊	◊	◊	•	◊	◊	H	L
		◊	•	•	•	•	◊	◊	•		H	L

Portion size: 100 g portion used for analysis. For reference: ½ cup = 113 g.

Glossary

ACIDULATED WATER Helps to prevent browning, or discoloration, of certain fruits and vegetables. Once peeled and cut, the fruit or vegetable can be dropped in water to which a mild acidic ingredient has been added (e.g., approximately 1 tbsp vinegar or 2 tbsp lemon or lime juice per 1¾ pts (1 L) water).

ALEURONE A protein-rich layer that surrounds the endosperm or food storage part of grass seeds. It is significant in the nutritional value of cereals; and is also important in processes, such as brewing, that harness chemical reactions occurring between the aleurone layer and the embryo on germination.

ALKALOIDS Organic chemicals, often toxic, that are derived mainly from plants. Many plants that contain alkaloids are edible after cooking or steeping in water. Other alkaloids are among the active principles in drugs such as morphine, nicotine, cocaine, and quinine.

AMINO ACIDS The building blocks of proteins essential to the normal function of living cells. Of the 20 known amino acids, 10 can be synthesized by humans, and the others (essential amino acids) must be sourced from food. Animal products are termed "complete" because they contain all the essential amino acids. Most vegetable foods are incomplete, but when combined with other plant foods, such as rice or lentils, supply the full complement.

ANNUAL A plant that completes its entire life cycle in one year—from germination to seed setting and death. Annuals that grow in temperate climates are classified as either summer annuals, which grow through the warmer months and die when winter arrives; or hardy annuals, which germinate from late summer, overwinter in a juvenile form, and then complete their life cycle when warmer weather arrives.

ANTIOXIDANT Substance that reduces or prevents damage in the human body caused by free radicals (natural by-products of biochemical processes). Vitamins A, C, and E, beta-carotene, and many other components of fruits and vegetables, such as the lycopene in tomatoes, have antioxidant properties.

AQUATIC A plant that spends most of its life cycle in water. Aquatics are classed as submerged, emergent, or floating. Submerged and emergent aquatics are anchored in the soil, but floating aquatics spend at least part of their life cycle in a fully free-floating form.

ARIL A fleshy appendage to a seed that develops as an outgrowth of the seed stalk. Often succulent or brightly colored, arils attract birds or animals that disperse the seed (e.g., the flesh of the lychee is an aril).

BERRY Botanically, a fleshy fruit that has seeds embedded without being surrounded by a fibrous or hard layer, such as tomatoes and blueberries. Commonly, the term is used to include fruits that botanically have a different structure, such as blackberries and mulberries.

BETA-CAROTENE Found in dark green and orange-yellow vegetables, it is the precursor of vitamin A. In general, the deeper or more intense the color of the vegetable, the greater the content of beta-carotene.

BIOFUEL Refined fuel, usually liquid or gas, made from organic sources. The source material may be animal or vegetable, and the processes involved may require digesters for animal waste to produce biogas, distillation for sugars and starches, or pressing and purification for oil extraction to produce biodiesel.

BIOTECHNOLOGY Technically, any biology-based technology used in agricultural, food, or medical sciences. Commonly, the term is used as a synonym for genetic manipulation or genetic engineering—though the term also applies to natural methods of selection and hybridization.

BRASSICA Members of the cabbage family (Brassicaceae) normally used as edible vegetables, such as cabbage, cauliflower, brussels sprouts, broccoli, and turnips.

CALYX (plural: calyces) The whorl attached to the receptacle of a flower made up of sepals. Mature fruits may carry remnants of the calyx at the blossom end (e.g., blueberries) or at the top of the fruit (e.g., the green ruff on a strawberry).

CARBOHYDRATE The principal source of energy for most humans, and supplying 50–65 percent of energy in an average Western diet. Carbohydrates are the basic constituents of plants and plant foods—in the form of sugars, starches, cellulose, and pectin.

CARPEL The fundamental female organ of a flower, usually composed of an ovary containing an ovule or ovules (embryonic seeds), and a narrower style tipped by a stigma, which receives pollen. Carpels may be single or multiple in one flower, and multiple carpels are often fused together. Fruits such as blackberries and mulberries develop from multiple-fused carpels.

CATKIN A flower spike, often pendulous, composed of many minute, usually wind-pollinated, flowers (e.g., hazel flowers). Catkins are usually single sex, and a plant may carry both male and female catkins or just those of a single sex.

CEREAL Edible grain seeds harvested from certain grasses grown as crops, especially wheat *(Triticum)*, barley *(Hordeum)*, oats *(Avena)*, rye *(Secale)*, maize *(Zea)*, and millet *(Setaria)*. Also used as the collective name for these crops.

CLONE Plants produced asexually (i.e., by cuttings, grafting, tissue culture, etc.) from a single parent to which they are genetically identical.

CROP ROTATION The practice of varying the crops grown each year so that the soil is not continually depleted of the same nutrients, and pathogens are not allowed to develop to excess. The process often incorporates the use of leguminous green manures to add nitrogen, and fallow years, when no crop is grown, to aid soil recovery.

CULTIVAR A contraction of cultivated variety. Cultivars are given distinguishing names, and modern cultivars must be given names of non-Latin form that are enclosed in single quotes and capitalized (e. g., 'Golden Delicious').

DEHULL The process of removing the outer coat from a seed. This can be achieved by removing the casing of a many-seeded fruit, such as taking peas from their pods; or by removing the tough coating of an individual seed, such as stripping the outer coat of sunflower seeds to expose the kernels.

DRUPE A fleshy fruit in which the seeds are separated from the outer flesh by a hard inner layer (the endocarp) composed of bony, woody, or fibrous tissue (e.g., plums or olives).

DRUPELET A tiny drupe—of which several or many combine to form the fruit. Drupelets usually result from the fertilization of a single carpel of a many-carpelled flower (e.g., blackberries).

EMBRYO The part of a seed present after fertilization, but before germination, that will develop into a new plant—distinct from the seed's endosperm, or food storage, and its epidermis, or seed coat.

ENDOCARP The innermost layer of a fruit wall, enclosing the seed or seeds—most readily distinguished in a drupe, when it is thick and often tough and fibrous (e.g., a mango), or hard and stony (e.g., a plum).

ENDOSPERM Tissue within a seed composed of starch and/or sugars, oils, or proteins providing a source of energy for the germinating embryo. Cereal grains and coconuts are among the many seed types with large amounts of endosperm.

FERMENT The process by which yeasts, molds, or bacteria produce desirable changes in foods—often in order to convert them into a form that can be stored (e.g., sauerkraut from fresh cabbage), but also to change them from an indigestible to a digestible form.

FINING AGENT Used to clarify wine and beer, and also to remove excess tannin or reduce levels of astringency or bitterness in wine. Common fining agents include bentonite, gelatine, egg white (albumen), casein, and isinglass (prepared from the swim bladders of certain fish).

FLAVONOIDS A group of color-producing chemicals typically found in plants. In the human body they act as antioxidants and also have anti-inflammatory properties. Good sources of flavonoids include citrus fruits, apricots, cherries, grapes, blackcurrants, broccoli, onions, tomatoes, as well as green tea and red wine.

FRUIT The seed-containing organ of any of the flowering plants—whether fleshy or dry. In culinary terms, fruit usually refers to those used for sweet dishes, while those used for savory dishes (e.g., tomatoes and peppers) are considered vegetables.

GAP ("GOOD AGRICULTURAL PRACTICES") Although not a very clearly defined term, the principle behind GAP is that of sustainability. To comply with a basic UN definition, GAP programs allow for continued production of safe nutritious crops that meet the cultural demands of the society producing them, while not depleting the environment in which they are produced. GAP may be subdivided into practices in particular areas of production, such as soil, water, and pest management.

GENETIC MODIFICATION/ MANIPULATION/ENGINEERING The use of laboratory-based genetic technologies—such as DNA manipulation and gene transfer—to produce cultivars with increased productivity, disease resistance, or pest resistance. This may involve altering a plant's existing genetic material or introducing new material from an outside source, including animals. While offering the hope of greatly increased production, it raises fears of restricted seed supplies and possible genetic contamination.

GHEE A form of clarified butter (pure butter oil without lactose and other milk solids), and one of the traditional fats of Indian cuisine. It tolerates higher temperatures than butter without burning.

GLUTEN Derived from glutenin and gliadin—two proteins in the endosperm of wheat and, to a lesser degree, other grains. Kneading dough bonds these proteins, yielding gluten that provides bread with its elasticity and chewiness.

GRAIN The small dry fruit of any grass, though the terms is more commonly used for grasses grown as food crops (i.e., cereals).

GROATS Food, also known as grits in North America, derived from cereal grains that have been crushed, rather than ground, after removal of the tough outer husk. Groats are typically produced from oats, and grits from maize (corn).

HARDY A plant that is able to survive and thrive in a hostile environment. Gardeners in colder climates have generally narrowed its meaning to "frost hardy." There are several recognized guides to plant hardiness, the best known being the USDA zone system.

HERB Technically, a plant with non-woody stems, such as most annuals. In horticultural and culinary terms, an aromatic plant whose leafy parts add flavor or aroma, rather than bulk, to a cooked dish or salad.

HYBRID The progeny of cross-pollination between different species, varieties, or cultivars—combining the genetic makeup of both. Botanical names of hybrids between 2 species are indicated by the multiplication sign "x" inserted before the epithet (e.g., *Citrus* x *latifolia*). A hybrid between 3 or more species, different cultivars, or between a cultivar and a species, is given a new cultivar name. Hybrids may be fertile or sterile but most often must be propagated asexually to remain true to type.

HYBRIDIZATION The process of producing new plants by crossing parent plants that have desirable characteristics and differing genetic backgrounds. Plants must generally be closely related (e.g., within the same genus) to hybridize successfully.

HYDROPONICS The technique of growing plants in an inert medium, such as pumice, through which a nutrient solution is passed. This soilless cultivation allows for easy harvesting and accurate monitoring of the plants' growth and nutrient uptake. Hydroponic cultivation is often carried out under cover and in combination with partial or full artificial lighting.

INULIN A group of fructose polymers (long chains of fructose molecules) that is naturally produced by certain plants as carbohydrate storage. Inulin is not broken down by the usual enzymes that attack starch and, as a result, does not cause an increase in blood sugar levels.

IRRADIATION The process of treating fruits and plant material with low levels of ionizing radiation to eliminate pathogens, and to delay ripening or prevent early sprouting. The process is effective but controversial due to the possibility of causing genetic mutations in the irradiated material.

JERKY Food made by drying strips of meat in the sun or by a fire, a practice of both Native Americans and early European settlers. The name derives from the Peruvian *charqui,* meaning dried meat.

LOBES Large projections along the margin of a leaf, generally measuring at least a third of the distance from the leaf's midrib to its outer edge.

NEUROTOXIN A substance, such as some alkaloids, that affect the nerve cells or the neurons of animals. Neurotoxins can cause paralysis leading to suffocation or heart failure.

NITROGEN-FIXING The ability of leguminous plants (members of the pea family), in combination with certain bacteria (e.g., *Rhizobium* and cyanobacteria), to absorb nitrogen from the atmosphere and combine it with hydrogen and oxygen to form simple inorganic molecules (e.g., ammonia and nitrous acid) that plants can absorb, and allow them to thrive in nitrogen-deficient soils. Legumes also fix nitrogen in small nodules on their roots and, if used as a green manure, release nitrogen into the soil as the plants decompose.

NIXTAMALIZATION A process in which the bound niacin in maize (corn) is converted to free niacin (vitamin B3), thus enhancing the nutritional value of the grain. This involves soaking and cooking the maize in an alkaline solution. Today, enzymes are often used to speed up the process.

NUT Botanically, a fruit that is not fleshy and does not split open when ripe. Commonly, it is an edible seed, larger than a grain, that can be eaten raw or with minimal preparation, such as roasting.

OMEGA-6 ESSENTIAL FATTY ACIDS Present in polyunsaturated fats such as maize (corn), soybean, sunflower, and canola oils, these essential fatty acids (EFAs) cannot be synthesized by the body and must be obtained from food. In Western diets, the disproportionate ratio of omega-6 to omega-3 fatty acids (typically found in cold-water, fatty fish) is associated with an increased risk of heart and other diseases.

ORGANIC Scientifically, a branch of chemistry interested in compounds in which carbon and hydrogen predominate (e.g., living matter). Horticulturally and agriculturally, produce grown entirely by natural techniques (without the use of manufactured chemical fertilizers or pesticides), based on the belief that such chemicals are harmful to the soil and to the humans and animals who consume the produce.

PECTIN A polysaccharide composed of long chains of sugar molecules, derived from the cell walls of fruits and vegetables. Fruits particularly rich in pectin include pome and citrus fruits, and redcurrants. Pectin is essential for jams and jellies to set.

PERENNIAL Botanically, any plant that lives for 3 or more years. Commonly, it describes a non-woody stemmed plant and an herbaceous plant that lives for at least 3 years and is most often propagated by division of the rootstock.

PERMACULTURE Originally a plan for sustainable agricultural production. It has evolved into a scheme for the integration of all aspects of human existence in an ecologically sound system that works for the benefit of all, while minimizing detrimental environmental effects. Many aspects of the design of permaculture systems are adapted from nature, such as arranging layers of crops to mimic the natural growth zones in a forest clearing.

PHOTOPERIOD SENSITIVITY The degree to which a plant's growth, flowering, and fruiting is affected by the length of the day. For example, to flower, some plants require shortening day lengths (e.g., the arrival of autumn), while others require lengthening days (e.g., the onset of spring).

POLLINATION The transferring of pollen between the male and female sexual organs of a plant, whether in the same flower or different flowers, or on different plants. Agents of pollination include wind, insects, and birds. Pollen can be deliberately transferred by humans to ensure fruit set or to create hybrids.

PROTEIN Protein is essential to the health of humans and animals, and is important in all biochemical processes in both plants and animals. Produced by plants, it is the basis of all food chains. Proteins are complex chains of amino acids in specific sequences; they are denatured by heat, improving their digestibility.

SEPAL One segment of the calyx of a flower. Sepals are usually thickened, green, and leafy, in contrast to the colored petals, but occasionally they are petal-like and similarly colored. They may be fused to one another —at least toward their bases—and often remain conspicuous on a fruit (e.g., a persimmon).

SPECIES The basic unit of plant classification, usually defined by the ability of individuals within the species to breed freely with one another over many generations without obvious change in their progeny. Normally, if a species breeds with another species the resulting progeny do not remain constant through the generations, or may be sterile. The scientific name of a species consists of the name of the genus to which it belongs, followed by a name referred to as the specific epithet, somewhat like a person's given name (e.g., *Prunus cerasus*).

SPICE A dried substance, obtained from rhizomes, roots, bark, flowers, fruit, or seeds, used to add aroma or flavor to a cooked food.

STAMEN The male reproductive organ of a flower—typically, a slender stem (filament) with a pollen-sac (anther) at its tip, which opens by a slit, or pore, to release pollen.

STAPLE FOOD The primary source of energy in traditional diets. Typically, these are plant foods, or derived from plant foods, including wheat and bread in the West, rice in much of Asia, and yams and other tubers in many Pacific cultures.

STONE CELLS Small gritty cells or groups of cells sometimes found in the soft tissues of fruits (e.g., the nashi pear is sometimes called the sand pear for its tendency toward grittiness due to stone cells).

STONE FRUIT The edible fruits of the plants in the genus *Prunus*—plums, cherries, peaches, apricots, nectarines, and almonds—all drupes with a single seed enclosed in a very hard ridged endocarp, or "stone."

TANNIN Natural compounds—often bitter and astringent on the palate—found in certain fruits (e.g., persimmons), in tea and chocolate, and most particularly in the skins, seeds, and stems of grapes. In red wines, tannin enhances aging potential but excess tannin in young wines is undesirable.

TENDRIL A small growth, usually developing in the leaf axil, sometimes from the leaf tip, that coils around twigs, wires, or other objects enabling a plant to climb (e.g., grape vines).

TERMINAL BUD The bud at the apex of a stem or inflorescence branch. Terminal leaf buds are often pinched out to encourage lateral branching and compact growth.

TILLAGE The process of working the soil in preparation for a crop. In home gardening this may simply be digging over and raking the soil to a fine tilth. In farming, it may involve plowing, discing, rolling, and drilling.

TRANSGENIC A plant that has been altered by genetic engineering, usually where DNA from another species has been combined with the plant's genes in vitro and then reinserted into the donor with the objective of expressing characteristics from the new DNA (such as pest resistance), and transferring these characteristics to the plant's progeny.

TUBER An underground or surface storage organ from which new plants will develop under suitable conditions. Many tubers, such as potatoes, sweet potatoes, cassava, and yams, are important food plants as they often survive prolonged storage.

UMBEL An inflorescence in which the individual flower stalks (pedicels) radiate from the end of the common stalk (peduncle). Umbelliferous plants include several common herbs and vegetables, such as fennel, dill, carrots, and parsnips.

VARIETY Technically, a subdivision of a species, of lower rank than a subspecies but higher than forma. The term commonly refers to cultivars and clones.

Index

Page numbers in *italics* indicate an illustration (at the most common names only). Page numbers in **bold** type indicate feature boxes.

Acknowledgments

The Publisher would like to thank Dannielle Doggett, David Kidd, Philippa Sandall, and Linda Vergnani for their help during the conceptualization process prior to production.

CAPTIONS FOR PRELIMINARY PAGES AND OPENERS

1 Star fruit; **2** (tl) A field of soybeans; (tr) Habenero chilies; (bl) Tomatoes; (br) Beans; **3** Rambutans; **5** A vegetable plot, including beetroot and brassicas; **7** Olives and olive oil for sale at a market in the south of France; **8–9** Chilies and peppers at a market stall; **10** Bags of pulses and grains; **11** A cherimoya; **12** (tl) Harvesting grapes in Germany; (tc) Peppercorns; (tr) Tomatoes; (bl) *Adam and Eve* by Lucas Cranach the Elder (1472–1553); (bc) Almonds growing on a tree; (br) Chestnuts in autumn; **13** (t) A nineteenth-century engraving of the Seville orange; (b) A mature head of barley; **58** (tl) Pink grapefruit; (tc) Onions; (tr) An ear of wheat; (bl) Lemon verbena; (bc) Peppercorns; (br) Coffee beans; **59** (t) Peanuts; (b) Sunflowers; **60–61** Gooseberries, strawberries, cherries, blackcurrants, and redcurrants; **138–139** Red and green bell peppers; **212–213** A field of barley; **230–231** Chestnuts; **248–249** Peppermint; **274–275** Cinnamon sticks; **302–303** Cocoa beans; **322–323** Harvested sugar cane; **334–335** A variety of pulses.

PHOTO CREDITS

AUS = Auscape International; CB = Corbis; GI = Getty Images; PL = Photolibrary

01c GI **002**bl GI/StockFood Creative, **02**br CB, **02**tl CB, **02**tr GI/Tim Boyle, **03**c GI/StockFood Creative, **05**c GI/Dorling Kindersley, **07**t GI/Robert Harding, **08-09**c GI, **10**c GI/Stone, **11**t GI/StockFood Creative, **12**bc GI/StockFood Creative, **12**bl The Art Archive/National Museum of Prague/Gianni Dagli Orti, **12**br GI/LOOK, **12**tc CB, **12**tl GI/LOOK, **12**tr GI/StockFood Creative, **13**b CB, **13**t The Art Archive/Eileen Tweedy, **14**b GI/Prehistoric, **14**t CB, **15**b The Art Archive/National Anthropological Museum Mexico/Gianni Dagli Orti, **15**t The Art Archive/Historiska Muséet Stockholm/Gianni Dagli Orti, **16**b CB, **16**t The Art Archive/Free Library Philadelphia, **17**b The Art Archive/Egyptian Museum Turin/Gianni Dagli Orti, **17**t GI/Inga Spence, **18**b GI/National Geographic, **18**t PL, **19**b PL, **19**t GI/Prehistoric, **20**b GI/National Geographic, **21**b GI/AFP/Getty Images, **21**tl GI/Bruno Morandi, **21**tr CB, **22**b The Art Archive/Egyptian Museum Cairo/Gianni Dagli Orti, **22**t CB, **23**b CB, **23**t CB, **24**b GI/Eisenhut & Mayer, **24**t GI/Tim Rand, **25**b CB, **25**t CB, **26**b CB, **26**t CB, **27**b GI/Food Photography Eising, **27**t CB, **28**l The Art Archive/Grand Masters' Palace La Valletta Malta/Gianni Dagli Orti, **29**b The Art Archive/Musée du Louvre Paris/Gianni Dagli Orti, **29**t The Art Archive/Private Collection/Gianni Dagli Orti, **30**b GI/AFP/Getty Images, **30**t PL, **31**b PL, **31**t The Art Archive/New York Public Library/Harper Collins Publishers, **32**l CB, **32**t CB, **33**b The Art Archive, **33**t PL, **34**b The Art Archive/Musée de la Marine Paris/Gianni Dagli Orti, **34**t CB, **35**b The Art Archive, **35**t CB, **36**b PL, **36**t CB, **37**b GI/John Turner, **37**t GI/Martin Page, **38**b The Art Archive/Museo Naval Madrid/Gianni Dagli Orti, **39**b GI/James P. Blair, **39**tl PL, **39**tr CB, **40**b CB, **40**t CB, **41**b PL, **41**t PL, **42**b GI/Michael Rosenfeld, **42**t CB, **43**b GI/Chinese School, **44**b CB, **44**t GI/Datacraft, **45**b CB, **45**t PL, **46**b CB, **46**t CB, **47**b CB, **47**t CB, **48**b CB, **49**b CB, **49**tl CB, **49**tr CB, **50**b CB, **50**t GI/Image Source, **51**b CB, **51**t CB, **52**b PL, **52**t CB, **53**b CB, **53**t CB, **54**b CB, **54**t PL, **55**b PL, **55**t GI/Time & Life Pictures/Getty Images, **56**b CB, **56**t PL, **57**b PL, **57**t PL, **58**bc IS/Michele Lugaresi, **58**bl GI/StockFood Creative, **58**br The Art Archive/Gianni Dagli Orti, **58**tc GI/Digital Vision, **58**tl GI/Visuals Unlimited, **58**tr GI/Nordic Photos, **59**bl GI, **59**tl CB, **60-61** GI/Klaus Hackenberg, **62**t GI/Frank Rothe, **63**b GI/Roger Phillips, **63**t GI/Ricardo de Vicq, **64**l PL, **65**b PL, **65**t GI/Sarah Cuttle, **66**b PL, **66**t PL, **67**t CB, **68**l CB, **69**b PL, **69**t PL, **70**b PL, **71**b GI/Jan Mammey/STOCK4B, **71**t CB, **72**b GI/Michael Rosenfeld, **73**b GI/Louise Lister, **73**t CB, **74**b GI/Simon Smith, **75**b CB, **75**t GI/Karl Newedel, **76**t Ozstock Images, **77**b CB, **77**t GI, **78**b CB, **78**t GI/Tom Grill, **79**b CB, **80-81** PL, **80**l CB, **81**r CB, **82**b GI, **82**t CB, **83**b AUS/Greg Harold, **84**t PL, **85**b PL, **85**t PL, **86**b The Art Archive/Museo di Roma Rome/Alfredo Dagli Orti, **87**b PL, **87**t PL, **88**t GI/Wolfgang Feiler, **89**l PL, **89**r GI, **90**t CB, **91**b PL, **91**t GI/Teubner, **92**t CB, **93**b GI/altrendo images, **93**t CB, **94**b CB, **94**t PL, **95**t GI/Linda Whitwam, **96**b PL, **96**t PL, **97**t Ian Maguire UF/IFAS/TREC & FLREC, **98**b The Art Archive/Eileen Tweedy, **98**t PL, **99**t Ozstock Images, **100**b PL, **100**t Global Publishing Australia, **101**t Ozstock Images, **102**b PL, **102**t PL, **103**b GI/Dorling Kindersley, **103**t Trade Winds Fruit, www.tradewindsfruit.com, **104**b GI/Teubner, **104**t CB, **105**t CB, **106**b CB, **106**t GI/Karl Newedel, **107**t CB, **108**t PL, **109**cr CB, **109**tl CB, **110**b Global Publishing Australia, **110**t PL, **111**b CB, **112**b PL, **112**t PL, **113**t Ozstock Images, **114**b PL, **114**t AUS/BIOS, **115**b PL, **116**b PL, **116**t PL, **117**t PL, **118**t PL, **119**b GI/Peter Anderson, **119**t The Art Archive/Royal Horticultural Society/Eileen Tweedy, **120**b PL, **121**b PL, **121**t PL, **122**t CB, **123**b PL, **123**t CB, **124**b PL, **124**t PL, **125**b PL, **126**c PL, **127**b GI/Barbel Buchner, **127**t PL, **128**t PL, **129**b PL, **129**t PL, **130**b PL, **131**b PL, **131**t GI, **132**t GI, **133**b CB, **133**t PL, **134**t PL, **135**b GI/Pete Oxford, **135**t PL, **136**b GI/Luis Veiga, **137**c GI/Dorling Kindersley, **137**t GI/Juan Silva, **138-139** GI/ColorBlind, **140**l GI, **140**t GI, **141**b GI/Panoramic Images, **142**b The Art Archive/Galleria Borghese Rome/Alfredo Dagli Orti, **142**t PL, **143**b PL, **144**bl PL, **144**br PL, **145**t PL, **146**b CB, **146**t CB, **147**b PL, **148**t CB, **149**b PL, **149**t PL, **150**b PL, **150**t PL, **151**b The Art Archive/Bibliothèque des Arts Décoratifs Paris/Gianni Dagli Orti, **151**t PL, **152**b PL, **153**b PL, **153**t PL, **154**b PL, **155**b The Art Archive/Private Collection/Marc Charmet, **155**t PL, **156**b CB, **157**b CB, **157**t CB, **158**b CB, **159** GI/Linda Lewis, **160**b GI/Roger Phillips, **160**t PL, **161**c PL, **162**b PL, **163**tl PL, **163**tr PL, **164**b CB, **165**b CB, **165**t GI/Michael Rosenfeld, **166**t PL, **167**b PL, **167**t PL, **168**b PL, **168**t PL, **169**t PL, **170**b The Art Archive, **170**t PL, **171**b PL, **172**b PL, **172**t PL, **173**b PL, **174**b CB, **174**t PL, **175**t CB, **176**l GI/Dorling Kindersley, **177**b GI/Teubner, **177**t PL, **178**t GI/Michael Rosenfeld, **179**b PL, **179**t PL, **180**b The Art Archive/Musée du Louvre Paris/Gianni Dagli Orti, **180**t PL, **181**t PL, **182**t PL, **183**b PL, **183**t PL, **184**b PL, **185**b PL, **185**t PL, **186**b PL, **186**t PL, **187**t PL, **188**b iStockphoto, **188**t PL, **189**b iStockphoto, **190**b PL, **191**b PL, **191**t GI/Michael Paul, **192**b GI/Dorling Kindersley, **193**b GI/Dave King, **193**t GI/Dave King, **194**t GI/Rita Maas, **195**b GI/Carl Tremblay, **195**t GI/Tom Grill, **196**t GI/Ulrike Schmid, **197**b CB, **197**t CB, **198**b GI/David Cavagnaro, **199**b GI/Michele Lamontagne, **199**t PL, **200**b CB, **200**t GI/Alan Richardson, **201**b PL, **202**t GI/Francesco Ruggeri, **203**b PL, **203**t GI/National Geographic, **204**b GI/Eric Anthony Johnson, **205**b PL, **205**t GI/John Kelly, **206**t GI/Altrendo Images, **207**b CB, **207**t PL, **208**t PL, **209**b GI/Eising, **209**t CB, **210**b PL, **211**b PL, **211**t PL, **212-213** GI/TG Stock, **214-15** GI/Michael Rosenfeld, **214**t PL, **215**t The Art Archive/Gianni Dagli Orti, **216**br GI/Ken Lucas, **216**cl GI/Paul M. Breeden, **217**t GI/Getty Images, **218**t PL, **219**b GI/Tohoku Color Agency, **219**t PL, **220**b GI/Christian Guy, **221**b PL, **221**t PL, **222**b CB, **222**t PL, **223**t GI/Frank Krahmer, **224**b CB, **224**t CB, **225**t PL, **226**b PL, **226**t GI/Kevin Morris, **227**b GI/Matthew Wakem, **227**t GI/ Ulrich Kerth, **228**t The Art Archive/Mireille Vautier, **229**b GI, **229**t PL, **230-231** GI/Jean Du Boisberranger, **232**b The Art Archive/Biblioteca Estense Modena/Alfredo Dagli Orti, **233**b PL, **233**t PL, **234**b PL, **235**b CB, **235**t GI/Monica Varella, **236**l CB, **237**b CB, **237**t GI/Inga Spence, **238**b PL, **238**t PL, **239**b Roger Leakey, Agroforestry and Novel Crops Unit, James Cook University, Cairns, Australia, **240**l CB, **240**t GI/Amy Neunsinger, **241**r PL, **242**t GI/Michelle Garrett, **243**b CB, **243**t GI/Marc O. Finley, **244**b CB, **244**t CB, **245**t GI, **246**l CB, **246**r GI, **247**t CB, **248-249** GI/Photolibrary, **250**t PL, **251**b GI/Lisa Romerein, **251**t CB, **252**t PL, **253**b GI/John Shipes, **253**t PL, **254**b CB, **255**b PL, **255**t PL, **256**b PL, **256**t CB, **257**t PL, **258**b The Art Archive/National Gallery London/Eileen Tweedy, **258**t PL, **259**r CB, **260**b CB, **261**b GI/Spencer Jones, **261**t GI/Anne Hyde, **262**b PL, **263**t GI/Peter Anderson, **264**b The Art Archive/Bibliothèque Nationale Paris, **264**t GI/DEA/C.DANI, **265**b PL, **266-267** PL, **266**b GI/Michael Rosenfeld, **267**b GI/John William Waterhouse, **268**tl CB, **268**tr PL, **269**b PL, **269**t PL, **270**t GI/DEA/B.DE CANDIA, **271**b PL, **271**t GI/Craig Knowles, **272**b The Art Archive/Marc Charmet , **272**t PL, **273**b PL, **274-275** GI/Ken Lucas, **276**l GI/Robert George Young, **277**cr The Art Archive/Eileen Tweedy, **277**tl The Art Archive/Bibliothèque des Arts Décoratifs Paris/Gianni Dagli Orti, **278**t GI/Kroeger/Gross, **279**b PL, **279**t GI/Mansell, **280**b PL, **281**b The Art Archive/Musée Archéologique Naples/Gianni Dagli Orti, **281**t PL, **282**b The Art Archive/Museo Correr Venice/Alfredo Dagli Orti, **282**t PL, **283**b PL, **284**b PL, **285**b The Art Archive/Musée des Beaux Arts Antwerp/Gianni Dagli Orti, **285**t PL, **286**b PL, **287**t GI/Neil Fletcher & Matthew Ward, **288**b PL, **289**b The Art Archive, **289**t GI/George Loun, **290**l PL, **291**b PL, **291**t PL, **292**l The Art Archive/Navy Historical Service Vincennes France/Gianni Dagli Orti, **293**b PL, **293**t PL, **294**b PL, **294**t PL, **295**t PL, **296**b The Art Archive/Navy Historical Service Vincennes France/Gianni Dagli Orti, **296**t PL, **297**b PL, **297**t PL, **298**b PL, **298**t PL, **299**b PL, **300**r PL, **301**b PL, **301**t The Art Archive/Bibliothèque Nationale Paris/Marc Charmet, **302-303** GI/James L Stanfield, **304**b CB, **304**t The Art Archive/Mireille Vautier, **305**b CB, **306**b GI/Francesca Yorke, **306**t CB, **307**b PL, **308**t PL, **309**b GI/Win Initiative, **309**t PL, **310**b GI, **311**b GI, **311**t PL, **312**t PL, **313**b CB, **313**t GI/Bodo A. Schieren, **314**b PL, **314**t PL, **315**r CB, **316**b PL, **316**t PL, **317**b GI/DEA/P. Puccinelli, **318**l PL, **319**b PL, **319**t PL, **320**b PL, **320**t PL, **321**t GI/James L. Stanfield, **322-323** GI, **324**b GI, **325**br GI/Gustavo Di Mario, **325**t PL, **326**r PL, **327**b PL, **327**t PL, **328**t GI/Travel Ink, **329**b GI/Jonathan Kantor, **329**t PL, **330**b GI/Christian Teubner, **330**t GI/Bridget Webber, **331**b GI/Travel Ink, **332**b GI/Peter Dazeley, **332**t GI/Teubner, **333**t PL, **334-335** GI/Amana Images.